Tumors of the Serosal Membranes

AFIP Atlases of Tumor and Non-Tumor Pathology

AMERICAN REGISTRY
OF PATHOLOGY
Publication & Education

ARP PRESS

Arlington, Virginia

Director of Publications: Amy Goldenberg, PhD
Production Editor: Dian S. Thomas
Technical Editor: Elizabeth Tomlinson
Copyeditor: Audrey Kahn

American Registry of Pathology
Arlington, Virginia 22203
www.arppress.org
ISBN 1-933477-19-9
978-1-933477-19-0

AFIP ATLASES OF TUMOR AND NON-TUMOR PATHOLOGY

Fifth Series
Fascicle 11

TUMORS OF THE SEROSAL MEMBRANES

by

Andrew G. Nicholson, DM, FRCPath
Consultant Histopathologist
Royal Brompton and Harefield Hospitals
Guy's and St Thomas' NHS Foundation Trust
Honorary Professor of Respiratory Pathology
National Heart and Lung Institute, Imperial College
London, United Kingdom

Aliya N. Husain, MD
Professor of Pathology
Pritzker School of Medicine
University of Chicago
Chicago, Illinois

Alain C. Borczuk, MD
Professor of Pathology
Weill Cornell Medicine
Department of Pathology and Laboratory Medicine
New York, New York

Published by the
American Registry of Pathology
Arlington, Virginia
2022

AFIP ATLASES OF TUMOR AND NON-TUMOR PATHOLOGY

EDITOR
Jason L. Hornick, MD, PhD
Professor of Pathology, Harvard Medical School
Director of Surgical Pathology and Immunohistochemistry
Brigham and Women's Hospital
Boston, Massachusetts

Manuscript reviewed by:
Lynette M. Sholl, MD

DEDICATIONS

To my ever lovely wife, Francesca, and all the members of the Nicholson and Garnham families who have been so supportive whilst writing this book, now including Tim.

Andrew G. Nicholson, DM, FRCPath

To my mentor, chief, and friend, Dr. Thomas Krausz, whose guidance, advice, and collaboration, especially in mesothelial lesions, are much appreciated.

Aliya N. Husain, MD

To my wife and family for their ongoing support; and to my aunt Jeanne, who passed away from pleural mesothelioma.

Alain C. Borczuk, MD

EDITOR'S NOTE

The Atlases of Tumor Pathology have a long and distinguished history. They were first conceived at a cancer research meeting held in St. Louis in September 1947, as an attempt to standardize the nomenclature of neoplastic diseases. The first series was sponsored by the National Academy of Sciences-National Research Council. The organization of this formidable effort was entrusted to the Subcommittee on Oncology of the Committee on Pathology, and Dr. Arthur Purdy Stout was the first editor-in-chief. Many of the illustrations were provided by the Medical Illustration Service of the Armed Forces Institute of Pathology (AFIP), the type was set by the Government Printing Office, and the final printing was done at the Armed Forces Institute of Pathology. The American Registry of Pathology (ARP) purchased the Fascicles from the Government Printing Office and sold them virtually at cost. Over a period of 20 years, approximately 15,000 copies each of nearly 40 Fascicles were produced. The worldwide impact of these publications over the years has largely surpassed the original goal. They quickly became among the most influential publications on tumor pathology, primarily because of their overall high quality, but also because their low cost made them easily accessible the world over to pathologists and other students of oncology.

Upon completion of the first series, the National Academy of Sciences-National Research Council handed further pursuit of the project over to the newly created Universities Associated for Research and Education in Pathology (UAREP). The Second Series was started, generously supported by grants from the AFIP, the National Cancer Institute, and the American Cancer Society. Dr. Harlan I. Firminger became the editor-in-chief and was succeeded by Dr. William H. Hartmann. The Second Series Fascicles were produced as bound volumes instead of loose leaflets. They featured a more comprehensive coverage of the subjects, to the extent that the Fascicles could no longer be regarded as "atlases" but rather as monographs describing and illustrating in detail the tumors and tumor-like conditions of the various organs and systems.

Dr. Juan Rosai was appointed as editor-in-chief of the Third Series, and Dr. Leslie Sobin became associate editor. A distinguished Editorial Advisory Board was also convened, and these outstanding pathologists and educators played a major role in the success of this series, the first publication of which appeared in 1991 and the last (number 32) in 2003.

The same organizational framework applied to the Fourth Series, meticulously edited by Dr. Steven Silverberg with Dr. Ronald DeLellis as the associate editor. With UAREP and AFIP no longer functioning, ARP remained the responsible organization. The Fourth Series volumes were hardbound with illustrations almost exclusively in color. There was also an increased emphasis on the cytopathologic (intraoperative, exfoliative, or fine needle aspiration) and molecular features that are important in diagnosis and prognosis. At the time of the Fourth Series, ARP also produced

Atlases of Non-Tumor Pathology; these volumes were numbered separately from the tumor volumes.

As in the prior series, the goal of the Fifth Series includes a continuous attempt to correlate, whenever possible, the nomenclature used in the Fascicles with that proposed by the World Health Organization Classification of Tumors, as well as to ensure a consistency of style. Including molecular diagnostics is more important than ever, as is the availability of an online component, a more nimble website (www.arppress.org), and a social media presence. Now in this series, the tumor and non-tumor volumes are combined and consecutively numbered as the Atlases of Tumor and Non-Tumor Pathology. Close cooperation between the various authors and their respective liaisons from the Editorial Board will continue to be emphasized in order to minimize unnecessary repetition and discrepancies in the text and illustrations.

Particular thanks are due to the members of the Editorial Advisory Board, our reviewers, the editorial and production staff, and the individual Fascicle authors for their ongoing efforts to ensure that this series is a worthy successor to the previous four.

Jason L. Hornick, MD, PhD

PREFACE

Since the 4th Series of the Fascicle was published, there have been considerable advances in our understanding of tumors within the serosal membranes, in particular the classification of mesothelial tumors and their genetics. There is no perfect classification system for tumors arising in these sites, and the previous edition classified tumors first into diffuse and localized forms and secondarily into benign and malignant variants. This 5th Series Fascicle aligns with the terminology recommended by the WHO in its most recent edition for pleural and pericardial tumors, with expansion to reflect serosal membranes at other sites and including recognition of mesothelioma in situ. This book, however, reflects our views beyond the WHO classification; the Fascicle template allows more scope for discussion in detail. There is also a chapter on tumor-like lesions to complement the tumor data within the WHO books.

The authors acknowledge all the members of the International Mesothelioma Pathology Group, organized by Francoise Galateau-Salle, which has provided a forum for advancing our knowledge in this field, as well as for debating and improving our understanding in controversial areas. We also acknowledge the International Mesothelioma Interest Group (IMIG) and the International Association for the Study of Lung Cancer (IASLC), as both societies provide platforms for discussion, especially in a multidisciplinary context.

We would also like to thank all the individuals who have provided materials, especially images, which are acknowledged in the legends.

Finally, we wish to thank the staff at ARP Press for their editorial assistance and for guiding us from our original vision to the bookshelf.

Andrew G. Nicholson, DM, FRCPath

Aliya N. Husain, MD

Alain C. Borczuk, MD

Permission to use copyrighted material has been granted by:

American Thoracic Society
American Review of Respiratory Disease 1975;111:17. For figure 1-4.

College of American Pathologists
Archives of Pathology and Laboratory Medicine 2018;142:89-108. For tables 3-4, 3-5, and 3-6.

Elsevier
The Developing Human: Clinically Oriented Embryology, 10th ed; 2015. For figures 1-5 and 1-6.

Wolters Kluwer Health, Inc.
American Journal of Surgical Pathology 2002;26:1200-1. For figures 1-7 above and 1-7 right.

CONTENTS

1. Anatomy, Development, and Normal Function of the Serosal Membranes 1
   - Gross Anatomy 1
     - Anatomic Sites 1
     - Blood Supply 1
     - Lymphatic Drainage 1
     - Innervation 2
   - Microscopic Anatomy 2
   - Development 5
   - Normal Function 8
2. Classification of Tumors of the Serosal Membranes 11
3. Diffuse Mesothelioma 13
   - General Features 13
     - Incidence 13
     - Epidemiology 13
     - Etiology 15
     - Incidence and Etiology at Sites Other than the Pleura 17
   - Clinical Findings 17
     - Pleura 17
     - Peritoneum 18
     - Pericardium and Tunica Vaginalis 18
   - Diagnostic Specimens 18
     - Pleura 18
     - Peritoneum, Pericardium, and Tunica Vaginalis 19
   - Gross Findings 19
     - Pleura 19
     - Peritoneum 19
     - Pericardium and Tunica Vaginalis 20
   - Microscopic Findings 20
     - Mesothelioma in Situ 23
     - Diffuse Mesothelioma: Epithelioid Mesothelioma 28
     - Sarcomatoid Mesothelioma 42
     - Biphasic Mesothelioma 47
   - Immunohistochemical Findings 48
     - Epithelioid Mesothelioma 50
     - Sarcomatoid Mesothelioma 50
   - Cytologic Findings 50
     - Cytologic Features of Malignancy 53
     - Reactive Atypia in Mesothelial Cells 53

Molecular Genetic Findings 62
Biomarkers in Mesothelioma 62
Molecular Markers 62
Ultrastructural Findings 64
Differential Diagnosis 65
Benign Versus Malignant Mesothelial Proliferations 65
Epithelioid Mesothelioma Versus Metastatic Carcinoma 72
Sarcomatoid Mesothelioma Versus Other Spindle Cell Malignancies 74
Other Differential Diagnoses 74
Treatment 77
Pleura 77
Peritoneum 79
Pericardium and Tunica Vaginalis 79
Prognosis 79
Survival Data 79
TNM Staging 80
Other Prognostic Factors 81
4. Other Mesothelial Tumors 97
Well-Differentiated Papillary Mesothelial Tumor 97
Localized Mesothelioma 99
Multicystic Mesothelioma 101
Adenomatoid Tumor 104
5. Mesenchymal Neoplasms 109
Solitary Fibrous Tumor 109
Epithelioid Hemangioendothelioma 114
Synovial Sarcoma 117
Desmoplastic Small Round Cell Tumor 119
Ewing Sarcoma 119
Other Primary Mesenchymal Neoplasms of Serosal Membranes 122
6. Lymphoproliferative Neoplasms 133
Primary Effusion Lymphoma 133
Diffuse Large B-Cell Lymphoma Associated with Chronic Inflammation 136
Erdheim-Chester Disease 136
7. Secondary Malignant Neoplasms 143
Pleural Disease 143
Sites of Origin 143
Mechanism of Serosal Involvement 143
Patterns of Involvement 143
Gross Findings 144
Microscopic Findings 145
Differential Diagnosis 149

Treatment and Prognosis 149
Peritoneum 149
Sites of Origin 149
Clinical Features 150
Microscopic Findings 150
Differential Diagnosis 152
Pericardium 153
8. Non-Neoplastic Entities and Entities of Uncertain or Overlapping Histogenesis 157
Neutrophilic Effusions and Empyema 157
Lymphocytic Effusions and Reactive Fibrous Pleuritis 157
Pleural Plaques 160
Mesothelial Hyperplasia 161
Cystic Tumor of the Atrioventricular Node 163
Eosinophilic Pleuritis 164
Granulomatous Pleuritis 165
Hemothorax 166
Rounded Atelectasis 166
Amyloidosis 168
IgG4-Related Disease 169
Endometriosis 169
Trophoblastic Implants 171
Walthard Rests 171
Gliomatosis Peritonei 171
Disseminated Peritoneal Leiomyomatosis 171
Splenosis 173
Melanosis 173
Chondroid and Osseous Metaplasia 173
Omental-Mesenteric Myxoid Hamartoma 173
Index 181

1 ANATOMY, DEVELOPMENT, AND NORMAL FUNCTION OF THE SEROSAL MEMBRANES

GROSS ANATOMY

Anatomic Sites

The serous membranes consist of the pleura, pericardium, peritoneum, and tunica vaginalis.

Pleura. The pleura forms a continuous layer over the thoracic structures. It is divided into visceral and parietal portions. The visceral pleura covers the lungs and interlobar fissures, with the latter dividing the lobes of the lung either completely or partially. The parietal pleura lines the thoracic wall, including the thoracic inlet, the lateral aspect of the mediastinum, and the thoracic surface of the diaphragm. It forms the suprapleural membrane, which is the thickened portion of endothoracic fascia extending over the cupola of the pleura (cervical pleura).

Peritoneum. The peritoneum forms a continuous layer over the abdominal structures, with the exception of the ostia of the oviducts. Approximately 10 percent of individuals have defects (or fenestrations) in the peritoneum overlying the liver, which is clinically relevant in relation to the spread of endometriosis to the right pleural space (1). The visceral peritoneum covers the intra-abdominal organs and their mesenteries. The parietal peritoneum lines the abdominal wall, the pelvis, the inferior surface of the diaphragm, and part of the anterior surfaces of the retroperitoneal viscera (2). The peritoneal cavity is further divided into the supracolic, right and left infracolic, and pelvic regions. The shape of the mesentery of the small intestine causes the right infracolic space to taper inferiorly, and the left to taper superiorly. This has clinical relevance since these partitions influence the distribution of fluid within the abdominal cavity.

Pericardium. The serosal membrane covers the inner surface of the pericardium on one side and the epicardial surface of the heart on the other.

Tunica Vaginalis. The tunica vaginalis is continuous with the lining of the peritoneal cavity and covers the outer surface of the contents of the testis.

Blood Supply

The parietal pleura derives its blood supply from the intercostal, internal mammary, and phrenic arteries. The visceral pleura is supplied by the bronchial arteries. Venous drainage runs parallel to the arterial supply. The peritoneum and mesentery are supplied by the splanchnic arterial vessels, although a limited portion is supplied by the lower intercostal and subcostal, lumbar, and iliac arteries. Venous drainage is by the splanchnic veins and the portal system.

Lymphatic Drainage

The potential space between the visceral and parietal pleurae is drained by an extensive lymphatic system. The anterior parietal pleura drains into the intercostal lymphatic network, while the lower parietal pleura drains into the retroperitoneal nodes in the region of the kidney and adrenal gland. The diaphragmatic pleura drains into the lower mediastinal lymph nodes. Visceral pleural lymphatics drain into the pulmonary hilar lymph nodes via intersegmental pathways (3). There is also an outward flow of lymph from the peripheral one third of the lung parenchyma toward the visceral pleura, which explains the spread of intrapulmonary tumor to the visceral pleura and contributes to the common occurrence of malignant pleural effusions in lung cancer. Lymph flow from the inner two thirds of the lung is toward the hilum.

In the normal state, the peritoneal cavity contains about 100 mL of serous fluid (4). The peritoneal lymphatics follow the blood vessels. Intestinal efferent lymphatics communicate with the intramuscular and subserosal plexuses. These channels unite in the mesentery to form collecting trunks, and the lymph passes through groups of lymph nodes and lymphoid aggregates

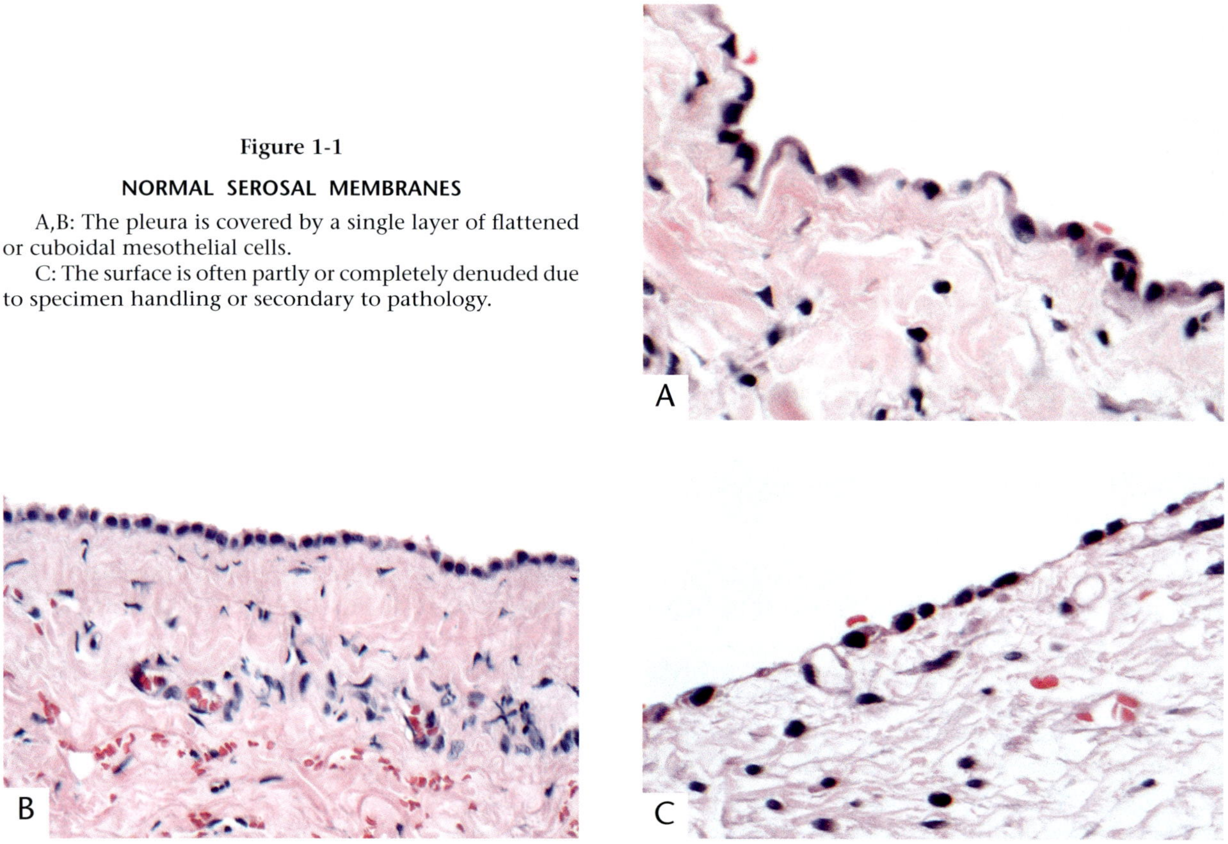

Figure 1-1

NORMAL SEROSAL MEMBRANES

A,B: The pleura is covered by a single layer of flattened or cuboidal mesothelial cells.

C: The surface is often partly or completely denuded due to specimen handling or secondary to pathology.

in the omentum, known as "milk spots." The afferent channels ultimately form an intestinal trunk that drains the region supplied by the superior mesenteric artery and empties into the cisterna chyli or the left lumbar trunk (5). Lymph from the left side of the colon, kidney, adrenal gland, gonad, and lower extremities is drained by the left and right lumbar lymphatic trunks (6).

There are extensive communication networks between the lymphatic vessels in the serosa on both sides of the diaphragm. Those on the right are larger and transport more fluid than those on the left. The peritoneal space drains to the celiac, superior mesenteric, and periportal lymph node groups first and continues via the thoracic duct to the mediastinal lymph nodes (7,8).

Innervation

The parietal pleura is innervated by sensory branches from the intercostal nerves. The visceral pleura is devoid of pain fibers but receives innervation from the vagus nerve and sympathetic trunks. The phrenic nerve supplies the central portion of the diaphragm. The nerves of the lower intercostal and abdominal wall supply the peripheral portion of the diaphragm (9).

The parietal peritoneum is innervated by the spinal nerves supplying the abdominal wall. These nerves include sensory pain fibers. The visceral peritoneum receives sympathetic nerve fibers only and is devoid of pain fibers.

MICROSCOPIC ANATOMY

The serosal membranes are lined by a single layer of flattened mesothelial cells that rest on a basal lamina (fig. 1-1A,B). These cells have abundant cytoplasm, centrally placed round nuclei, and a single small nucleolus. They range in diameter from 16 to 40 µm. The mesothelium lining the visceral pleura is characterized by more abundant microvilli than that of the parietal pleura. Mesothelial cells of the parietal pleura are separated by gaps to communicate with the underlying lymphatic vessels (10,11). In surgical specimens, mesothelial cells are often cuboidal and either partly or completely denuded (fig. 1-1C), secondary to pathologic conditions or handling.

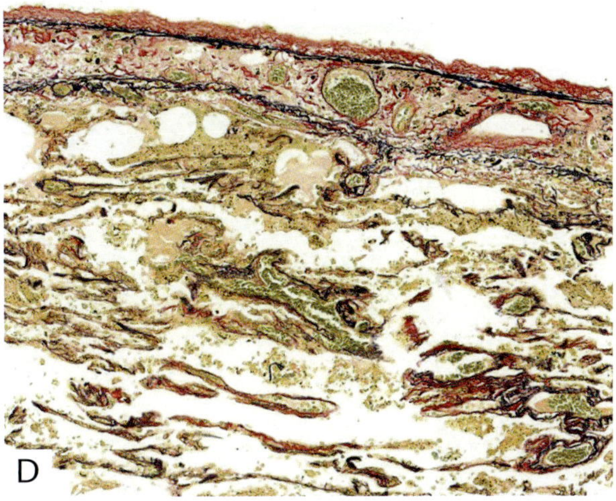

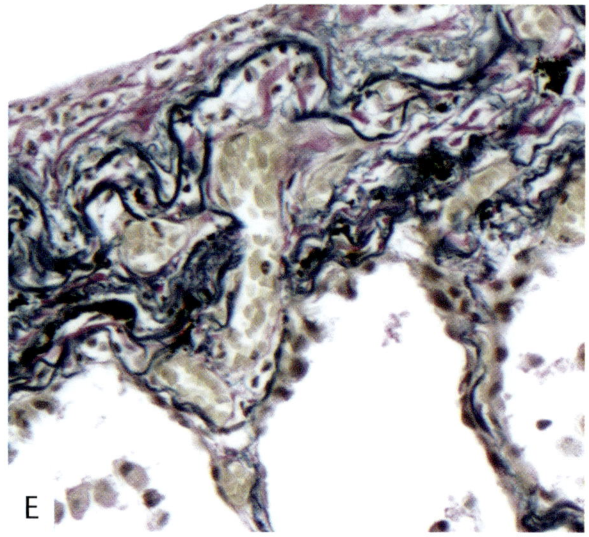

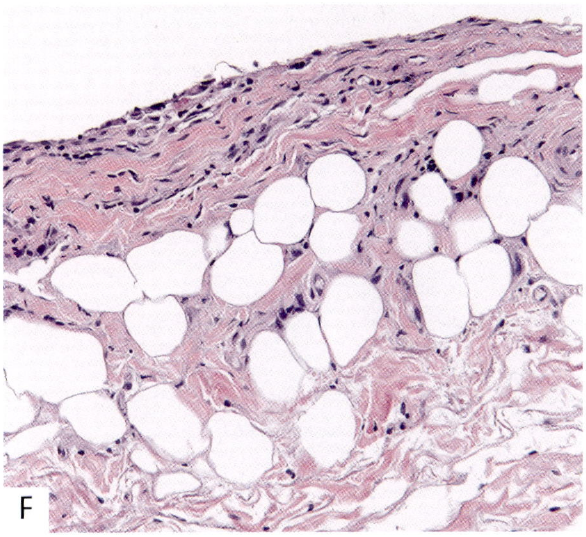

Figure 1-1, continued

D: Normal visceral pleura has a thin layer of fibrosis deep to the mesothelial cells, with two elastin layers often visible and separated by a further layer of fibrosis (elastin van Gieson [EVG] stain).

E: The elastin layers can become less defined, with more disordered elastosis, in chronic disease, interspersed by a thin layer of connective tissue (EVG stain).

F,G: The parietal pleura contains a less well-defined elastin layer (G: EVG stain).

H: The pericardium comprises a single layer of mesothelium with an underlying dense fibrous layer.

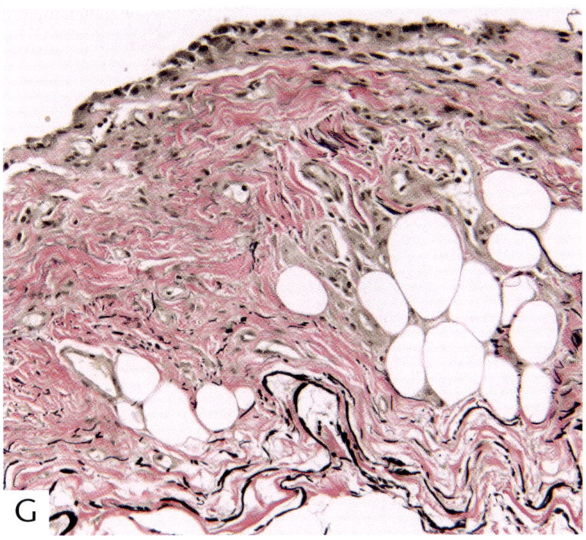

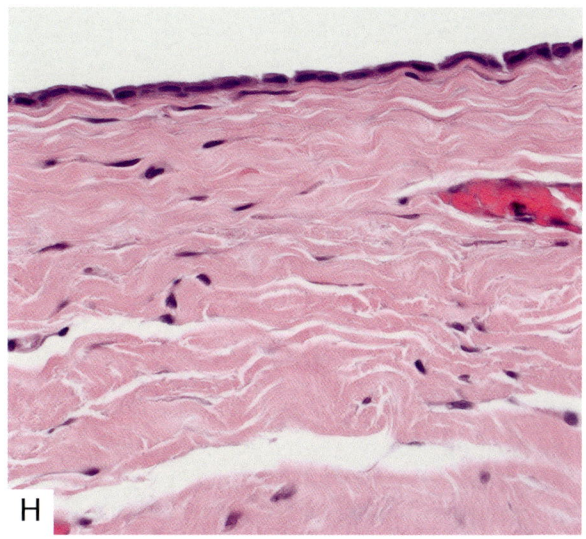

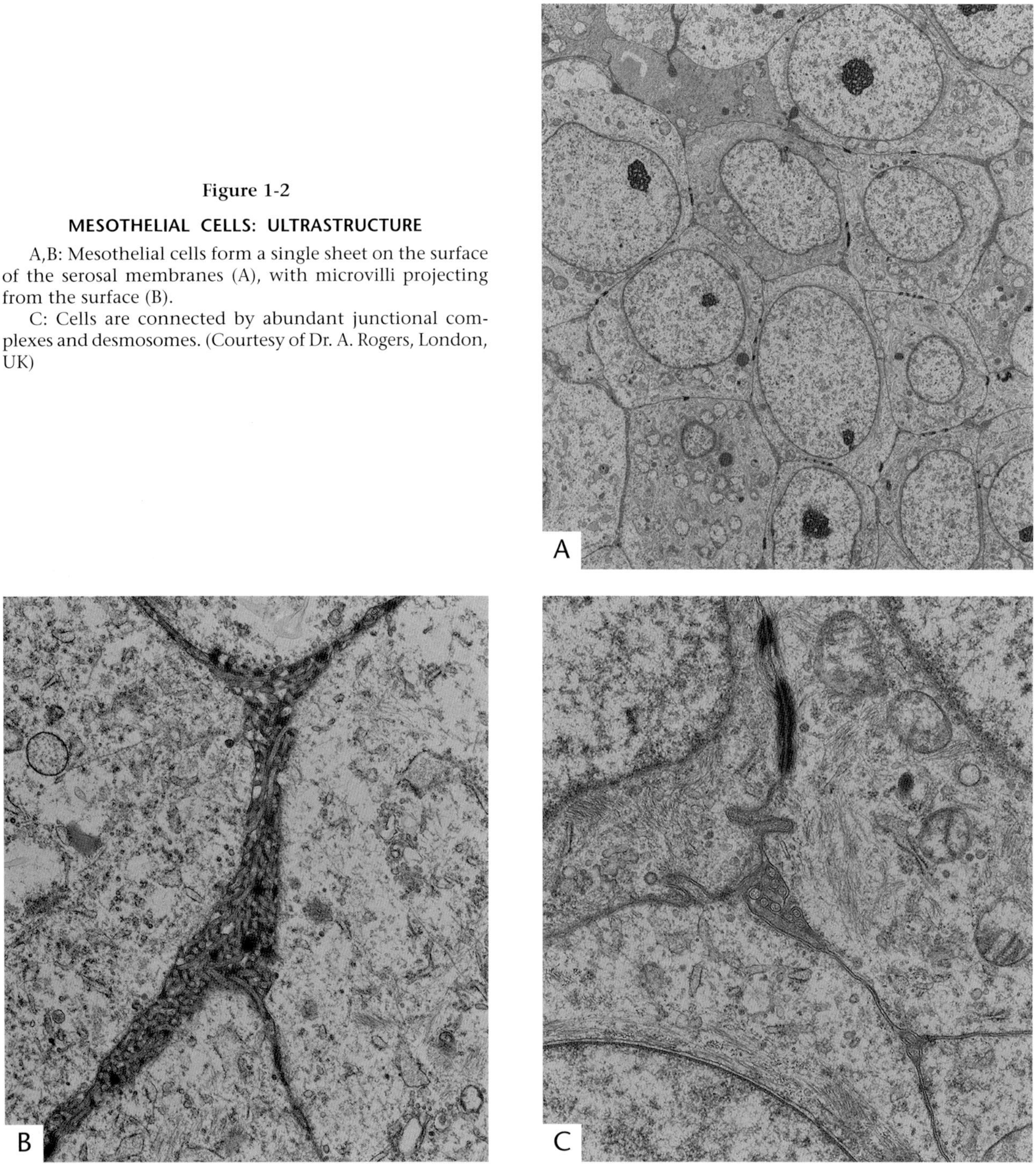

Figure 1-2

MESOTHELIAL CELLS: ULTRASTRUCTURE

A,B: Mesothelial cells form a single sheet on the surface of the serosal membranes (A), with microvilli projecting from the surface (B).

C: Cells are connected by abundant junctional complexes and desmosomes. (Courtesy of Dr. A. Rogers, London, UK)

The visceral pleura consists of five layers, which may vary in presence and thickness. Mesothelial cells and basement membrane overlie a thin layer of fibrous tissue, with two layers of elastin separated by a further layer of fibrosis (fig. 1-1D). This can become disordered in chronic inflammatory conditions (fig. 1-1E). The parietal pleura contains a less well-defined elastin layer (fig. 1-1F,G). The pericardium consists of a single layer of mesothelium with an underlying dense fibrous layer (fig. 1-1H).

Ultrastructurally, mesothelial cells have moderately dense cytoplasm that contains ribosomes, rough endoplasmic reticulum,

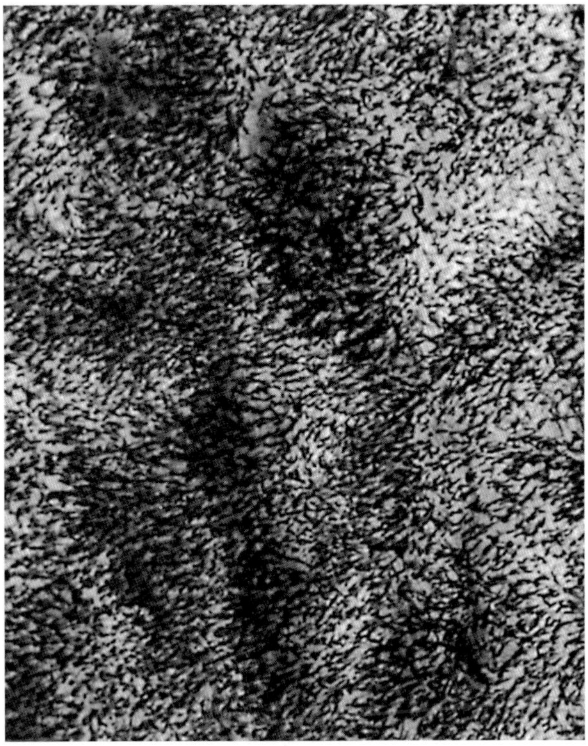

Figure 1-3

MESOTHELIAL CELLS: ULTRASTRUCTURE

A scanning electron microscopic view of visceral rat pleura shows the profusion of microvilli on the surface.

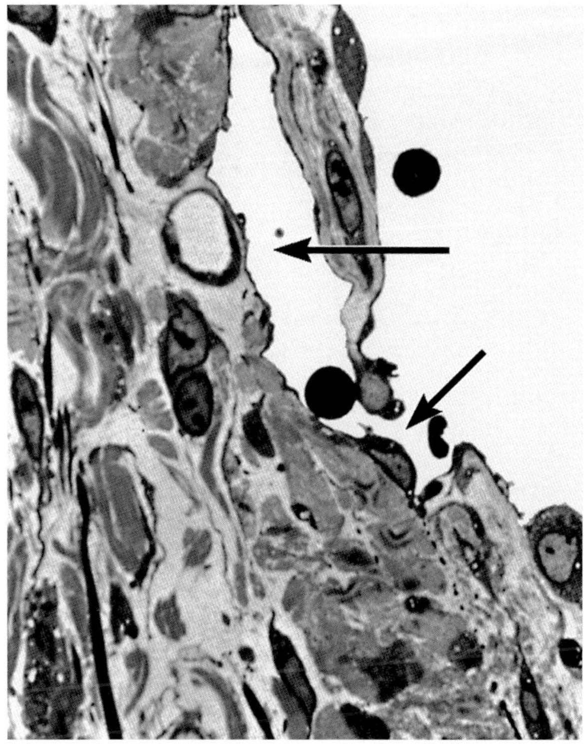

Figure 1-4

STOMA MESOTHELIAL SURFACE

A lymphatic lacuna (long arrow) opens onto the mesothelial surface via a narrow stoma (short arrow). (Copyright © 2022 American Thoracic Society. All rights reserved. Cite: Wang NS/1975/The preformed stomas connecting the pleural cavity and the lymphatics in the parietal pleura/ Am Rev Respir Dis/111/17/Figure 9. The American Review of Respiratory Disease* is an official journal of the American Thoracic Society. *Now titled the American Journal of Respiratory and Critical Care Medicine.)

perinuclear tonofilaments, and moderate numbers of mitochondria (fig. 1-2). Micropinocytotic vesicles may be observed in the cell membrane. Adjacent cells are connected by relatively abundant junctional complexes and desmosomes; however, these junctions are discontinuous to permit ready diffusion of molecules between cells (12). Studies suggest pores of different sizes are present (13,14).

The most prominent and characteristic ultrastructural feature of mesothelial cells is the presence of long, slender, surface microvilli (figs. 1-2, 1-3). These microvilli measure up to 3.0 µm in length and 0.1 µm in diameter. Microvilli may be longer around diaphragmatic stomata (15) and are most abundant on the surface of organs that actively move (11,16). It is hypothesized that the microvilli serve to entrap hyaluronate, which reduces friction at these sites.

A thin basal layer lies underneath the mesothelial cells, separating the cells from a connective tissue layer that consists of variable amounts of collagen and elastic fibers, fibroblast-like cells, capillaries, and lymphatics. Adipocytes may also be present. Where mesothelial cells come into contact with terminal lymphatic vessels, the basal lamina is attenuated or missing, and there are gaps between cells. In the parietal pleura, these communications take the form of stomata, which allow direct continuity between mesothelium and lymphatic endothelium (fig. 1-4) (11).

DEVELOPMENT

The intraembryonic coelomic cavity is the forerunner of the body cavities that eventually become lined by the pleural, peritoneal, and pericardial mesothelia. This cavity originates within the lateral and cardiogenic mesoderms as isolated

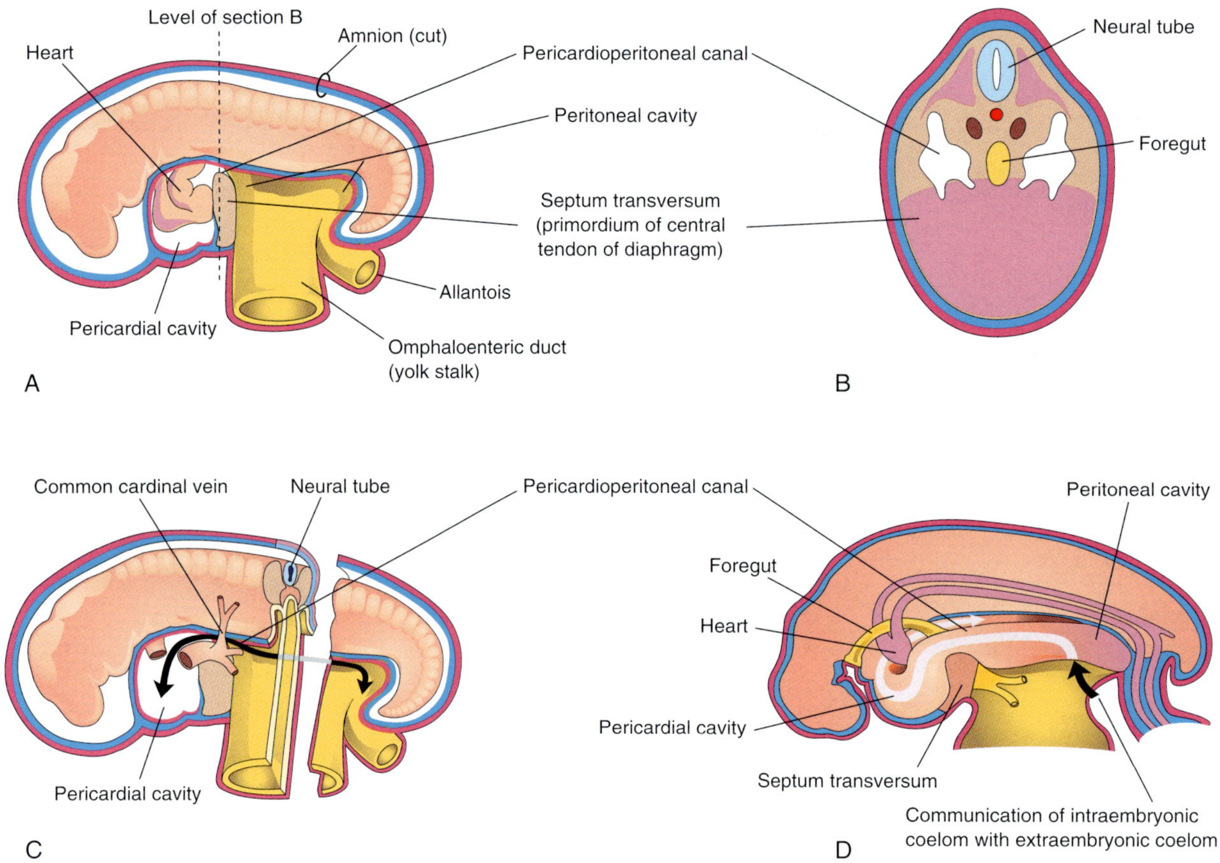

Figure 1-5

EMBRYO AT APPROXIMATELY 24 DAYS

A: The lateral wall of the pericardial cavity has been removed to show the primordial heart.

B: Transverse section of the embryo illustrates the relationship of the pericardioperitoneal canals to the septum transversum (primordium of central tendon of diaphragm) and the foregut.

C: Lateral view of the embryo with the heart removed. The embryo has also been sectioned transversely to show the continuity of intraembryonic and extraembryonic coeloms or body cavities (arrow).

D: The pericardioperitoneal canals arise from the dorsal wall of the pericardial cavity and pass on each side of the foregut to join the peritoneal cavity. The arrow shows the communication of the extraembryonic coelom with the intraembryonic coelom and the continuity of the intraembryonic coelom at this stage. (Fig. 8-4 from Moore KL. The developing human: clinically orientated embryology, 10th ed. Philadelphia: WB Saunders; 2015.)

spaces that coalesce to form a horseshoe-shaped cavity. The curve of this cavity is the future pericardial cavity and the lateral parts are the future pleural and peritoneal cavities (17). The intraembryonic coelom divides the lateral mesoderm into the somatic and splanchnic layers that form the linings of the primitive coelomic cavity.

As development continues, the intraembryonic portion of the coelom closes off ventrally from the extraembryonic part. The area around the umbilicus is the last to close (fig. 1-5). The original right and left coelomic cavities advance toward each other, meeting in the midline. The lateral halves of the pericardial cavity are the first part of the intraembryonic coelom to form by the splitting of the mesoderm and the first part completed by the union of the cavities. Layers of splanchnic mesoderm couple in the mid-portion of the embryo to form the various mesenteries. Mesoderm ventral to the gut attenuates, and the cavities coalesce.

The septum transversum partitions the single coelomic space into thoracic and abdominal portions; it is joined laterally by the pleuro-

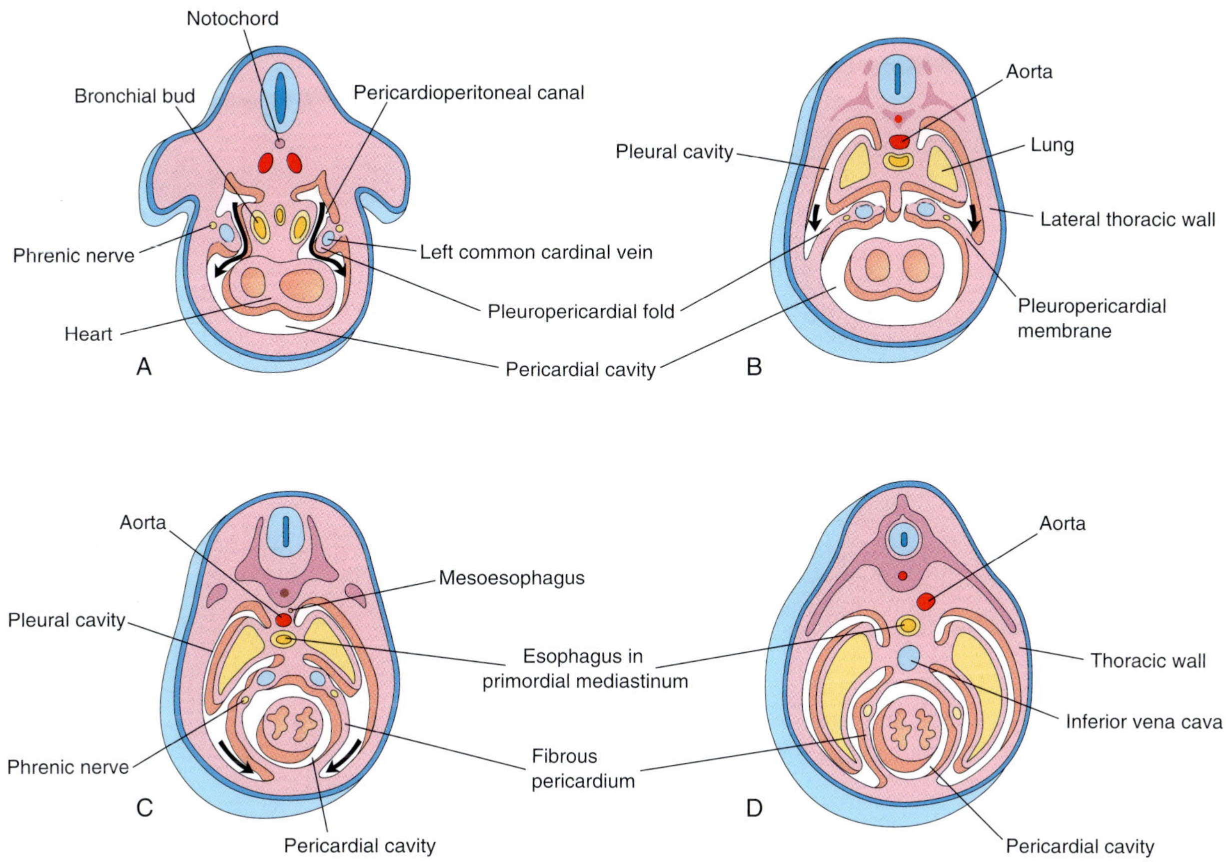

Figure 1-6

STAGES IN THE DEVELOPMENT OF THE PLEURAL CAVITIES

Transverse sections through embryos cranial to the septum transversum illustrate successive stages in the separation of the pleural cavities from the pericardial cavity. Growth and development of the lungs, expansion of the pleural cavities, and formation of the fibrous pericardium are also shown.

A: At 5 weeks, there is communication between the pericardioperitoneal canals and the pericardial cavity (arrows).

B: At 6 weeks, the pleural cavities expand into the body wall (arrows).

C: At 7 weeks, the pleural cavities expand ventrally around the heart. The pleuropericardial membranes are now fused in the median plane, with the mesoderm ventral to the esophagus.

D: At 8 weeks, the lungs and pleural cavities continue to expand and form the fibrous pericardium and thoracic wall. (Fig. 8-5 from Moore KL. The developing human: clinically orientated embryology, 10th ed. Philadelphia: WB Saunders; 2015.)

peritoneal folds that arise from the body wall. Bilateral pleuropericardial folds then partition the pleural from the pericardial cavities and eventually join the septum transversum (17,18). The septum transversum arises in the cervical area of the embryo, moving caudally to the level of the first lumbar vertebra by the second month of gestation. As it descends, it carries nerve fibers from the phrenic nerves. The lungs also begin their development in a cephalad location, arising as buds from the endoderm that project into the early pleuropericardial folds from the mesodermal bridge on each side (fig. 1-6).

The pleural and pericardial cavities first complete separation on the right side. On the left side, a communication between these cavities usually persists (18). The diaphragm derives from the septum transversum (which forms the central tendon), the two pleuroperitoneal membranes, the mesentery of the esophagus, and the dorsal and lateral body walls (which form the muscular portion).

The parietal pleura, pericardium, and peritoneum are derived from somatic mesoderm, whereas the visceral membranes are derived from splanchnic mesoderm. These membranes consist

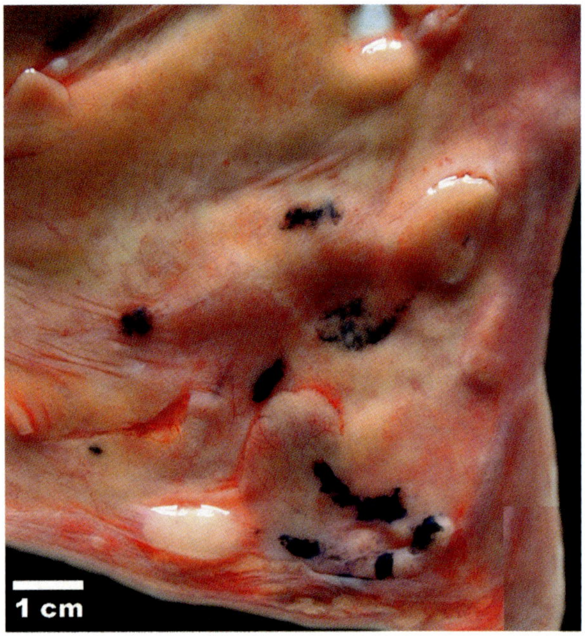

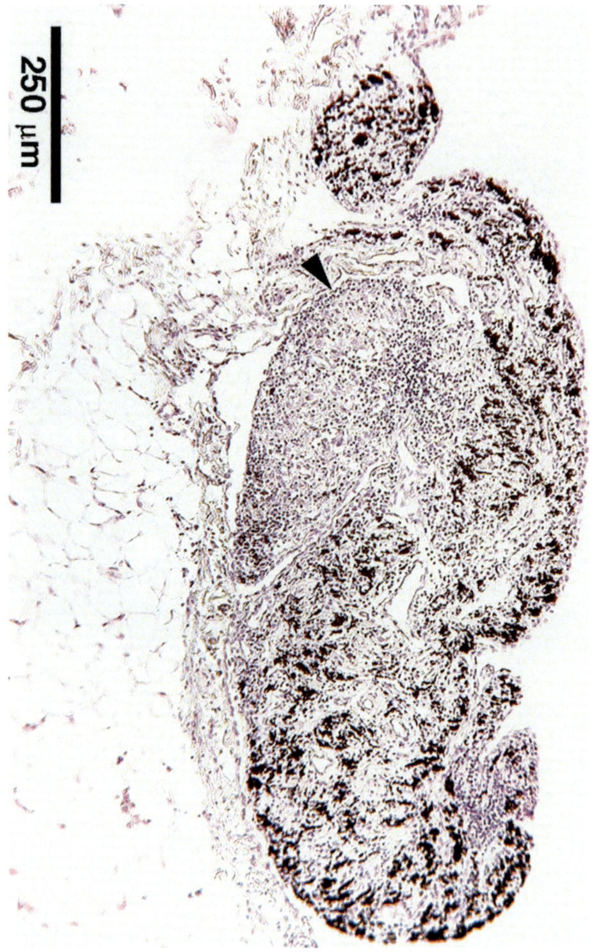

Figure 1-7

BLACK SPOTS ON THE PARIETAL PLEURA

Above: Pigmented spots observed at autopsy in the parietal pleura range from 3 to 10 mm in diameter.

Right: Hematoxylin and eosin (H&E)-stained section of a pigmented spot. The arrow indicates the area with a chronic inflammatory reaction. (Figs. 1D and 2A from Mitchev K, Dumortier P, De Vuyst P. 'Black spots' and parietal pleural plaques on the parietal pleura of 150 urban necrotic cases. Am J Surg Pathol 2002;26:1200-1.)

of a layer of persisting mesenchymal tissue and a layer of flattened surface cells that become the mesothelium. These layers retain the potential to differentiate in a number of ways (18).

NORMAL FUNCTION

The purpose of the serosal lining is to provide an essentially frictionless surface over which the viscera can move as they perform their contractile and expansile functions. A thin layer of liquid that is rich in hyaluronate, a glycoaminoglycan lubricant, coats the serosal surfaces, and these surfaces are actively involved in fluid transport (19,20). The visceral and pleural surfaces are separated by a distance of 5 to 10 µm. This space is occupied by hyaluronate and cells, including pleural macrophages (21).

Fluid moves from the bloodstream to the pleural space and vice versa by means of hydrostatic and colloidal osmotic pressures in the serosal capillaries. Translocation of fluid from the arterial ends of the capillaries in the parietal pleura with reabsorption in the venous ends of the capillaries in the visceral pleura is favored, since the hydrostatic pressure in the systemic circulation supplying the former is greater than the pressure in the pulmonary circulation (22).

The lymphatic vessels of the parietal pleura reabsorb 10 to 20 percent of the pleural fluid. Larger molecules (e.g., proteins) enter directly into the lymphatic channels. Particulate matter and cells are also taken up by the parietal pleura, principally through the stomata of the lower mediastinal and costal areas (fig. 1-7) (20,22–24). Particulate matter in the peritoneal cavity is reabsorbed through diaphragmatic lymphatic vessels. The flow of lymph is influenced by muscular activity during respiration.

Obstruction of lymphatics may result in the accumulation of fluid in the pleural, pericardial,

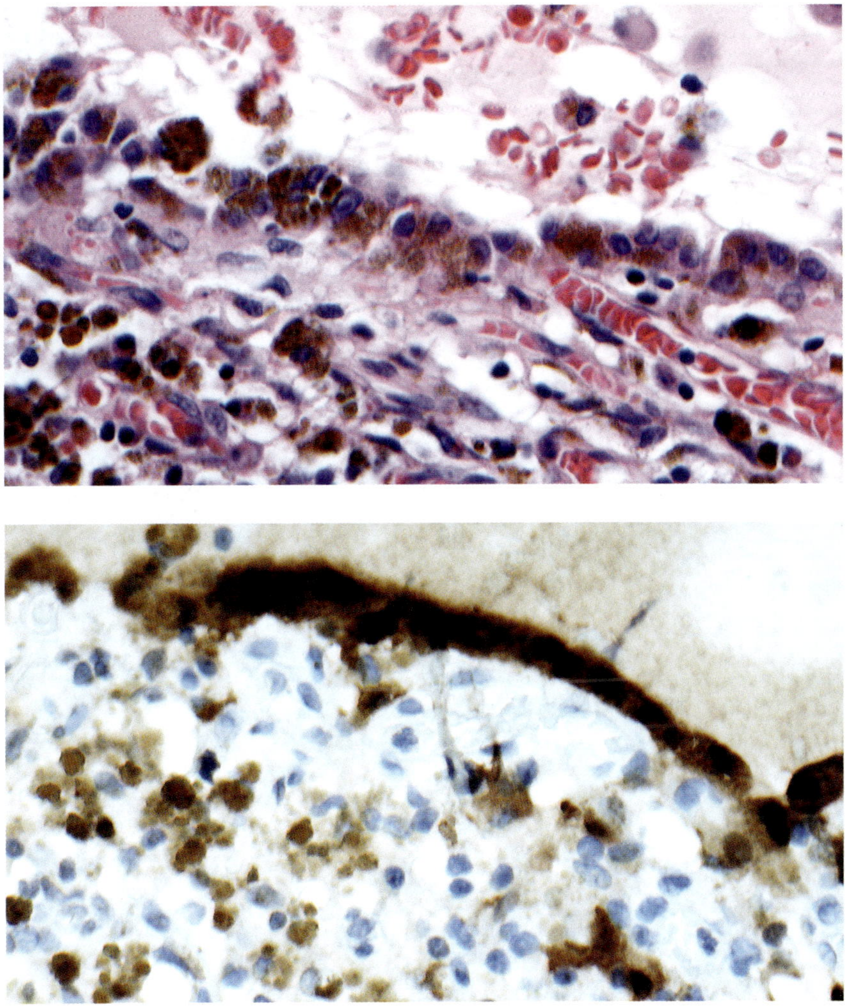

Figure 1-8

MESOTHELIAL CELLS: PHAGOCYTOSIS

Top: Hemosiderin pigment is seen within the mesothelial cells on the surface of chronically inflamed pleura.

Bottom: Calretinin staining confirms mesothelial origin.

or peritoneal spaces. Lymph flow can be effectively blocked by tumor, fibrosis, or particulate matter. The proteinaceous content of accumulated fluid can augment the retention of additional fluid by means of its oncotic pressure. Pleural thickening (e.g., diffuse pleural fibrosis) may seal lymphatic vessels and thus prevent the absorption of fluid. Overload of the lymphatic drainage system in the peritoneal cavity may result in retention of fluid in the thorax (particularly on the right), since ascitic fluid drains through the diaphragm into the pleural cavity.

The relative importance of the various compartments within the mesothelial cell for fluid transport is incompletely understood and somewhat controversial. Fluid and particle transmission may occur through pinocytotic vesicles or directly through the cytosol. Long surface microvilli increase the surface area and thus promote fluid transport. Pinocytosis may be more important for the equalization of protein and salt composition of the serous fluid. Some have suggested that intercellular clefts are the major sites of solid and fluid transport across the peritoneal membrane (12), with pores of varying sizes having different functions (13,14). This has relevance clinically since reactive mesothelial cells have the capacity to accumulate hemosiderin within their cytoplasm (fig. 1-8).

There is also increasing evidence that mesothelial cells show a remarkable degree of plasticity. They have the potential to generate other mesenchymal-derived cell lines (25), behave as vascular progenitor (so-called MesoT) cells (26), and regulate homeostasis (27).

REFERENCES

1. Joseph J, Sahn SA. Thoracic endometriosis syndrome: new observations from an analysis of 110 cases. Am J Med 1996;100:164-70.
2. Bender MD, Ockner RK. Diseases of the peritoneum, mesentery and diaphragm. In: Seidsinger MH, Fordtran JS, eds. Gastrointestinal disease: pathophysiology, diagnosis and management, 2nd ed. Philadelphia: Saunders; 1978:1947-8.
3. Fourdrain A, Lafitte S, Iquille J, et al. Lymphatic drainage of lung segments in the visceral pleura: a cadaveric study. Surg Radiol Anat 2018;40:15-9.
4. Flessner MF. Peritoneal transport physiology: insights from basic research. J Am Soc Nephrol 1991;2:122-35.
5. Haagensen CD. The lymphatics in cancer. Philadelphia: Saunders; 1972.
6. Standring S, ed. Gray's anatomy: the anatomical basis of clinical practice, 41st ed. New York: Elsevier; 2016.
7. Parungo CP, Soybel DI, Colson YL, et al. Lymphatic drainage of the peritoneal space: a pattern dependent on bowel lymphatics. Ann Surg Oncol 2007;14:286-98.
8. Abernethy NJ, Chin W, Hay JB, Rodela H, Oreopoulos D, Johnston MG. Lymphatic drainage of the peritoneal cavity in sheep. Am J Physiol 1991;260(Pt 2):F353-8.
9. Baum GL, Crapo JD. Baum's textbook of pulmonary diseases, 7th ed. Philadelphia: Lippincott Williams & Wilkins; 2004.
10. Wang NS. The preformed stomas connecting the pleural cavity and the lymphatics in the parietal pleura. Am Rev Respir Dis 1975;111:12-20.
11. Wang NS. The regional difference of pleural mesothelial cells in rabbits. Am Rev Respir Dis 1974;110:623-33.
12. Cotran RS, Karnovsky MJ. Ultrastructural studies on the permeability of the mesothelium to horseradish peroxidase. J Cell Biol 1968;37:123-37.
13. Devuyst O, Rippe B. Water transport across the peritoneal membrane. Kidney Int 2014;85:750-8.
14. Devuyst O, Margetts PJ, Topley N. The pathophysiology of the peritoneal membrane. J Am Soc Nephrol 2010;21:1077-85.
15. Li J. Ultrastructural study on the pleural stomata in human. Funct Dev Morphol 1993;3:277-80.
16. Andrews PM, Porter KR. The ultrastructural morphology and possible functional significance of mesothelial microvilli. Anat Rec 1973;177:409-26.
17. Moore KL, Persaud TV, Torchia MG. The developing human: clinically orientated embryology, 10th ed. Philadelphia: Elsevier; 2016.
18. Sadler TW. Langman's medical embryology, 12th ed. Philadelphia: Wolters Kluwer Health/Lipincott, Wilkins & Williams; 2012.
19. Mills SE, ed. Histology for pathologists, 5th ed. Philadelphia: Wolters Kluwer; 2020.
20. Pistolesi M, Miniati M, Giuntini C. Pleural liquid and solute exchange. Am Rev Respir Dis 1989; 140:825-47.
21. Choe N, Tanaka S, Xia W, Hemenway DR, Roggli VL, Kagan E. Pleural macrophage recruitment and activation in asbestos-induced pleural injury. Environ Health Perspect 1997;105(Suppl 5): 1257-60.
22. Black LF. The pleural space and pleural fluid. Mayo Clin Proc 1972;47:493-506.
23. Albertine KH, Wiener-Kronish JP, Staub NC. The structure of the parietal pleura and its relationship to pleural liquid dynamics in sheep. Anat Rec 1984;208:401-9.
24. Mitchev K, Dumortier P, De Vuyst P. 'Black Spots' and hyaline pleural plaques on the parietal pleura of 150 urban necropsy cases. Am J Surg Pathol 2002;26:1198-206.
25. Gotloib L, Gotloib LC, Khrizman V. The use of peritoneal mesothelium as a potential source of adult stem cells. Int J Artif Organs 2007;30:501-12.
26. Colunga T, Hayworth M, Kress S, et al. Human pluripotent stem cell-derived multipotent vascular progenitors of the mesothelium lineage have utility in tissue engineering and repair. Cell Rep 2019;26:2566-79.
27. do Amaral R, Arcanjo KD, El-Cheikh MC, de Oliveira FL. The peritoneum: health, disease, and perspectives regarding tissue engineering and cell therapies. Cells Tissues Organs 2017;204:211-7.

2 CLASSIFICATION OF TUMORS OF THE SEROSAL MEMBRANES

There is no ideal classification system for tumors arising in the serosal membranes (1). The histogenesis of certain lesions remains obscure (e.g., solitary fibrous tumor), serosal membranes are affected by diseases with varying frequencies reflected by the organs which they abut, and some tumors have the same histogenesis but are primarily classified on the basis of being localized rather than diffuse (e.g., localized versus diffuse mesothelioma). Furthermore, mesothelial cells show remarkable morphologic plasticity, having the capacity to be both epithelioid (rounded) and sarcomatoid (spindle shaped), as well as undergoing differentiation to other mesenchymal elements (e.g., osteosarcomatous differentiation within mesothelioma). For these reasons, the previous edition of this Fascicle primarily classified tumors first into diffuse and localized forms and secondarily into benign and malignant variants (1). Since then, the World Health Organization (WHO) has published two further classification systems for mesothelial tumors arising in the pleura, the most recent taking into account the recognition of mesothelioma in situ (Table 2-1) (2,3).

This book uses the terminology recommended by the WHO in its most recent edition for pleural and pericardial tumors, although expanded to reflect serosal membranes at other sites. Mesenchymal, hematolymphoid, and secondary tumors are discussed in separate chapters; also described are both non-neoplastic lesions and lesions of uncertain histogenesis (see chapter 8), since these commonly are in the differential diagnosis of neoplastic disease. Cytologic findings are presented in relation to individual entities, where appropriate.

Table 2-1

WORLD HEALTH ORGANIZATION (WHO) CLASSIFICATION OF MESOTHELIAL TUMORS OF THE PLEURA AND PERICARDIUM, WITH SUBTYPES[a]

Tumors	Morphology Code[b]
Benign and preinvasive mesothelial tumors	
Well-differentiated papillary mesothelial tumor	9052/1
Adenomatoid tumor	9054/0
Mesothelioma in situ	9050/2
Mesothelioma	
Diffuse mesothelioma	9050/3
Epithelioid mesothelioma	9052/3
Sarcomatoid mesothelioma	9051/3
Biphasic mesothelioma	9053/3
Localized mesothelioma	9050/3
Epithelioid mesothelioma	9052/3
Sarcomatoid mesothelioma	9051/3
Biphasic mesothelioma	9053/3

[a]WHO Classification of Tumours Editorial Board. Thoracic tumours, 5th ed. Lyon: IARC; 2021.

[b]Morphology code of the International Classification of Diseases for Oncology (ICD-O) (6) and the Systematized Nomenclature of Medicine (http://snomed.org). Behavior is coded /0 for benign tumors, /3 for malignant tumors, and /1 for borderline or uncertain behavior.

REFERENCES

1. Churg A, Cagle PT, Roggli VL. Tumors of the serosal membranes. AFIP Atlas of Tumor Pathology, 4th Series, Fascicle 3. Washington, DC: American Registry of Pathology; 2006.
2. Travis WD, Brambilla E, Burke AP, Marx A, Nicholson AG, eds. WHO classification of tumours of the lung, pleura, thymus and heart, 4th ed. Lyons: IARC; 2015.
3. WHO Classification of Tumours Editorial Board. Thoracic tumours, 5th ed. Lyon: IARC; 2021.

3 DIFFUSE MESOTHELIOMA

Diffuse mesothelioma is a malignant tumor originating from mesothelial cells that diffusely infiltrates and covers the serosal surfaces. Diffuse mesothelioma should be differentiated from localized mesothelioma (see chapter 4), which has a different clinical behavior (1). Tumors are either purely epithelioid, purely sarcomatoid, or biphasic. The epithelioid cells have rounded or epithelial morphology and the sarcomatoid cells have a mesenchymal or spindle cell morphology. Tumors are classified as biphasic when both cell types are present. The term *diffuse malignant mesothelioma* in relation to any of the major subtypes is acceptable. Epithelial should not be used instead of epithelioid, sarcomatous should not be used instead of sarcomatoid, and the terms mixed mesothelioma, mixed epithelioid and sarcomatoid mesothelioma, and mixed epithelial and sarcomatous mesothelioma should be avoided for biphasic mesothelioma.

GENERAL FEATURES

Incidence

The incidence of diffuse mesothelioma is low compared to solid tumors such as lung carcinoma. However, rates of occurrence vary regionally based on age, gender, and occupational exposure. World Health Organization (WHO) data spanning the period 1994 to 2008 (2) show an overall rate of 6.2 per million, with an age-adjusted rate of 4.9 per million and a 3.6 to 1.0 ratio of men to women. Data from 2018 reported just over 30,000 new cases and nearly 26,000 mesothelioma deaths worldwide, with an incidence of disease higher in Western countries (3). In a recent review of mesothelioma data in the United States, the rates of disease were decreasing (4). However, age subgroups, such as patients older than 65 and older than 85, had increasing rates across all races studied. Several other studies support increasing rates among older patients (over 65 and over 85) but decreasing rates in younger patients, likely reflecting an exposure history preceding the curtailment of widespread industrial asbestos use (4–6).

Epidemiology

Global Variations. Rates of mesothelioma vary internationally (Table 3-1). Among countries with high incidence rates, such as Turkey, the male to female ratio is only 1.36 to 1.0, with an average age of 61.7, for a rate of 23 per million. A projected decrease in the next 10 years is hoped for, based on migration patterns away from regions with high asbestos and erionite exposure (7). In Belgium, high use of asbestos resulted in a rate of mesothelioma in men of 29 per million, with a rate much higher in men than women (8).

Mesothelioma rates also vary over time, as reductions in asbestos exposure have an effect: for example, in New South Wales, it is estimated that the maximum rate was reached by 2010 (9); in other countries, the maximum rate has yet to be reached. In Brazil (10), a continued increase in disease rates reflect a lag in the peak use of asbestos. In Ireland, the rate of mesothelioma, while low at 5 per million, increased from 1994 to 2009. In China (11), the rate of mesothelioma is 15 per million, with a male to female ratio of 1.26 to 1.0, but the per annum rate continues to increase by 2.5 percent. The type of fiber involved may also be different between countries: for example, in the United Kingdom, amosite fibers are the most commonly found (12).

Occupation. Among the occupations with the highest disease association and mortality are insulation workers and ship builders; lower but significant rates are seen in pipe fitters and layers, plumbers, chemical technicians and engineers, metal workers, electricians, sailors, US navy veterans, and as an industry, construction (12,13). These rates also vary with time as occupational levels change within countries (14).

Table 3-1

RATES OF MESOTHELIOMA (PER MILLION), AGE ADJUSTED AND BY GENDER[a]

Country	Male Rate	Male ASR[b]	Female Rate	Female ASR[b]
USA	16	9	5	3
United Kingdom	70	34	14	6
Australia	52	29	11	6
Netherlands	55	28	8	4
Belgium	40	20	9	4
Germany (N. Rhine)	38	16	8	3
Ukraine	7	5	5	3
Canada (Ontario)	21	12	5	2
Japan (Osaka)	27	11	7	2
Denmark	36	19	7	3
Austria	20	11	6	3
Norway	28	15	5	3
Rep. Korea	3	2	2	1
Italy (Milan)	48	21	24	7
Croatia	26	15	6	3
Ireland	15	10	2	2
Israel	9	7	3	2
France (Loire)	45	24	11	4
Switzerland (Zurich)	41	22	8	4
China (Beijing)	6	3	4	2
Turkey (Izmir)	9	8	7	6
Turkey (Eskisehir)	38	32	38	29
Italy (Reggio Emilia)	56	26	14	5
Uruguay	5	3	2	1

[a]Data from references 14a and 14b.
[b]ASR = Age standardized rate.

Age. Diffuse pleural mesothelioma generally presents in patients over 50 years of age, and is more common in men. Patients presenting with mesothelioma at 40 years of age or younger have nearly equivalent rates of pleural and peritoneal disease as well as an even gender ratio (15). These patients have a tendency toward better survival (see Prognosis).

Mesothelioma Arising in Women. Most data report on men with mesothelioma but there are recent reports on the epidemiology of mesothelioma in women. In Italy, there is a gender ratio (F/M) of 0.38 and 0.70 for pleural and peritoneal cases, respectively. Most cases in women are related to para-occupational exposure (16), but occupational exposures occur in the chemical and plastic industry, and mainly in the nonasbestos textile sector. Gender ratio is inversely correlated with the mortality rate (17).

Genetic Predisposition. In 2010, somatic mutations in BRCA1-associated protein 1 gene (*BAP1*) were found to be associated with metastatic uveal melanoma (18), with one mutation found to be in the germline. Soon after, a familial cancer syndrome was identified (19), and mesothelioma was identified as a tumor type within that syndrome (20). Further study of these patients has suggested onset at an earlier age (21,22), more frequent peritoneal origin, and improved survival. Median ages of onset are lower in null than missense variants (22). Patients with *BAP1* germline mutations have lower asbestos exposure than sporadic mesothelioma control patients (23). Patients with lower asbestos exposure have an increased number of pathogenic truncating mutations in genes related to DNA repair pathways, raising the possibility of other contributing factors (24). Familial cases exist that are independent of *BAP1* mutations, but predisposition needs to be weighed in conjunction with potential common exposures (25). A *CDKN2A* germline mutation has also been associated with familial mesothelioma and melanoma (26). Mesotheliomas are reported in patients with neurofibromatosis type 2 (NF2) (27) and Lynch syndrome (28), although it is likely that exposure to asbestos or additional mutational steps are required for tumor development.

Other genetic factors may also influence susceptibility to the development of mesothelioma. For example, familial clusters have been described (29). It has been postulated that genetic effects may exert their influence, in part, from increased sensitivity to the oncogenic effects of asbestos, but it is difficult to eliminate the effect of common occupational and environmental exposures in these clusters. For example, when environmentally exposed to erionite, mesothelioma was more prevalent in certain families compared to others (30). Also, one report of mesotheliomas in Italy determined possible single nucleotide polymorphisms associated with asbestos susceptibility (31). Nevertheless, the impact on latency and disease incidence as well the contribution of dose and fiber type on potential subgroups of susceptible individuals remain to be determined.

Etiology

Asbestos Exposure. The major risk factor for the development of mesothelioma is asbestos exposure, and historically this has been occupational exposure. The attributable risk of pleural mesothelioma is considered to be high in men (88 percent), with a rate of 58 percent for peritoneal disease. In women, the attributable risk is lower, at 23 percent for both pleural and peritoneal disease (32).

Asbestos comprises six naturally occurring silicate minerals that have characteristic physical qualities: long and thin fibrous crystals (Table 3-2). Initially, the minerals were known by their colors (blue, brown, white, green), but are now divided into two classes, serpentine (curly) and amphibole (needle-like). The serpentine class consists of the chrysotile type; the amphibole class contains five types: amosite, crocidolite, tremolite, anthophyllite, and actinolite. Minerals such as richteriite and winchite, which have similar qualities, are classified as asbestiform fibers.

Table 3-2

ASBESTOS TYPES AND MINERALS SIMILAR TO ASBESTOS[a]

Serpentine	
Chrysotile	Antigonite[a]
Amphiboles	
Amosite	Winchite
Actinolite	Richterite
Anthophyllite	Fluoro-edenite[a]
Crocidolite	
Tremolite	
Zeolite	
Erionite	

[a]Data from references 32a, 32b, 64, and 68.

The identification of mesothelioma risk among asbestos-exposed individuals is the result of a landmark study from the North Western Cape province of South Africa (33). In this study, many cases were occupational, but some patients were just residents of the region. Subsequent reports from Queen's University of Belfast (34) and London Hospital (35) supported the association with asbestos. In the United States, a study of insulation workers showed an increased risk of mesothelioma in occupations handling asbestos (36). In these series, both pleural and peritoneal diseases were reported. In a Canadian series, McDonald (37) also established that the risk of asbestos exposure and mesothelioma was mainly secondary to occupational use. There was also a lack of smoking association. Amosite, another amphibole fiber, was implicated soon after (38).

Fiber Type and Mesothelioma. There is continued debate over fiber type exposure and mesothelioma. In a recent review of 665 cases (39), the relative risk of mesothelioma after amphibole asbestos exposure was statistically significant, x21, while the relative risk with chrysotile asbestos, x3.8, was not statistically significant.

Fiber length and the impact on biological clearance appear to be critical in the effect of exposure on subsequent mesothelioma risk.

Chrysotile asbestos is more rapidly cleared, in as little as weeks to months (40,41), while amphiboles, such as crocidolite, can take nearly a decade to clear (42,43). Studies from Canada, the United States, and South Africa, as well as studies performed in the automotive mechanic industry (44,45), do not support chrysotile exposure as causative of mesothelioma (46,47). However, it is acknowledged that heavy exposure to chrysotile may be associated with an increased risk (48). It is also of concern that amphibole contamination of chrysotile asbestos is difficult to eliminate, especially in the context of heavy exposure. The importance of amphibole contamination for increasing the risk of mesothelioma is shown in vermiculite from Libby, Montana, which had winchite and richterite (amphibole fibers) contamination (49).

Key studies have focused on the rates of mesothelioma from nearly 11,000 asbestos mine and mill workers in Canada from the towns of Asbestos and Thetford. The degree of amphibole exposure differed among these workers, but importantly, tremolite was found in the lungs of workers from Thetford and crocidolite from workers in the town of Asbestos (50), indicating that amphibole contamination of chrysotile was important in the development of mesothelioma. Deaths from mesothelioma were associated with tremolite exposure (51): overall mesothelioma risk from chrysotile exposure was very low, unless tremolite contaminated (52). Studies among automotive mechanics and brake lining workers (53), as well as studies of environmental chrysotile exposure (54), show a low to absent risk of mesothelioma. Indeed, frequent amphibole contamination in high-dose chrysotile exposure may represent the major public health issue in the use of chrysotile asbestos and mesothelioma risk (55).

Latency. The latency of exposure to the development of mesothelioma is prolonged, from 20 to 40 years (56), and may be independent of dose (57,58). Population data indicate that the rate of mesothelioma increased from 1999 to 2015, but the age-adjusted rate decreased, with the largest increase in patients 85 years or older (13). There may be a plateau at 40 years, after which the incidence no longer increases (59).

While environmental and household exposures in children may account for mesothelioma in adulthood, there is limited data. Exposure to blue asbestos in childhood was associated with elevated mesotheliomas in adults in a study from Western Australia (60) and a Danish study suggested that environmental exposure during childhood significantly increased the risk of mesothelioma in later life (61).

Environmental Exposure. While occupational exposure as causation for mesothelioma is well established, environmental exposure has been a more controversial area, despite being suggested in the original South African study (33). Living near contaminated sites and household exposure have both been suggested as risk factors (62,63). The International Association for the Study of Lung Cancer (IASLC) and the International Mesothelioma Interest Group (IMIG) have suggested that criteria for environmental exposure in a population include: equal male to female ratio, equal rate of peritoneal and pleural disease, and age under 55 years (64–67). Some amphibole fibers encountered in the environment (winchite, actinolite, richterite) and erionite may explain increased numbers of cases among these groups. However, some of these fiber types are either not classified or are nonasbestos fibers. In one series, living within 2000 meters of asbestos-related industries accounted for a statistically significant increased RR of mesothelioma of 11.5. While occupational exposure remains the most important cause of mesothelioma, environmental exposure may contribute to 5 to 10 percent of cases, depending on region and source of exposure (67).

Other Causes. Zeolite fibers, such as erionite, are strongly associated with the development of mesothelioma, particularly in the Cappadocia region of Turkey (68), as well as in focal areas of Central Mexico (69). The risk of mesothelioma appears to be independent of cigarette smoking (70). No increase in risk was found in talc millers and miners or in patients having talc pleurodesis (71).

Postradiation. Studies of radiation exposure and mesothelioma shows patients are younger and have improved survival. Latency is also seen, with an average 20- to 25-year interval between radiation and mesothelioma, with a wide range (72–75). This includes occupational exposure incurred in the mining of uranium (75a) as well as radiation administered for medical reasons. In larger series of patients treated with radiation for breast cancer (76,77), prostate cancer (78), non-Hodgkin lymphoma (79), and Hodgkin disease (80), mesothelioma risk was increased, with a notable 20-fold increase in risk after treatment for Hodgkin disease (81).

Incidence and Etiology at Sites Other than the Pleura

Peritoneum. The rate of peritoneal mesothelioma is low, with estimated rates from 0.5 to 3.0 per million in men and 0.2 to 2.0 per million in women (82,83). No relation to chrysotile asbestos exposure has been reported. Somewhat paradoxically, while peritoneal mesothelioma has a lower attribution to asbestos exposure, asbestos-associated cases have among the highest exposure loads (84,85). Attribution to non-asbestos causes is supported by a greater frequency in women, even in areas where pleural mesothelioma has a high male to female ratio (9,17,86). Either *BAP1* mutation or allelic loss is seen in peritoneal mesotheliomas, with 57 percent of tumors showing complete loss (87).

Pericardium and Tunica Vaginalis. The rate of pericardial mesothelioma is very low, estimated at 0.08 to 0.1 per million (Surveillance, Epidemiology, and End Results [SEER] data). There is a male predominance, with an average age of 47 years, with pediatric cases also reported. Although the number of cases is small, no asbestos association was seen in pericardial or tunica vaginalis mesothelioma when SEER regions with high rates of pleural mesothelioma were studied (88–90). More recent studies, however, report a stronger association between asbestos exposure and tunica vaginalis mesothelioma (91,92).

CLINICAL FINDINGS

Pleura

The median age at presentation of diffuse pleural mesothelioma is around 60 years, although the rate is increasing in older individuals, particularly among populations exposed to amphibole asbestos. In this group, it is more common in men. Thoracic mesotheliomas occasionally occur in younger patients, including children and teenagers, and these tumors are histologically similar to those of adults. The gender ratio is closer to 1 to 1 in younger patients and prognosis may be better in this group (15).

The clinical presentation often reflects symptoms related to pleural effusions (93), which include dyspnea, cough, and chest pain. Fatigue, night sweats, fever, and weight loss also occur. Dull chest pain, associated with pleural effusion, can become severe localized chest pain when chest wall invasion occurs. Decreased breath sounds on physical examination and, eventually, palpable nodules are potential physical examination findings. Rare presentations include pneumothorax (94,95), lymphadenopathy (96), and diffuse pulmonary spread mimicking interstitial lung disease or military tuberculosis (97).

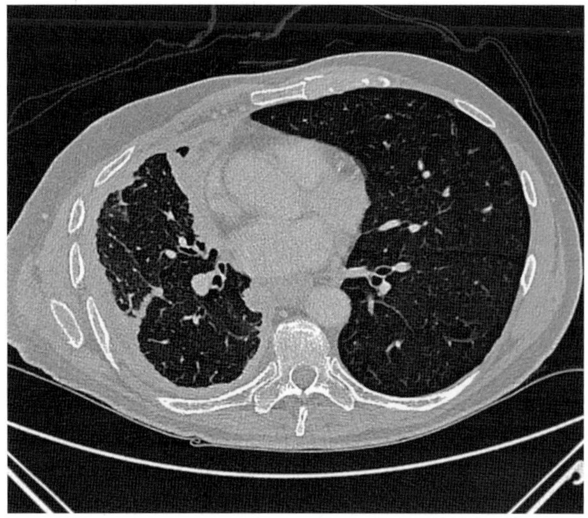

Figure 3-1

DIFFUSE MESOTHELIOMA: PLEURA

Computerized tomography (CT) shows right-sided diffuse pleural thickening with focal invasion into the lung and loss of lung volume.

Pleural effusion is common. Chest X ray (98) may reveal the effusion; pleural plaques and lung fibrosis (due to asbestosis) may be seen. Computerized tomography (CT) remains the standard of care imaging modality and readily demonstrates the diffuse, usually unilateral, pleural thickening and loss of lung volume (fig. 3-1). In some cases, mass lesions predominate, mimicking lung carcinoma (fig. 3-2). Magnetic resonance imaging (MRI) and positron emission tomography (PET) imaging may aid in diagnosis, staging, and prognostication (99–101). There have been recent recommendations for the standardization of reporting and integration into a multidisciplinary approach (99,102). Evaluation of therapeutic response remains a challenge, but computerized volume estimates

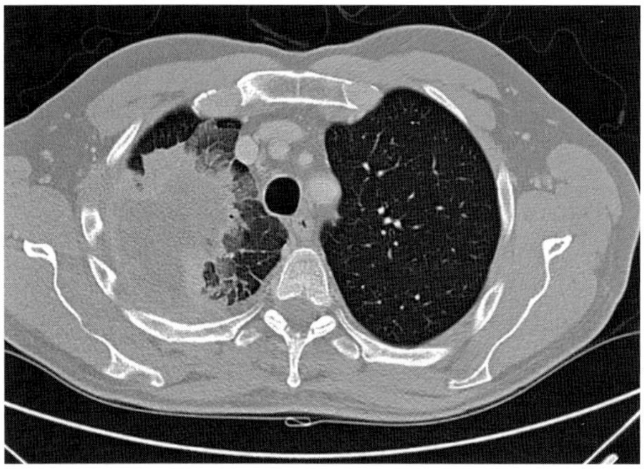

Figure 3-2

DIFFUSE MESOTHELIOMA: PLEURA

CT from a patient with sarcomatoid mesothelioma shows a right-sided mass that involves the lung, pleura, and chest wall.

may provide clinically relevant response and prognostic information (103,104).

Paraneoplastic syndromes are uncommon, but anti-Ma (neuronal nuclear) opsoclonus (105) and anti-Yo cerebellar degeneration have been described (106). Rare reports of membranous nephropathy (107), syndrome of inappropriate antidiuretic hormone (108), and thrombocythemia/thrombosis (109) have also been reported.

Peritoneum

Peritoneal mesothelioma is uncommon, but like its pleural counterpart, can be a treatment refractory and fatal disease (110). Median age of presentation is younger than for thoracic disease and a larger proportion of patients are women (86,111). Its presentation is variable, including patients with ascites ("wet type") or multiple nodules without ascites ("dry type"). Because of the abdominal location, intestinal obstruction can be a presentation of disease. Presentation within inguinal or umbilical hernias has also been reported (112). Occasionally, disease extends out of the abdomen into the chest, resulting in bi-cavity disease.

Pericardium and Tunica Vaginalis

Pericardial mesothelioma is rare (90). Patients present with dyspnea, effusion (113), tamponade, constrictive pericarditis (114), myocardial invasion (115), arrythmia (116), or atrial thrombus (117).

Mesotheliomas of the tunica vaginalis present with scrotal enlargement due to a mass or hydrocele; a testicular tumor is often the presumed diagnosis. Patients may also present with inguinal hernia (91,118–120).

DIAGNOSTIC SPECIMENS

The specific pathologic features present in a given specimen depend on the type and size of the specimen being examined. Cases undergoing maximal cytoreductive surgery (extrapleural pneumonectomy [EPP] or extended pleural decortication [EPD]) (121) should be distinguished from smaller samples, with the former being pathologically staged and the latter clinically staged (99). There is little evidence on what is the optimal number of biopsy samples required to ensure that there is adequate representation of the tumor, although expert consensus is that ideally at least three sites should be sampled, when feasible, including any identified as of interest on imaging, such as areas with high FDG uptake. This is because areas of invasion may be focal and morphology may vary considerably (99,122). Standardization of sampling will hopefully lead to more robust reporting, especially for clinical trials (123), and concordance between biopsies and resections has already been shown to be better when there are three compared to fewer biopsies, and also to be better for sarcomatoid versus epithelioid subtypes (124). Indeed, the more samples taken, the higher the rate of biphasic tumor diagnosis (125–127).

Pleura

In the pleura, diagnostic specimens range from effusions (see Cytologic Findings) and needle core biopsies, to thoracoscopic/surgical biopsy, to pleurectomy to extrapleural pneumonectomy, and finally to postmortem examination. Biopsies are mainly undertaken either via percutaneous biopsy, video-assisted thoracoscopy (VATS), or open surgical biopsy; the last two have the advantage of visualization of the pleural cavity. Thoracoscopy may be undertaken under local anesthetic (128,129). Unsurprisingly, VATS/open biopsies have a higher diagnostic rate than needle biopsies (90 to 100 percent), with a CT-guided biopsy needle having greater sensitivity (80 to 90 percent) than to a closed biopsy (about 50 percent) (129–134).

Taking full-thickness samples also helps in diagnosis as invasion of subpleural fat is a strong diagnostic feature for malignancy. If samples are taken from the visceral pleura and include lung, then elastin van Gieson staining is very useful in helping assess whether mesothelial cells have infiltrated the peripheral lung. If pleurodesis is being considered, then surgeons may send a frozen section to confirm malignancy, although additional samples should be sent to ensure tissue is fixed and processed to obtain optimal morphology. It is also worth orientating full-thickness samples so that sections can be undertaken perpendicular to the surface since this aids in the assessment of spindle cell proliferations. With increasing interest in molecular testing, pathologists should consider blocking multiple biopsies into more than one cassette to ensure sufficient tissue for both diagnosis and any potential treatment-related testing. Collection of frozen tissue for research should also be considered (99).

For larger specimens, such as extended pleural decortications and extrapleural pneumonectomies, there should be a thorough sampling of tumor deposits. Anatomic structures relevant to T staging should be assessed to determine the extent of disease spread. If lung tissue is present, this should also be sampled and assessed for the presence of asbestos as this may have relevance for the patient in terms of any potential litigation.

Peritoneum, Pericardium, and Tunica Vaginalis

Peritoneal samples vary from ascites (see Cytologic Findings), to biopsies and debulking of tumors. Rare localized mesotheliomas also occur at this site (1) and these may be completely resected. Most pericardial cases are diagnosed on biopsy although a few localized cases are resected (90). Cases from the tunica vaginalis are typically diagnosed on biopsy and, if localized, undergo resection. Assessment of lymph nodes is important (91,119,120).

GROSS FINDINGS

Pleura

Most mesotheliomas are diffuse, a term that implies multifocal but usually unilateral disease, in contrast to localized mesotheliomas, which occur rarely (see chapter 4). Mesotheliomas of the pleura likely begin as surface proliferations. The earliest lesions of mesothelioma, now viewed as mesothelioma in situ, may not be apparent grossly (135–137), although some cases have pleural effusions. Most cases, however, will show some macroscopic abnormality. As the disease progresses, multiple small nodules are formed which typically involve several areas on one side of the chest (fig. 3-3). These then grow over the surface to form plaque-like tumors and masses (fig. 3-4), eventually encasing the whole lung and forming a rind that may be several centimeters thick (fig. 3-5). The tumor is typically white to gray, varies from gelatinous to fibrous, and may have grossly visible cystic areas filled with mucoid-like material. It frequently extends along the fissures between the lobes of the lungs (fig. 3-6) and grows into the lung along interlobular septa. Eventually, the tumor grows into the underlying tissues including the lung, diaphragm, mediastinum and its contents (fig. 3-7), and chest wall. It can also grow into the lung parenchyma and along airways; thus it may be seen on a transbronchial biopsy, where it needs to be distinguished from lung carcinoma.

A dominant mass, giving the impression of a localized tumor, needs to be distinguished from localized mesothelioma (138). Mesotheliomas have the propensity to grow along prior biopsy tracks, through the chest wall, and into the subcutis (fig. 3-8). They can also grow through the diaphragm and into the peritoneal cavity, directly invading mediastinal fat, pericardium, and even the contralateral lung and pleura. Lymph node metastases, involvement of the contralateral side (fig. 3-9), and extrathoracic metastases (fig. 3-10) occur late in the disease. An autopsy study of 318 patients with pleural mesothelioma showed extrapleural dissemination in 88 percent and lymph node metastases in 53 percent (139). Tumor dissemination in extrathoracic sites was seen in 55 percent, including involvement of liver (32 percent), spleen (11 percent), thyroid gland (6 percent), and brain (3 percent).

Peritoneum

Depending on the extent of disease at presentation, tumors typically show variably thick cake-like thickening of the peritoneal and omental surfaces (fig. 3-11). Necrosis and hemorrhage may be present (fig. 3-12). Tumors may also be myxoid (fig. 3-13). Early disease may be hard to

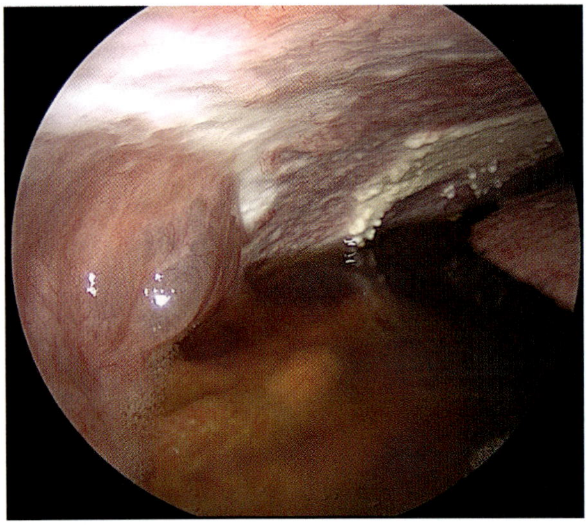

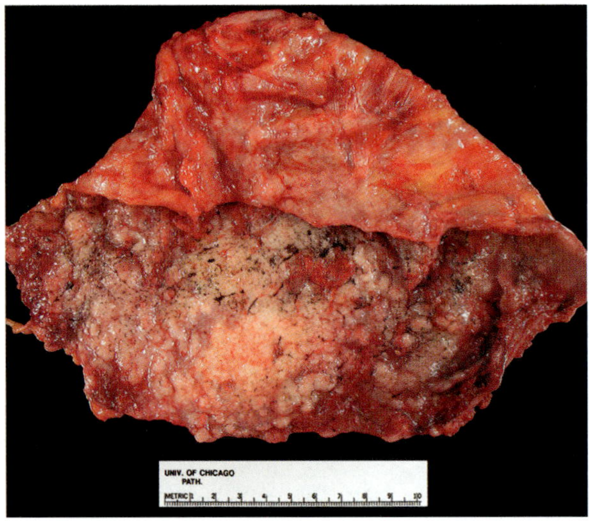

Figure 3-3

DIFFUSE MESOTHELIOMA: PLEURA

Left: Thoracoscopic appearance of diffuse mesothelioma involving the parietal pleura shows multiple yellow-white nodules as well as diffuse growth. Normal pleura is on the left. A large pleural effusion is present. (Courtesy of Dr. M. Ferguson, Chicago, IL)

Right: Pleurectomy specimen shows extensive tumor studding by small nodules.

visualize macroscopically and widespread sampling should be considered in suspicious cases.

Pericardium and Tunica Vaginalis

Most cases show diffuse involvement of the pericardium, although a few localized masses are described (90). Patients with tunica vaginalis involvement typically present with a scrotal mass, with or without a hydrocele. Disseminated disease at presentation is common (119).

MICROSCOPIC FINDINGS

Based on the microscopic appearance, mesotheliomas are classified into epithelioid, sarcomatoid, and biphasic subtypes (140,140a), classifications that are important for pathologic diagnosis, differential diagnosis, prognosis, and treatment. There is strong interobserver reproducibility for subtyping pleural mesotheliomas into these three types (141–143). Within each subtype, a variety of architectural patterns, cytologic features, and stromal features may be displayed (99). These need to be recognized by the pathologist in order to make a diagnosis and select appropriate ancillary tests for respective differential diagnoses; in some cases, subtype recognition contributes to prognosis and clinical behavior (Table 3-3). For EPD and EPP specimens (those that are pathologically staged), the percentage of each architectural pattern and prognostically relevant cytologic and stromal features should be documented. For smaller specimens (those that are clinically staged), the presence of patterns and features alone should be documented but not quantified (99). In many cases, the diagnosis of malignancy is obvious on hematoxylin and eosin (H&E)-stained slides, however, due to the morphologic overlap with other tumor types, the fact that serosal membranes are a common site for metastatic disease, and the medicolegal implications, a panel of immunohistochemical stains is routinely performed to make the diagnosis of mesothelioma.

The extent of mesothelioma infiltration depends on the stage of the disease at presentation, and ranges from mesothelioma in situ, through superficial infiltration of the fibrous pleural stroma (fig. 3-14), to involvement of the chest wall, including skeletal muscle (fig. 3-15). There is often a fibrotic/desmoplastic response, within which epithelioid tumor cells become entrapped and may resemble carcinomatous infiltration (fig. 3-16) or sarcomatoid mesothelioma (fig. 3-17). Tumor-infiltrating

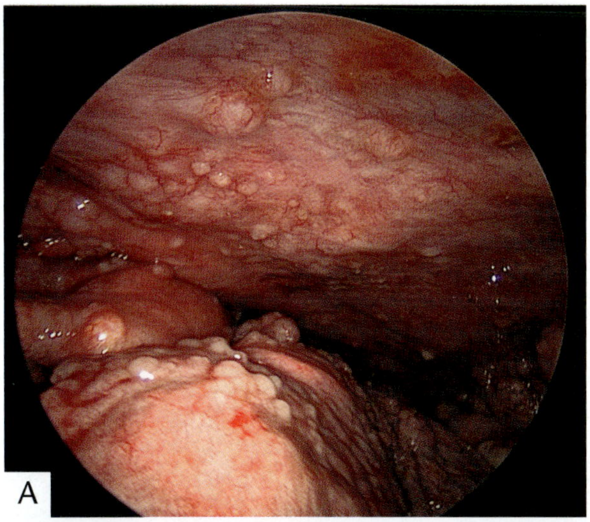

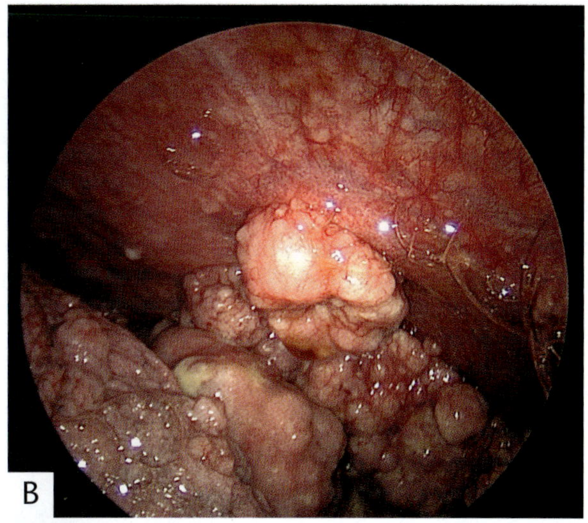

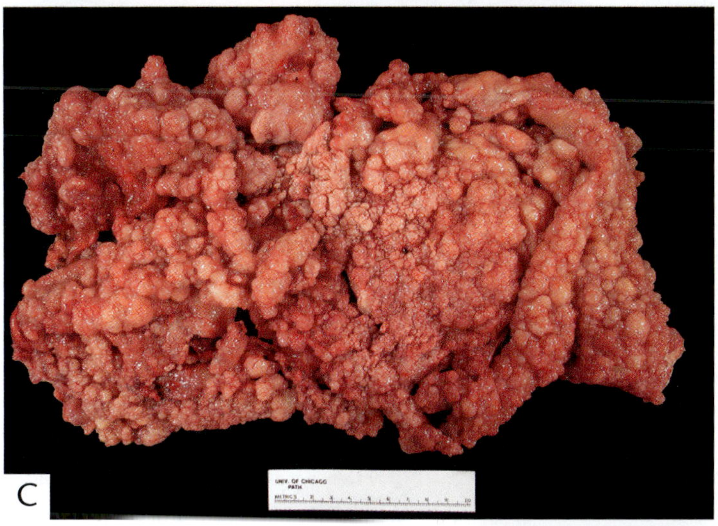

Figure 3-4

DIFFUSE MESOTHELIOMA: PLEURA

A: Thoracoscopic appearance of more extensive growth of diffuse mesothelioma involving parietal and visceral pleura. Multiple yellow-white nodules as well as plaque-like growth are seen. (Courtesy of Dr. M. Ferguson, Chicago, IL)

B: Thoracoscopic appearance of diffuse mesothelioma forming a large mass as well as multiple yellow-white nodules. (Courtesy of Dr. M. Ferguson, Chicago, IL)

C: Pleurectomy specimen shows extensive tumor studding by small nodules.

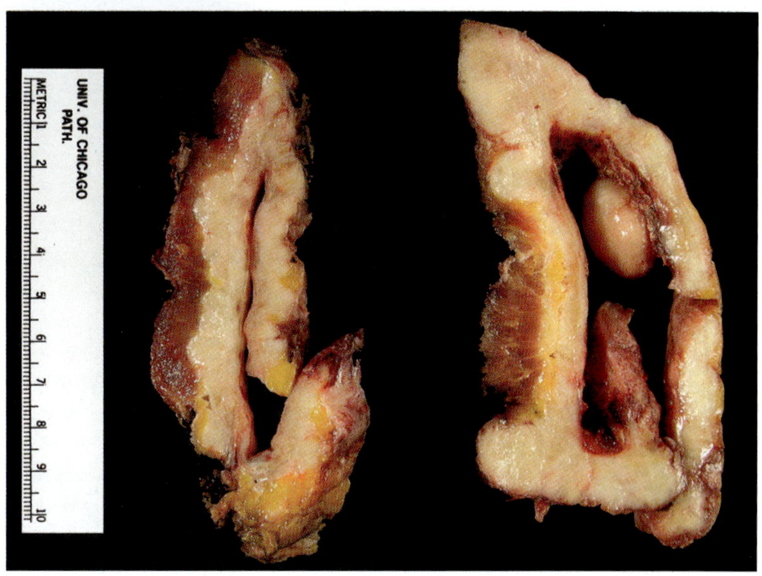

Figure 3-5

DIFFUSE MESOTHELIOMA: PLEURA

Cut surface of pleurectomy specimen shows tumor forming a rind. The tumor is gray-tan to white.

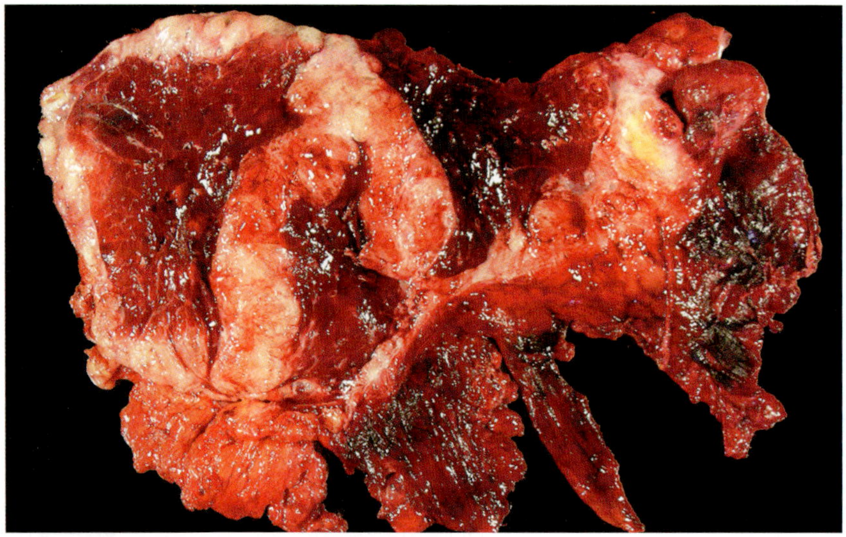

Figure 3-6

DIFFUSE MESOTHELIOMA: PLEURA

Extrapleural pneumonectomy specimen shows diffuse mesothelioma encasing the lung and growing into a major fissure.

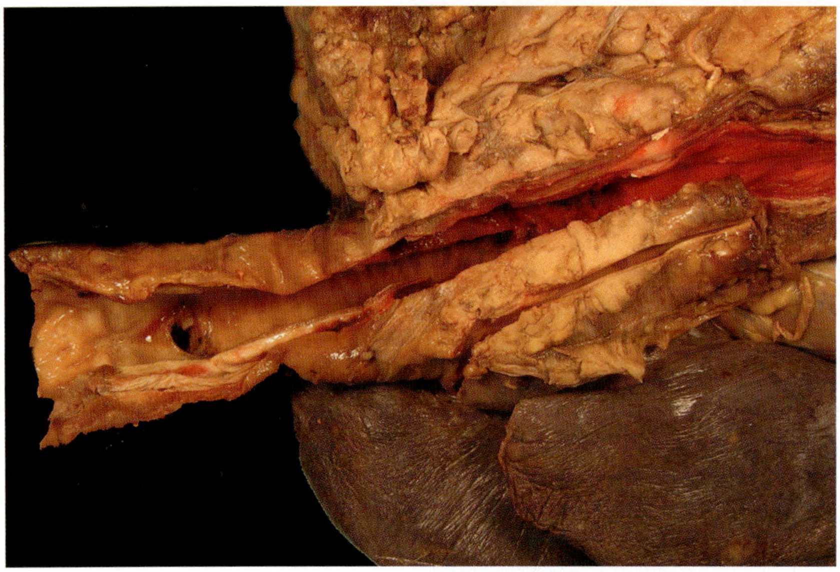

Figure 3-7

DIFFUSE MESOTHELIOMA: PLEURA

Postmortem specimen shows right-sided diffuse mesothelioma growing into the mediastinum and encasing the trachea, which has been opened posteriorly.

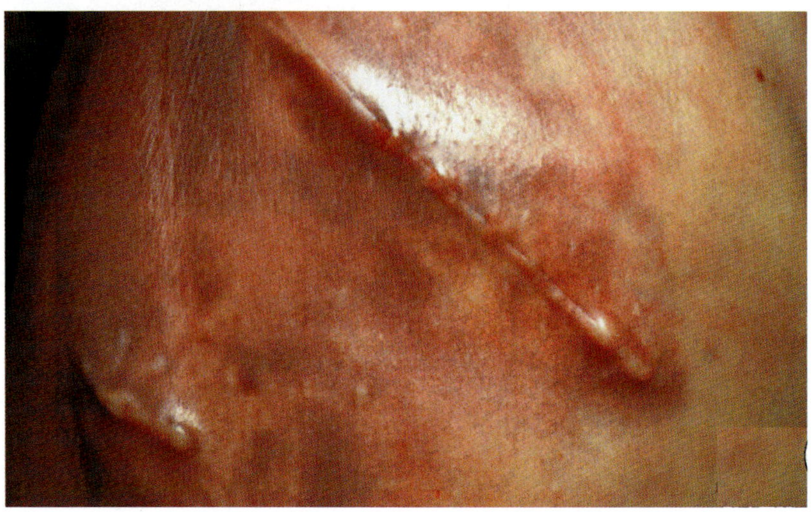

Figure 3-8

DIFFUSE MESOTHELIOMA: PLEURA

The tumor extends along a needle tract and also infiltrates a surgical scar. (Courtesy of Dr. D. Henderson, Adelaide, Australia)

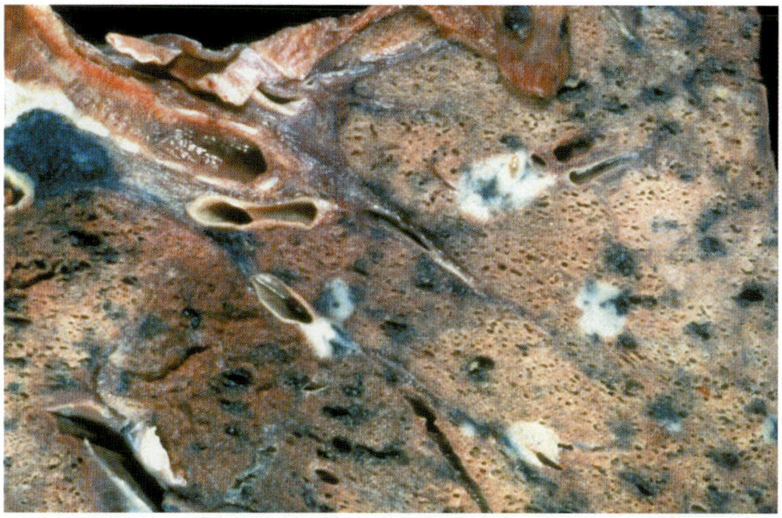

Figure 3-9

DIFFUSE MESOTHELIOMA: PLEURA

At autopsy, small tumor nodules are present in the contralateral lung, as well as diffuse pleural thickening.

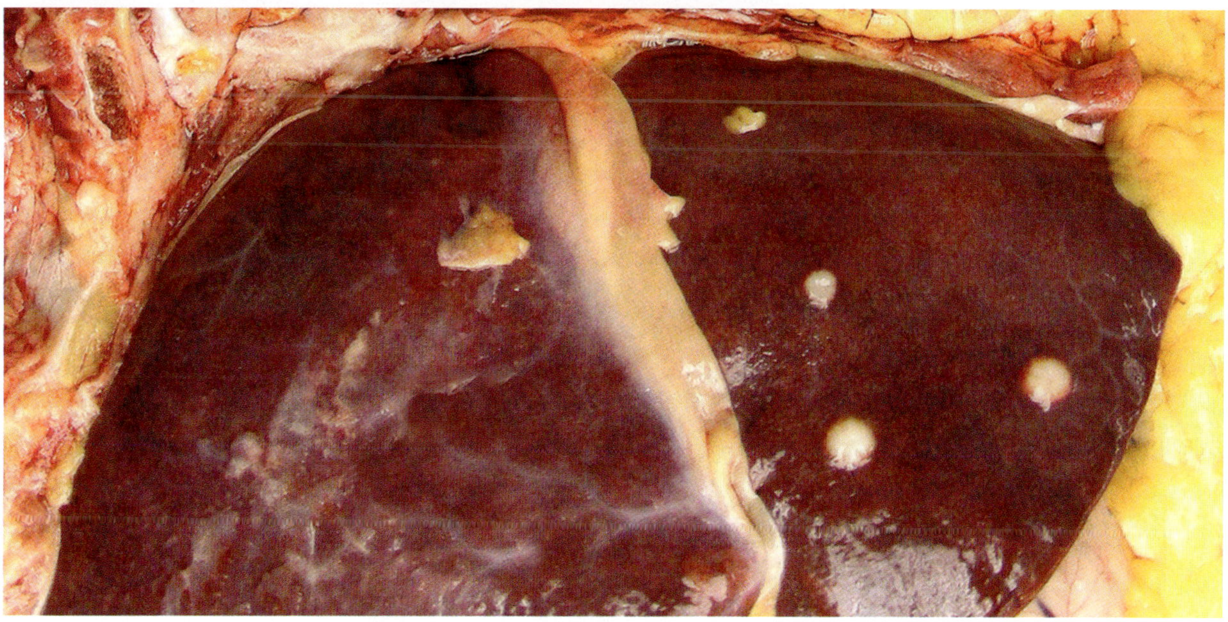

Figure 3-10

DIFFUSE MESOTHELIOMA: PLEURA

Metastatic tumor nodules are seen on the surface of the liver at postmortem examination.

inflammatory cells are the rule rather than the exception; most tumors have varying amounts of lymphocytes and macrophages within and adjacent to them, while others have more neutrophils (fig. 3-18).

Mesothelioma in Situ

The concept of *mesothelioma in situ* was first proposed based on examination of seven cases without grossly visible tumor in which there was a single layer or small papillary projections of cytologically atypical mesothelial cells on the pleural surface associated with microscopically invasive mesothelioma (144). This concept seemed logical since so many other tumors have an in situ phase. However, it could not be determined whether the surface cells really were mesothelioma in situ or represented surface

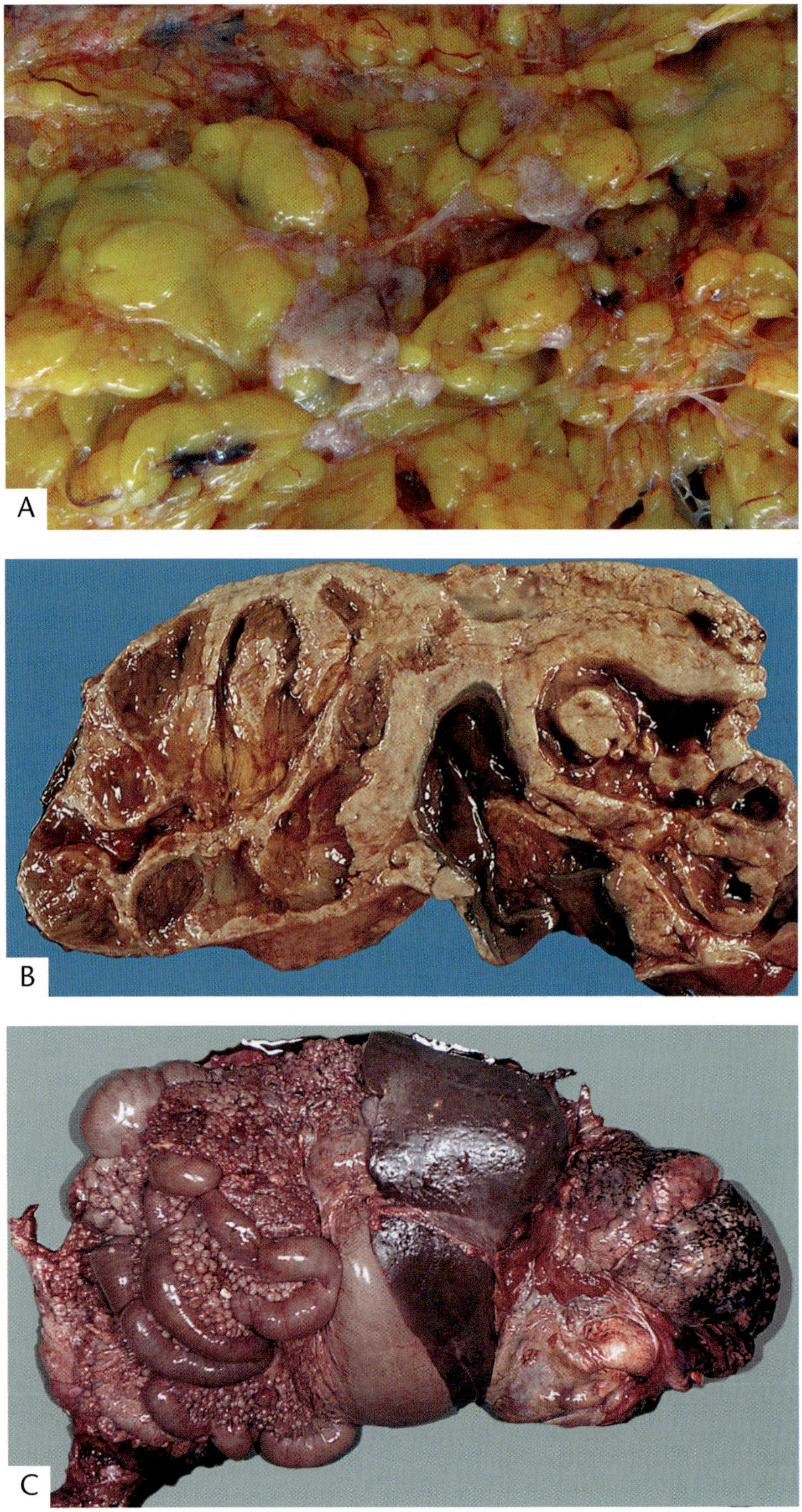

Figure 3-11

DIFFUSE MESOTHELIOMA: PERITONEUM

A: Mesenteric fat is covered by multiple foci of mesothelioma.

B: This diffuse tumor encases loops of bowel.

C: This tumor appears as multiple small nodules. Note the liver metastases.

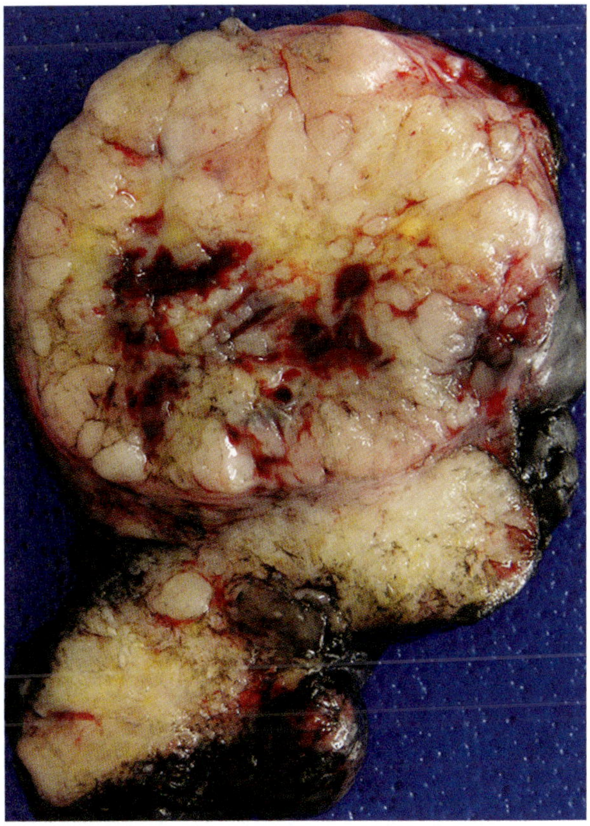

Figure 3-12

DIFFUSE MESOTHELIOMA: PERITONEUM

The tumor is nodular and hemorrhagic.

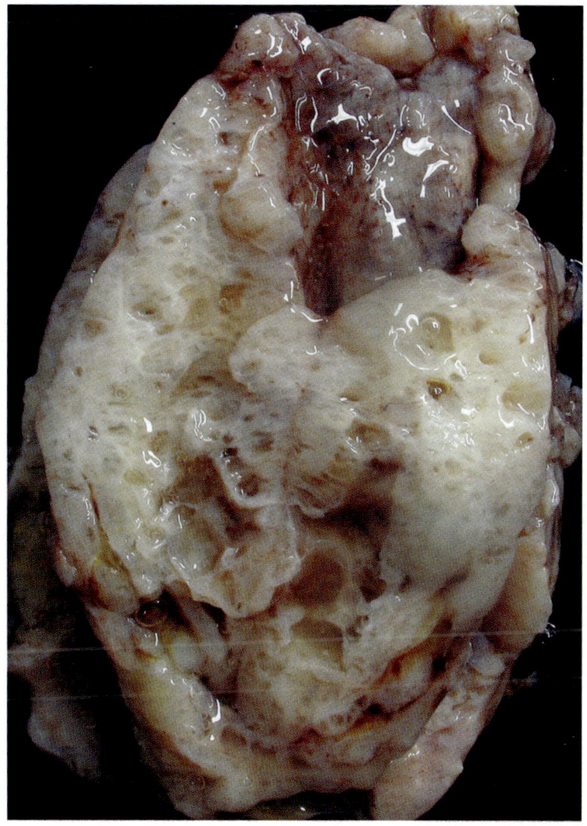

Figure 3-13

DIFFUSE MESOTHELIOMA: PERITONEUM

The tumor is strikingly myxoid.

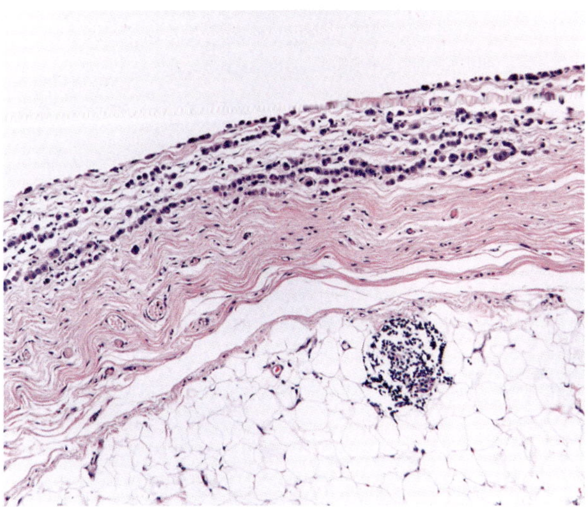

Figure 3-14

EPITHELIOID MESOTHELIOMA

Early infiltration under the surface is seen in this epithelioid mesothelioma.

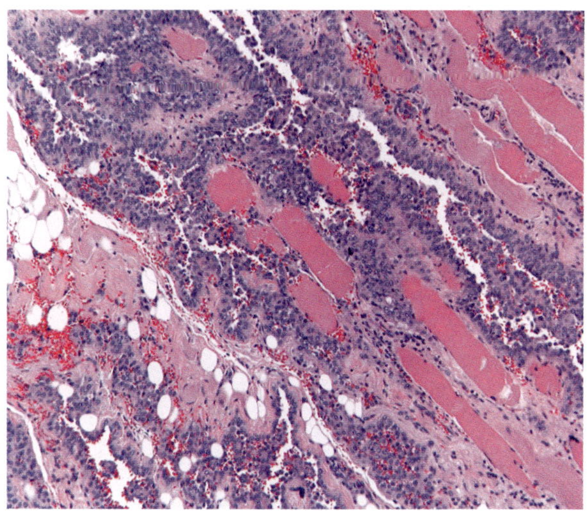

Figure 3-15

EPITHELIOID MESOTHELIOMA

Tumor infiltrates through fat into skeletal muscle.

Table 3-3

HISTOLOGIC PATTERNS AND FEATURES OF MESOTHELIOMA[a]

Epithelioid mesothelioma
- Architectural patterns (give percentages for EPD/EPP[b] and document patterns present for all other samples)
  - Tubulopapillary
  - Trabecular
  - Adenomatoid/microcystic
  - Solid
  - Micropapillary
- Cytologic features (give percentages for EPD/EPP. For all other samples, state "with...features present"
  - Pleomorphic[c,d]
  - Rhabdoid[c]
  - Lymphohistiocytoid[d,e]
  - Deciduoid
  - Small cell
  - Clear cell
  - Signet ring
- Stromal features (give percentages for EPD/EPP. For all other samples, state "with . . . features present"
  - Myxoid[e]
  - Desquamative interstitial pneumonia-like (DIP-like)

Sarcomatoid mesothelioma
- Cytologic features (give percentages for EPD/EPP. For all other samples, state "with . . . features present"
  - Transitional pattern[c,d]
  - Pleomorphic[c,d]
  - Lymphohistiocytoid[d,e]
- Stromal features (give percentages for EPD/EPP. For all other samples, state "with . . . features present"
  - Desmoplastic
  - With heterologous differentiation

Biphasic mesothelioma
- For EPD/EPP, any combination of patterns of epithelioid and sarcomatoid mesothelioma with at least 10% of each component. For all other samples, a diagnosis of "biphasic" can be made regardless of percentages of each component, but reposts should include a comment indicating the percentages of each component in the sample.

[a]Updated from reference 99. Some architectural patterns and cytologic and stromal features are important for prognosis while some are included only for clarity to avoid pathology misdiagnoses. When generating reports, please note that multiple architectural patterns and cytologic and stromal features may be present in a tumor and all patterns/features seen in a tumor should be included in the report.
[b]EPD = extended pleurectomy/decortication; EPP = extrapleural pneumonectomy.
[c]Cytologic and stromal features associated with a worse prognosis.
[d]Classification of transitional, pleomorphic, and lymphohistiocytoid cytologic features is problematic due to limited data. The current consensus is that pleomorphic and lymphohistiocytoid cytologic features can be seen in both epithelioid and sarcomatoid subtypes, but transitional cytologic features should be viewed solely as a cytologic variant of sarcomatoid subtype.
[e]Cytologic and stromal features associated with a better prognosis.

growth of an invasive mesothelioma. Since then, tests based on genomic analysis have become widely accepted and available which have greatly improved the ability to separate benign lesions from mesothelioma (145–153). Subsequently, mesothelioma in situ has been reported in cases with a mesothelial proliferation limited to the surface but with evidence of BAP1 loss on immunohistochemistry and/or homozygous deletion of *CDKN2A* (*p16*) by fluorescence in situ hybridization (FISH) (135–137). In the largest series, all seven patients developed invasive disease 12 to 92 months after biopsy, with three patients still free of invasive disease at 12, 57, and 120 months (136).

Although recognition of mesothelioma in situ has important ramifications in the pursuit of the earliest phase of the disease, it is important to ensure that the diagnosis is made in a reliable manner. A diagnosis can only be made in a combination with clinical, imaging, histologic, and molecular data (99), with no gross or microscopic evidence of invasive tumor. Microscopy shows a mesothelial proliferation limited to the serosal surface with either BAP1 loss and/or *CDKN2A/p16* homozygous deletion (fig. 3-19). It has also

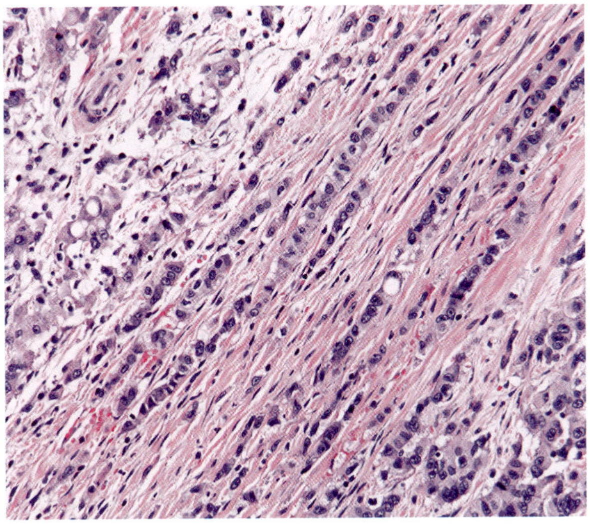

Figure 3-16

EPITHELIOID MESOTHELIOMA

The single file infiltration resembles breast carcinoma.

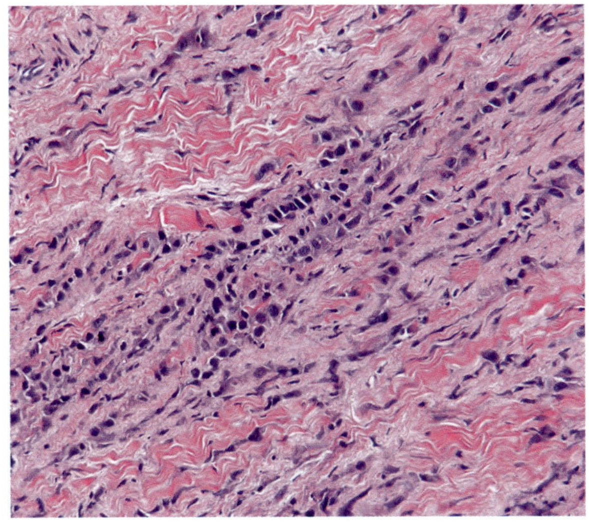

Figure 3-17

EPITHELIOID MESOTHELIOMA

Infiltrating tumor cells are compressed between collagen fibers making it difficult to differentiate from sarcomatoid morphology.

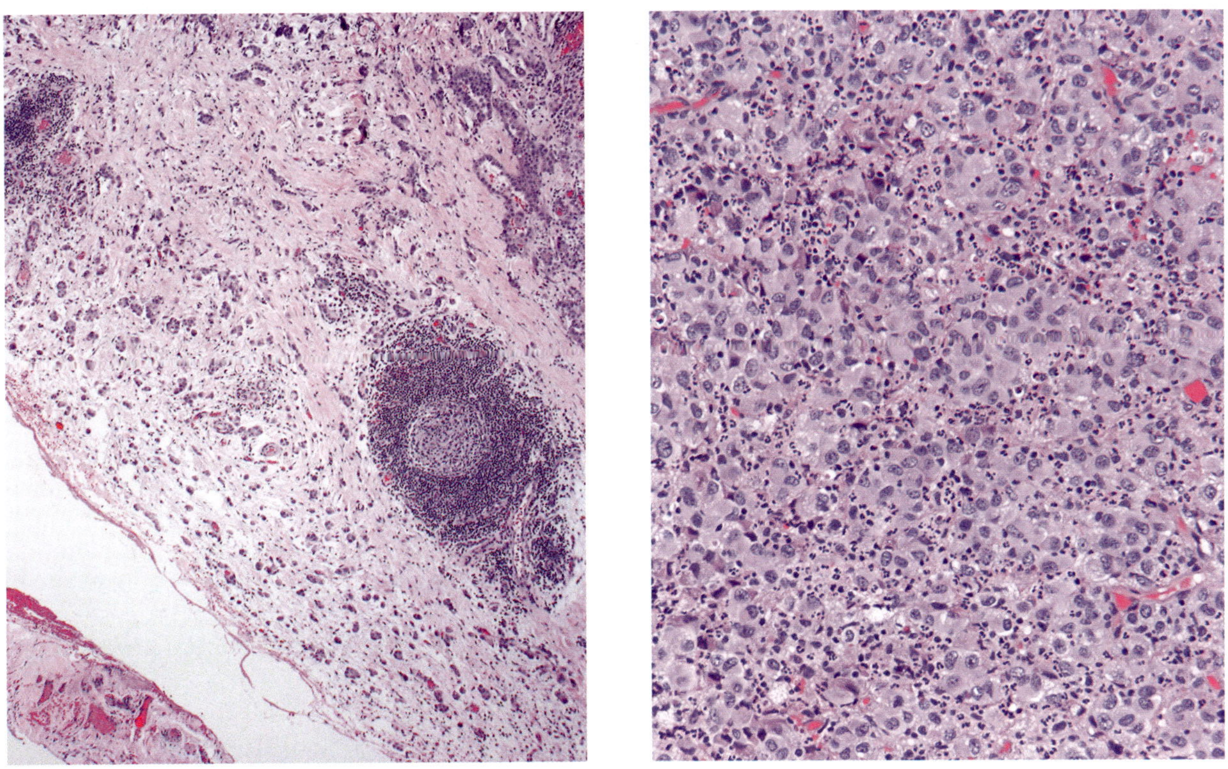

Figure 3-18

EPITHELIOID MESOTHELIOMA

Left: Lymphoid infiltrates are present both adjacent to and within the tumor. A germinal center is prominent. Right: This tumor shows marked intratumoral neutrophilic infiltration.

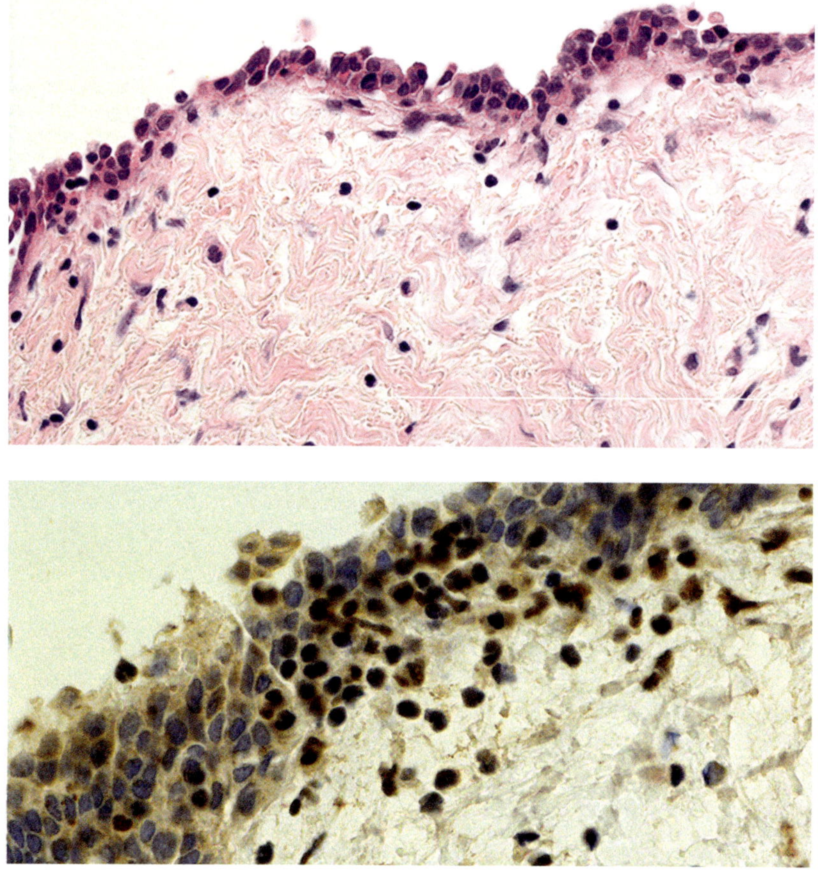

Figure 3-19

MESOTHELIOMA IN SITU

Top: The pleural surface is covered by a mildly atypical mesothelial proliferation. There is no invasion and no evidence of a mass lesion at surgery or on CT imaging.

Bottom: Mesothelial cells show loss of BAP1 nuclear staining.

been recommended that testing only be done in laboratories with experience using validated tests for the above molecular abnormalities (99). Recognition of either BAP1 loss and/or *CDKN2A/p16* homozygous deletion in cytologic material without a mass lesion, should prompt histologic sampling to confirm a lack of invasion.

There is no evidence to determine which group of patients may benefit from assessment for a molecular abnormality. In the setting of an unexplained pleural effusion, current consensus supports the use of testing to detect molecular abnormalities in mesothelial cells only in those patients where there is clinical suspicion of malignancy (99). Accurate categorization of these cases may prove to be important for identifying mesotheliomas at an earlier stage.

Diffuse Mesothelioma: Epithelioid Mesothelioma

Most *epithelioid mesotheliomas* consist of relatively monotonous and bland rounded tumor cells, with abundant cytoplasm and pale nuclei with easily seen nucleoli (fig. 3-20). The nuclear to cytoplasmic ratio is low when compared to most carcinomas, although many tumors show significant intratumoral architectural and stromal heterogeneity (fig. 3-21). The cytologic morphology, however, tends to be the same across different architectural patterns within the same tumor, although there are exceptions in which there is considerable cytologic variation, which has implications for grading. In any pattern, there may be discohesive cell groups and single cells within the stroma (fig. 3-22). Psammoma bodies are occasionally present (fig. 3-23). The literature contains numerous publications on variations in morphology, some related to potential misdiagnosis and others with prognostic significance, and it is useful to categorize them in terms of architectural patterns, cytologic features, and stromal variations (Table 3-3).

Architectural Patterns of Epithelioid Mesothelioma. Epithelioid mesotheliomas exhibit a wide variety of architectural patterns, and a combination of two or more patterns in different proportions is much more likely to be seen in any given case with adequate sampling, since pure patterns are rare (fig. 3-21). Small single biopsies, therefore, may not be representative of the tumor as a whole, hence the recommendations for standardization of adequate sampling (99,122,123).

Tubulopapillary is a common pattern in which there are variable amounts of papillary and tubular (gland-like) structures (fig. 3-24). It is rare for a tumor to be purely tubular or papillary. The papillae often branch and form complex structures (fig. 3-25). They have fibrovascular cores covered by a single layer of cuboidal to flattened tumor cells, with some areas having more than one layer of cells (fig. 3-26). The infiltrating component has both papillae and tubules, which are formed by similar cells. It is important to differentiate this pattern from well-differentiated papillary mesothelial tumors (WDPMT) which have a cytologically bland surface proliferation and papillary architecture (see figs. 4-2, 4-3) (154).

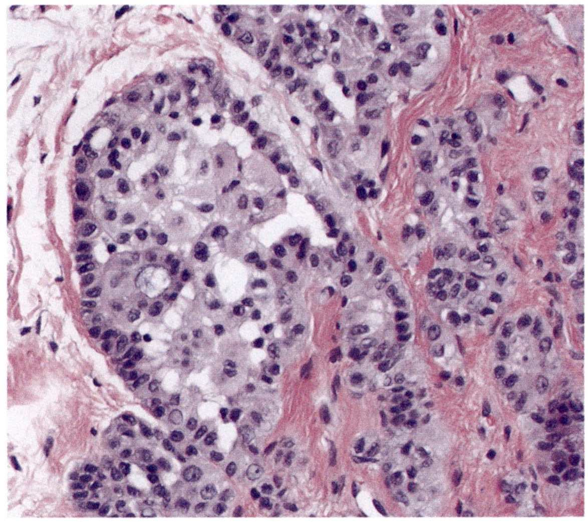

Figure 3-20

EPITHELIOID MESOTHELIOMA

Typical histologic appearance of epithelioid mesothelioma with large, rounded cells containing pale eosinophilic cytoplasm.

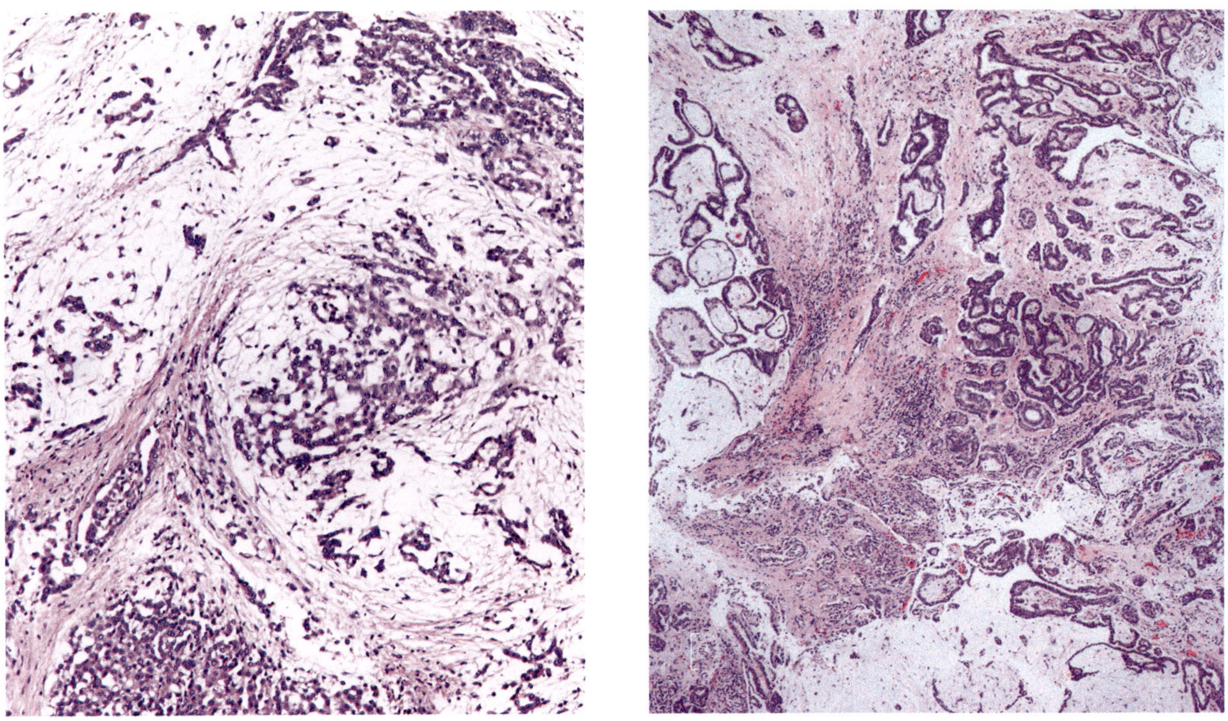

Figure 3-21

EPITHELIOID MESOTHELIOMA

Left: Multiple patterns are present in this epithelioid mesothelioma, including trabecular, single cell, and adenomatoid/microcystic. The stroma is notably myxoid.

Right: In this case, there are areas showing papillary and adenomatoid architecture, as well as focal micropapillary clusters in myxoid stroma (bottom).

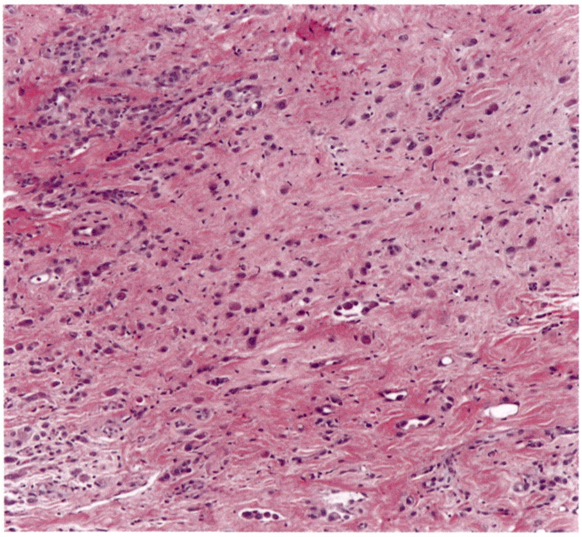

Figure 3-22

EPITHELIOID MESOTHELIOMA

Areas of single cell infiltration may be present.

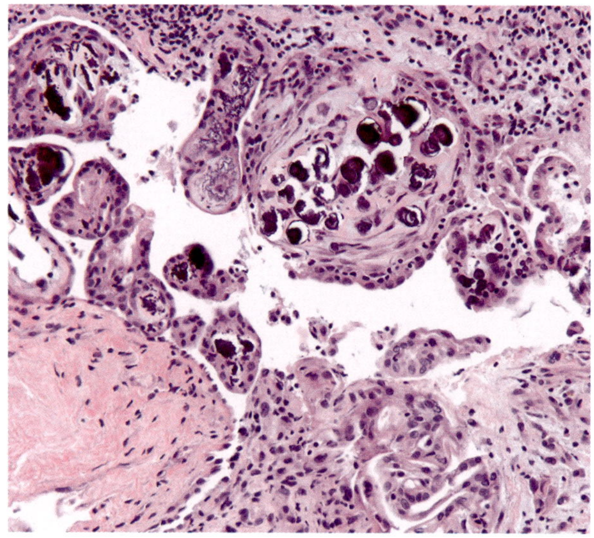

Figure 3-23

EPITHELIOID MESOTHELIOMA

Psammoma bodies are present in this pleural mesothelioma with papillary pattern.

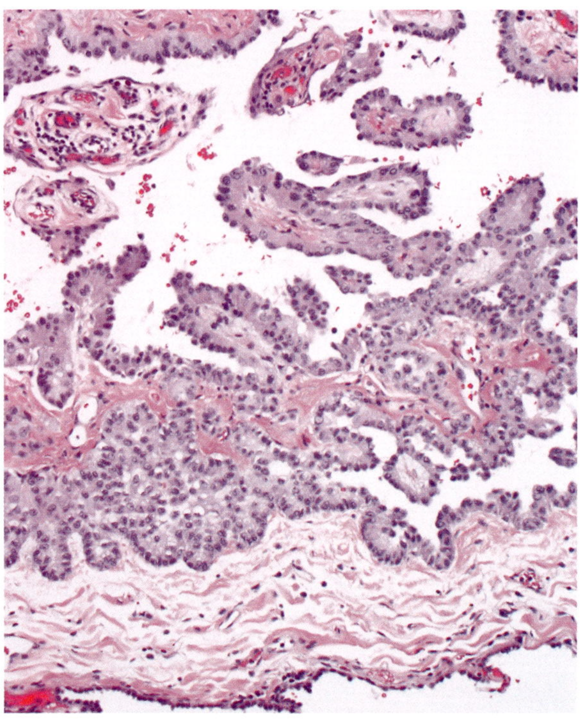

Figure 3-24

EPITHELIOID MESOTHELIOMA: TUBULOPAPILLARY PATTERN

The tumor forms tubular structures and spaces into which project papillae with fibrovascular cores. Benign-appearing mesothelium is seen at the bottom.

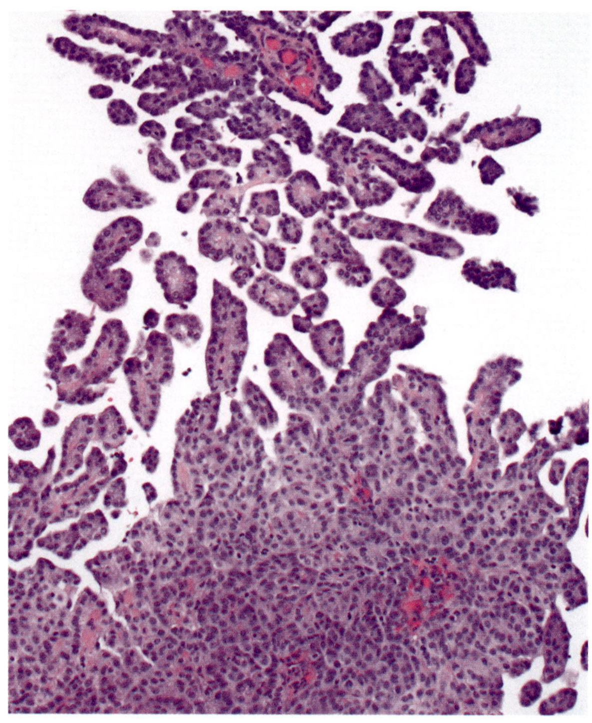

Figure 3-25

EPITHELIOID MESOTHELIOMA: PAPILLARY PATTERN

Complex branching pattern is present at the top and more solid growth is seen at the bottom.

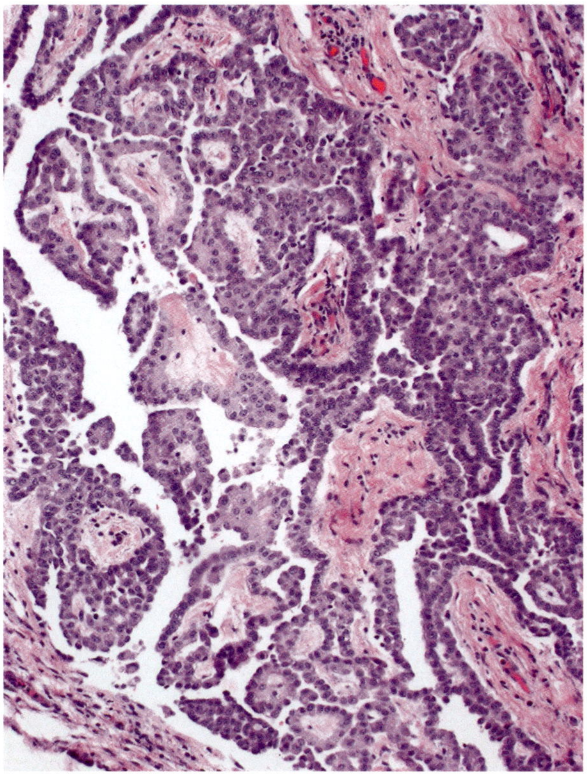

Figure 3-26

EPITHELIOID MESOTHELIOMA: PAPILLARY PATTERN

The papillae are covered by single to multiple layers of tumor cells.

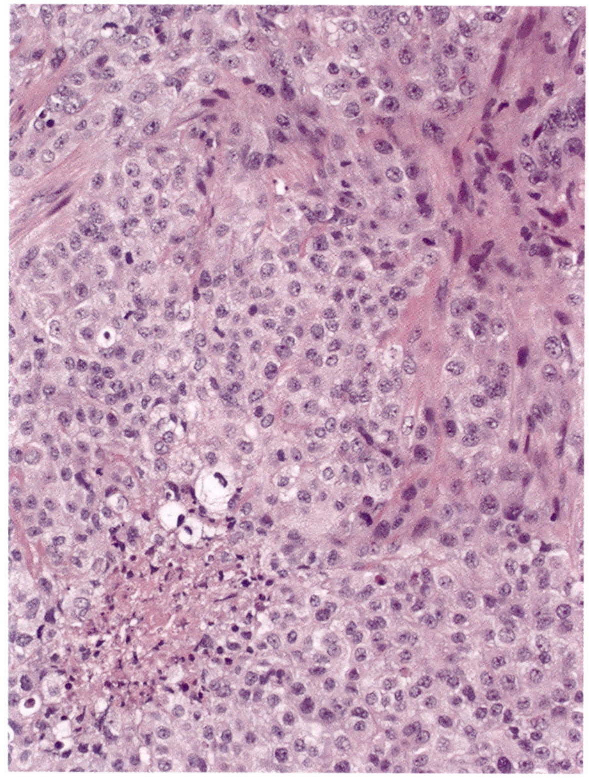

Figure 3-27

EPITHELIOID MESOTHELIOMA: SOLID PATTERN

There are sheets of epithelioid tumor cells with pale cytoplasm, round to oval nuclei, small to medium-sized nucleoli, and numerous mitotic figures. An area of necrosis is seen at lower left.

The *solid growth pattern* is very common; it has large nests or sheets of tumor cells that are typically uniform, although higher-grade tumors can occur. The cells may be round, oval, or polygonal, with abundant eosinophilic cytoplasm and round, vesicular nuclei often with prominent nucleoli (fig. 3-27). Necrosis is more common in the solid pattern, as is variation in nuclear atypia and mitoses. This pattern is associated with a poor prognosis compared to the tubulopapillary pattern (155,156).

The *trabecular pattern* is often seen together with other patterns, such as solid, acinar/tubular, and papillary. Epithelioid cells form cords that are 1 to 2 cell layers thick (fig. 3-28).

The *adenomatoid* (or *acinar*) and *microcystic patterns* are somewhat overlapping, essentially encompassing a glandular architectural pattern. Adenomatoid (acinar) tumors consist of elongated or branching gland-like lumens lined by relatively bland tumor cells (fig. 3-29), although some cases show more atypia and prominent nucleoli (fig. 3-30). Microcystic areas consists of bland, flat to cuboidal cells lining smaller gland-like structures (fig. 3-31). In some cases the acinar structures are more dilated (fig. 3-32). This architectural pattern is often seen in association with a prominent myxoid stroma (figs. 3-21, 3-33) and has a significantly better 2-year survival rate compared to other patterns of epithelioid mesothelioma (157). Some tumors resemble salivary gland tumors and have been described as having an adenoid cystic pattern, although this term is not recommended (fig. 3-34).

The *micropapillary pattern* is often a minor component of the tumor and is recognized by the fact that the papillary projections are small and have no fibrovascular core (fig. 3-35). Most often the tumor cells have low-grade morphology but some show significant pleomorphism (fig. 3-36). It is

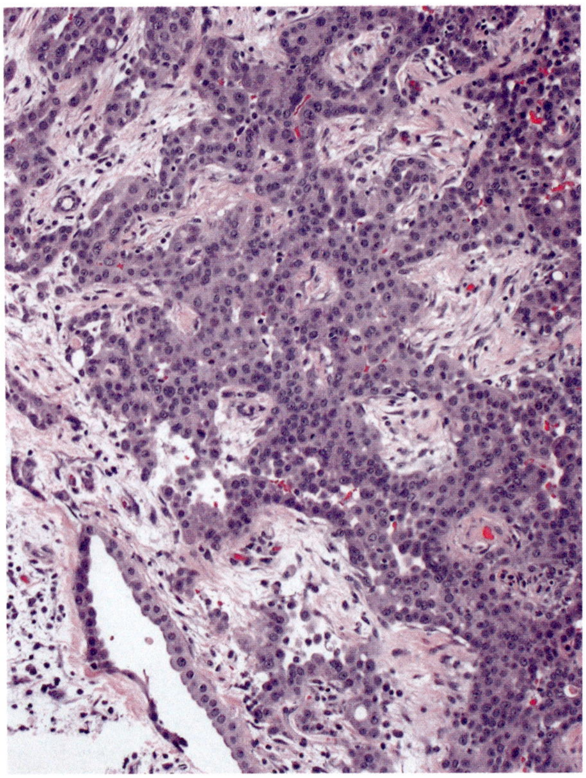

Figure 3-28

EPITHELIOID MESOTHELIOMA: TRABECULAR PATTERN

The tumor cells form thick cords invading loose fibromyxoid stroma.

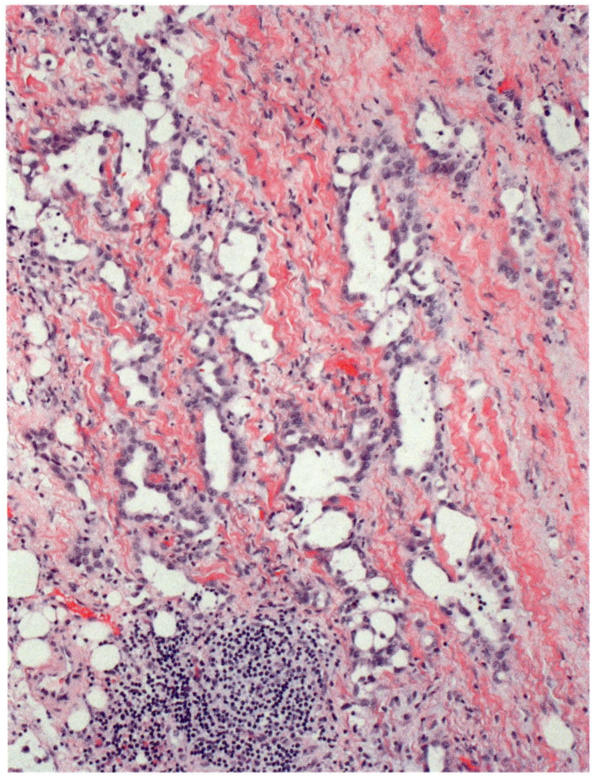

Figure 3-29

EPITHELIOID MESOTHELIOMA: ADENOMATOID (ACINAR)/MICROCYSTIC PATTERN

There are gland-like structures formed by bland mesothelial cells, infiltrating fibrous tissue.

Figure 3-30

EPITHELIOID MESOTHELIOMA: ADENOMATOID (ACINAR)/ MICROCYSTIC

There is marked nuclear pleomorphism in this tumor, with enlarged nuclei and prominent nucleoli.

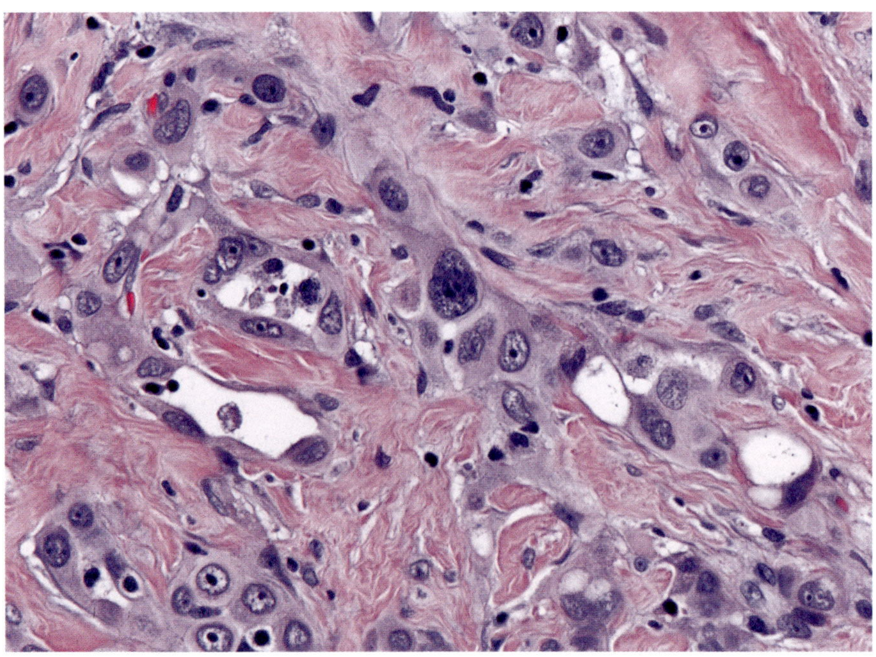

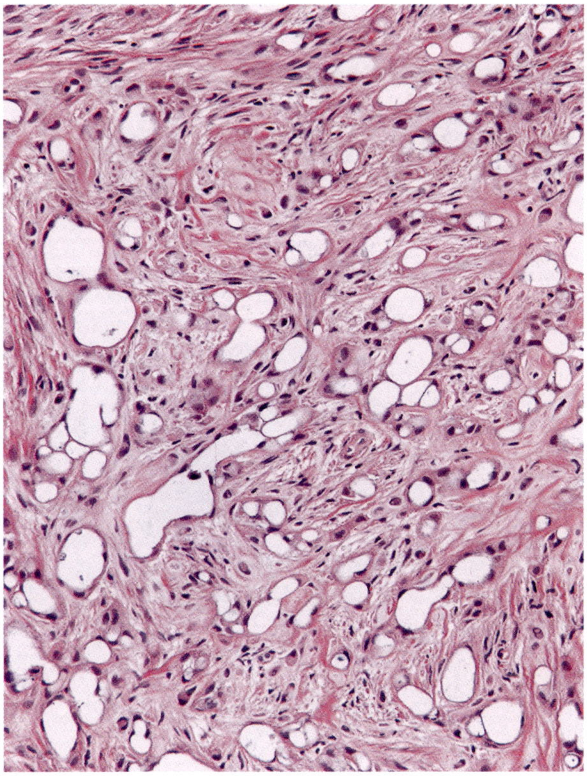

Figure 3-31

EPITHELIOID MESOTHELIOMA: ADENOMATOID (ACINAR)/MICROCYSTIC

The tumor cells form small gland-like spaces, which are lined by bland, flat to low cuboidal cells.

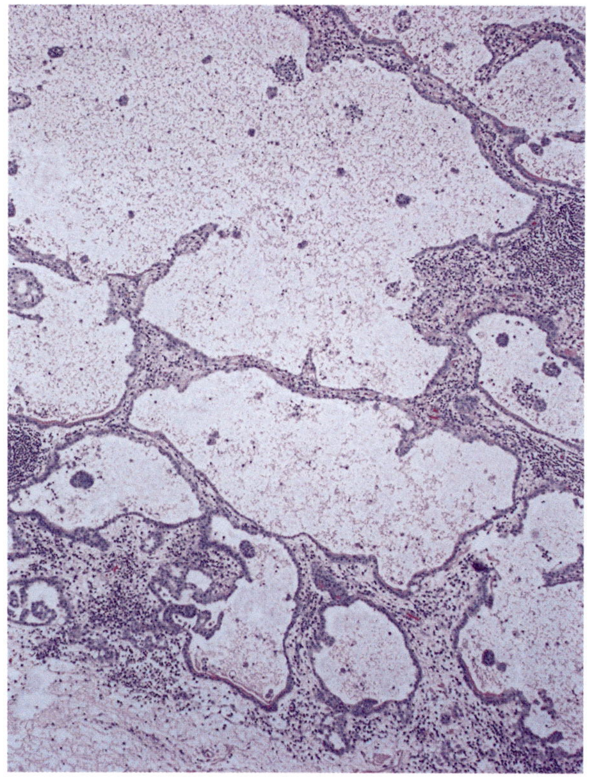

Figure 3-32

EPITHELIOID MESOTHELIOMA: ADENOMATOID (ACINAR)/MICROCYSTIC

Cyst-like spaces are formed by the bland tumor cells. There is amorphous material within the cysts.

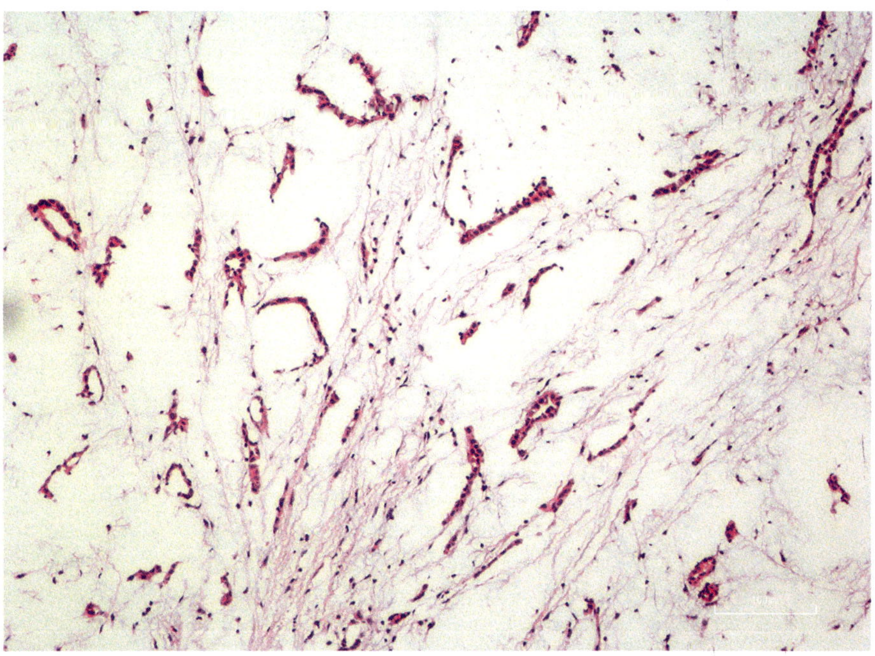

Figure 3-33

EPITHELIOID MESOTHELIOMA: ADENOMATOID (ACINAR)/MICROCYSTIC

There is abundant myxoid stroma, within which the bland tumor cells are floating, forming small cyst-like spaces and nests.

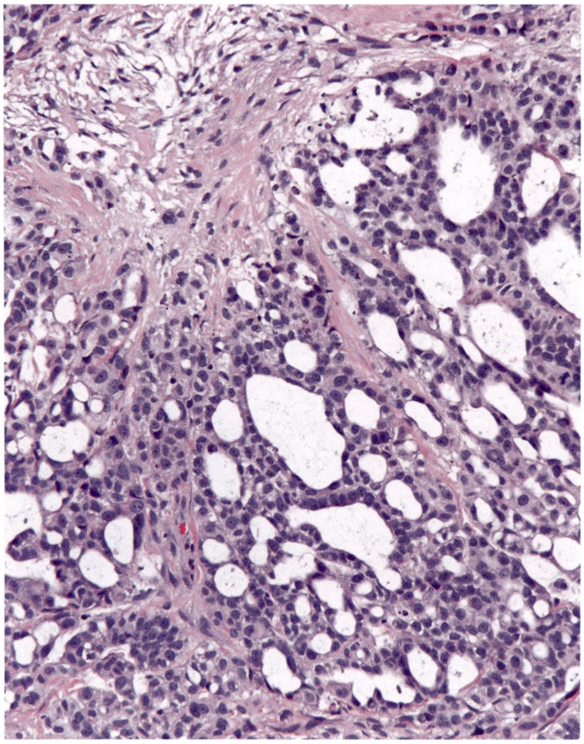

Figure 3-34

EPITHELIOID MESOTHELIOMA: ADENOMATOID (ACINAR)/MICROCYSTIC

The tumor cells have very low-grade morphology and show a cribriform pattern mixed with solid nests, reminiscent of adenoid cystic carcinoma.

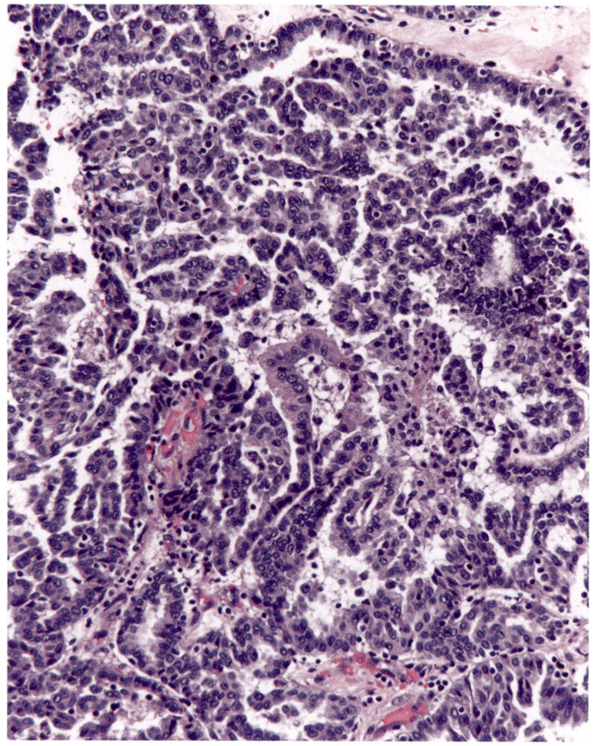

Figure 3-35

EPITHELIOID MESOTHELIOMA: MICROPAPILLARY AND PAPILLARY PATTERN

The tumor cells form multiple small papillae, most of which do not have fibrovascular cores, while others do. Mixed patterns are common.

important to recognize and report this pattern since it is associated with a higher incidence of lymphatic invasion (fig. 3-37) and lymph node metastasis (156).

Cytologic Variants of Epithelioid Mesothelioma. As well as the numerous architectural patterns seen in epithelioid mesotheliomas, individual cytologic variations from the classic epithelioid cell are common. Recognition of these is important since some have prognostic significance and others are more common in carcinomas arising at other sites and it is important for pathologists to recognize them as acceptable morphologies for mesothelioma. Most cases are easily resolved by application of appropriate immunohistochemical markers.

Signet ring morphology can be seen in any epithelioid pattern (158). This carries no prognostic significance but should not be misdiagnosed as metastatic adenocarcinoma (fig. 3-38).

Clear cell morphology consists of mesothelial cells with clear cytoplasm, which raises the differential diagnosis of metastatic renal cell carcinoma, nonsmall cell lung carcinomas with clear cell morphology, clear cell melanoma, and other clear cell tumors (fig. 3-39). Some mesotheliomas have lipidized cytoplasm, within which multiple small vacuoles are present (fig. 3-40).

Deciduoid morphology is a rare variant of mesothelioma that is composed of large, polygonal or ovoid cells with well-defined cell borders, dense eosinophilic cytoplasm, and single or multiple nuclei (fig. 3-41). Among mesotheliomas with deciduoid features, only those with high-grade features (wide variation in size and shape, marked nuclear atypia, and high mitotic activity) show a worse prognosis (fig. 3-42) (159–161).

Small cell morphology is rare in mesotheliomas and is usually combined with other patterns,

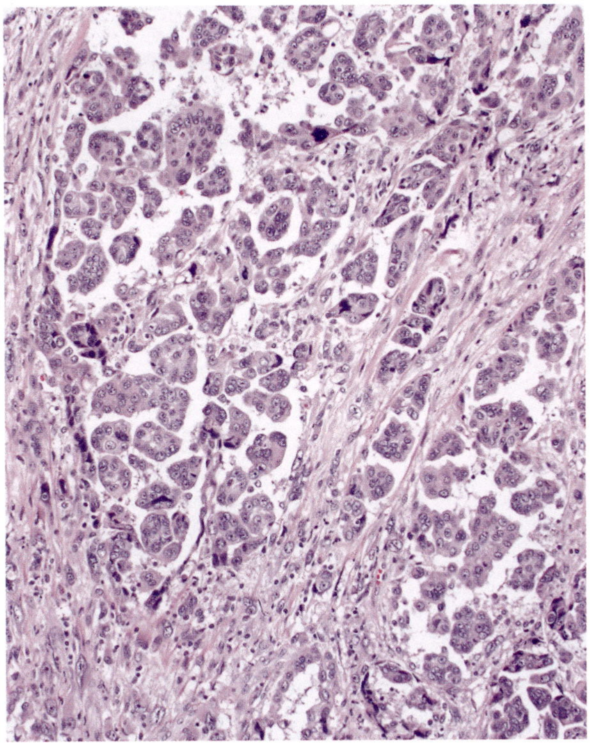

Figure 3-36

EPITHELIOID MESOTHELIOMA: MICROPAPILLARY PATTERN

The tumor forms micropapillary projections into spaces within the stroma. The tumor cells in this case are pleomorphic, with tumor giant cells and prominent nucleoli.

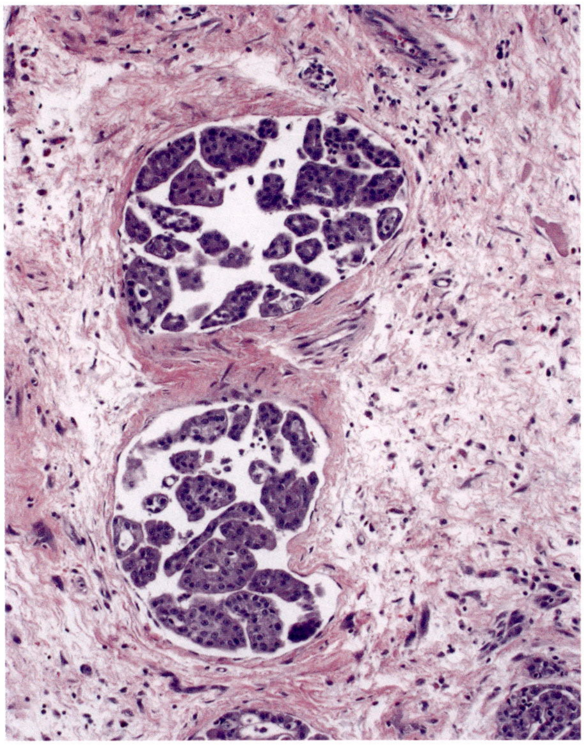

Figure 3-37

EPITHELIOID MESOTHELIOMA: MICROPAPILLARY PATTERN

Micropapillary clusters of mesothelioma cells are seen within dilated lymphatics.

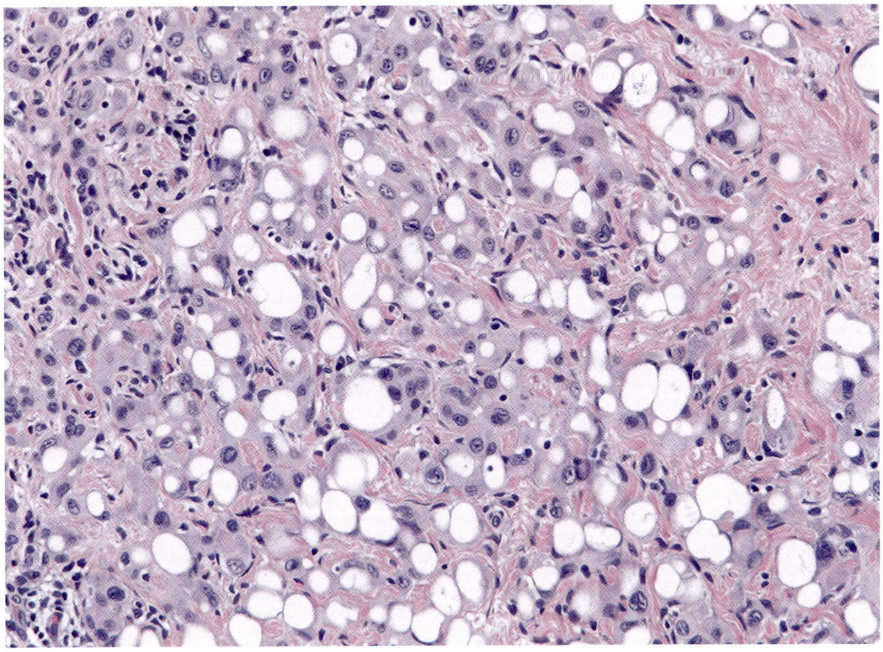

Figure 3-38

EPITHELIOID MESOTHELIOMA: WITH SIGNET RING CELLS

The tumor cells have a clear intracytoplasmic vacuole, giving them a signet ring cell appearance. This pattern puts signet ring cell carcinoma in the differential diagnosis.

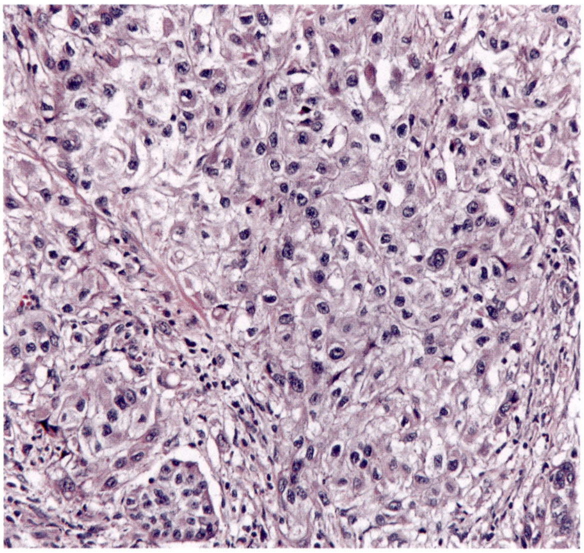

Figure 3-39

EPITHELIOID MESOTHELIOMA: WITH CLEAR CELLS

The cytoplasm of the tumor cells is variably clear.

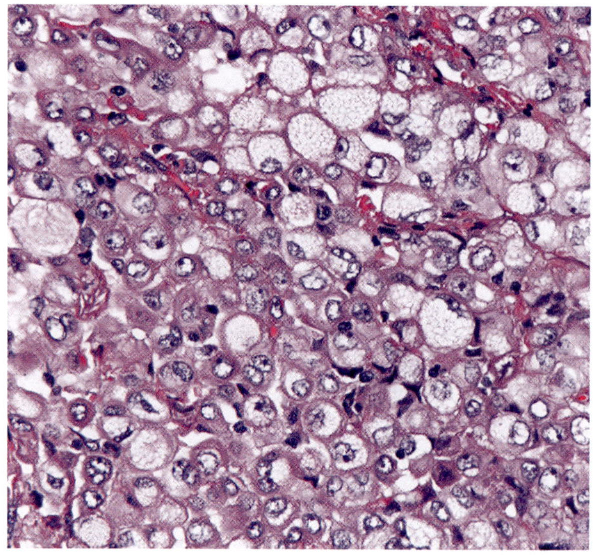

Figure 3-40

EPITHELIOID MESOTHELIOMA: LIPIDIZED

The tumor cells have microvacuolated cytoplasm.

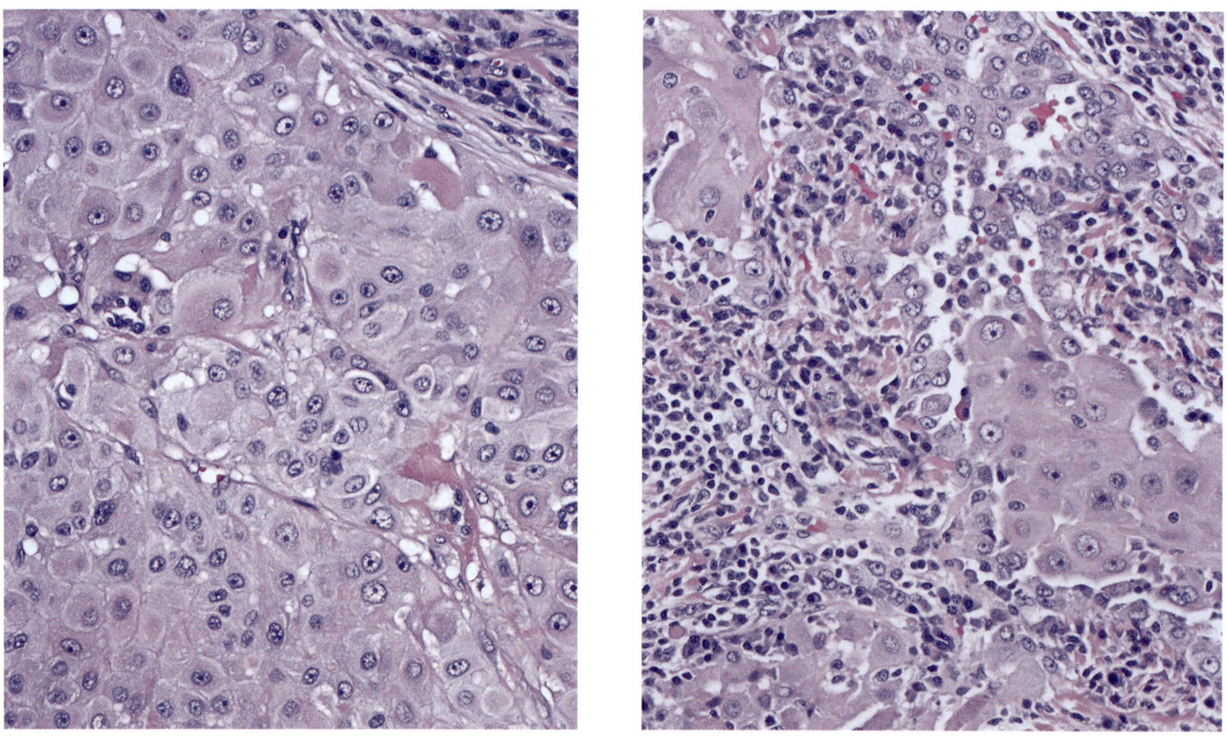

Figure 3-41

EPITHELIOID MESOTHELIOMA: DECIDUOID

Left: The tumor cells are large and polygonal, with well-defined cell borders, dense eosinophilic cytoplasm, and a low nuclear to cytoplasmic ratio.

Right: A different field from the same case shows that not all cells have deciduoid morphology. There is a prominent inflammatory infiltrate composed of lymphocytes and plasma cells.

forming only 15 to 20 percent of the tumor when resection specimens are examined. The small cells grow in solid nests of closely packed cells, often exhibiting a well-defined cell membrane and a high nuclear to cytoplasmic ratio (fig. 3-43). Mitotic activity is generally low and there is no karyorrhexis or hematoxyphilic staining of blood vessels (Azzopardi effect) as is seen in small cell carcinoma of the lung (162).

Rhabdoid morphology is another rare variant, in which tumors often form discohesive sheets of cells with prominent perinuclear eosinophilic inclusions (fig. 3-44) (163). This cytologic feature may also be seen in SMARCA4-deficient malignancies as well as in rhabdomyomatous tumors but, again, immunohistochemistry should allow identification of a mesothelial phenotype. SMARCA4 is rarely lost in mesotheliomas (164).

Cytologic Variants Seen in Both Epithelioid and Sarcomatoid Mesotheliomas. There remains debate about whether certain cytologic variants, namely pleomorphic,

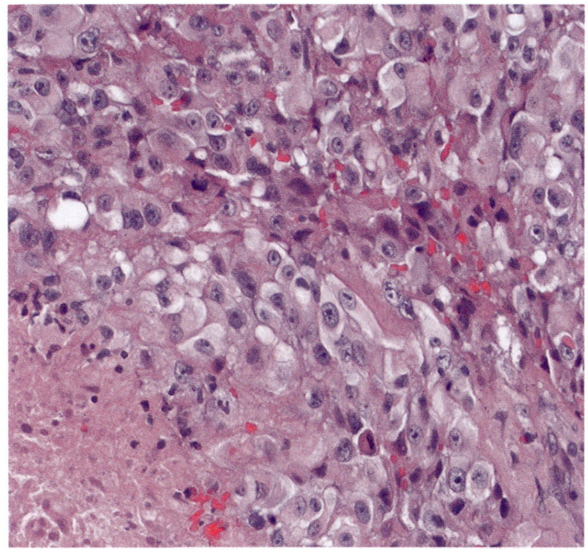

Figure 3-42

EPITHELIOID MESOTHELIOMA: DECIDUOID

This mesothelioma shows high-grade features such as nuclear atypia, mitotic figures, and necrosis.

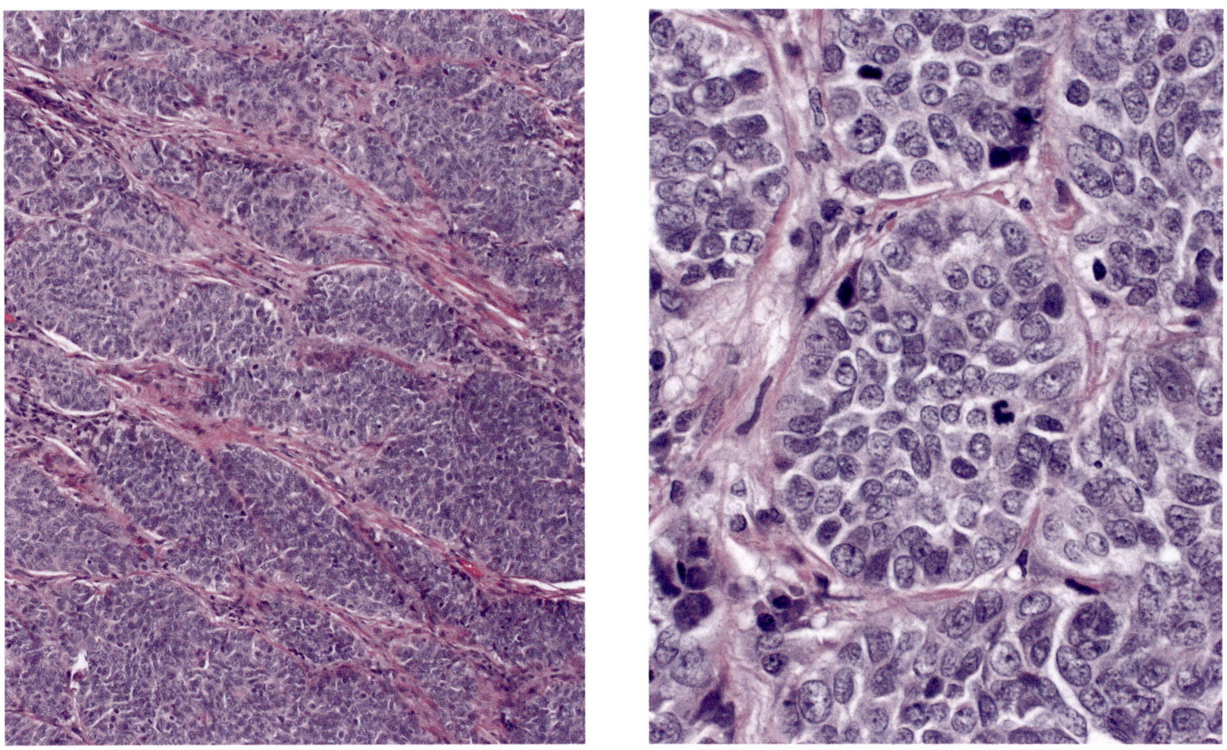

Figure 3-43

EPITHELIOID MESOTHELIOMA: SMALL CELL

Left: Solid nests of small cells are seen at medium power.
Right: At high power, nests of closely packed tumor cells show nucleoli and a few mitoses, suggesting small cell carcinoma.

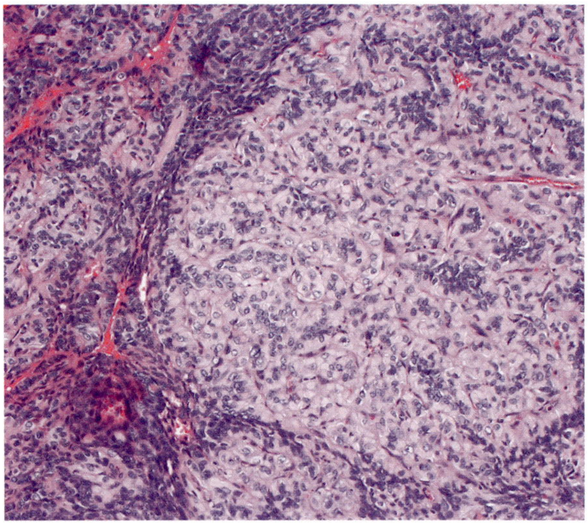

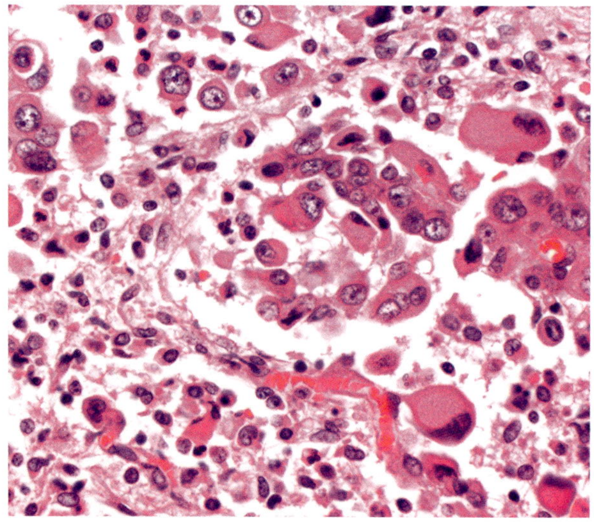

Figure 3-44

EPITHELIOID MESOTHELIOMA: RHABDOID

Left: Areas with rhabdoid features with pale eosinophilic inclusions in the cytoplasm are seen. Right: Perinuclear eosinophilic inclusions are focally present.

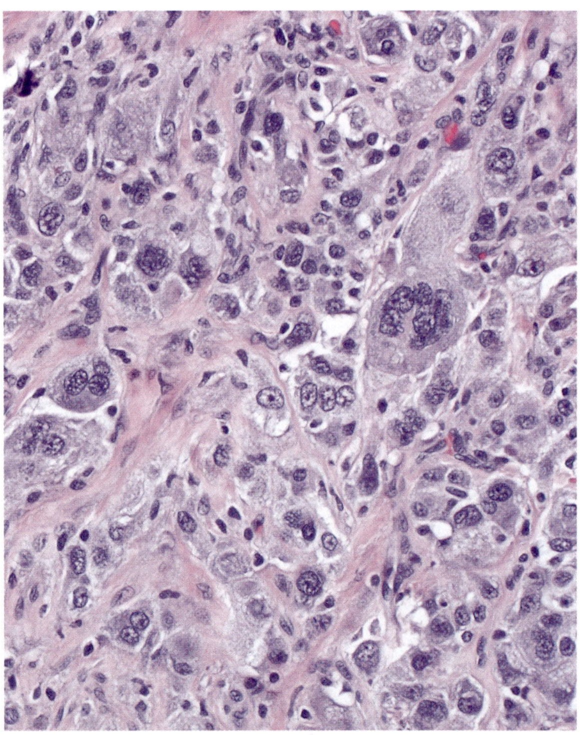

Figure 3-45

EPITHELIOID MESOTHELIOMA: PLEOMORPHIC

There is marked pleomorphism of tumor cells with multiple tumor giant cells, hyperchromasia, and bizarre nucleoli. Compare the nuclear size to benign stromal and lymphocytic nuclei.

lymphohistiocytoid, and transitional, should be classified under epithelioid or sarcomatoid subtype, or both (143,156,157,165).

Pleomorphic cytologic features are defined as those tumors in which 10 percent or more of their cells are anaplastic or form tumor giant cells (fig. 3-45) (156). The nuclei in these areas are much larger and more hyperchromatic than those seen in lower-grade tumors and there are many mitoses, including atypical ones (fig. 3-46). Lymph node metastases are frequent and these tumors have a highly aggressive clinical course, with median survival similar to that of sarcomatoid mesothelioma (156,166). However, most cases with pleomorphic areas arise on a background of otherwise epithelioid morphology (167).

Transitional cytologic features are defined as solid sheets of ovoid cells that overlap between epithelioid and sarcomatoid morphologies, although without overt sarcomatoid (fascicular, patternless) architecture (fig. 3-47) (143, 165,168). Most published data show that mesotheliomas with transitional areas have a prognosis closer to sarcomatoid than epithelioid subtypes, so the consensus view is that this pattern should be regarded solely as a variant of the sarcomatoid subtype (Table 3-3).

Lymphohistiocytoid mesotheliomas have a heavy lymphoid infiltrate obscuring polygonal

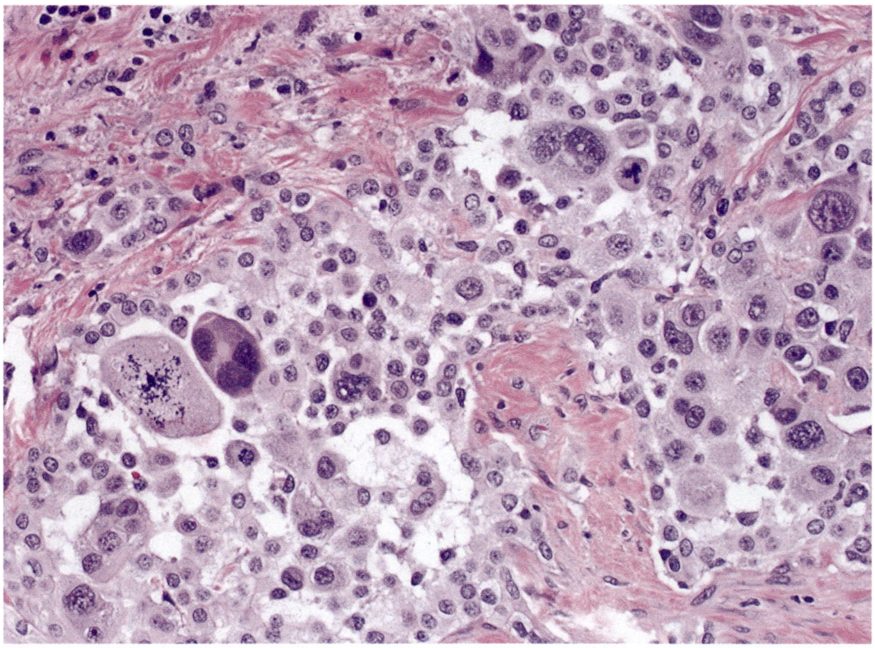

Figure 3-46

EPITHELIOID MESOTHELIOMA: PLEOMORPHIC

Atypical mitotic figures are present.

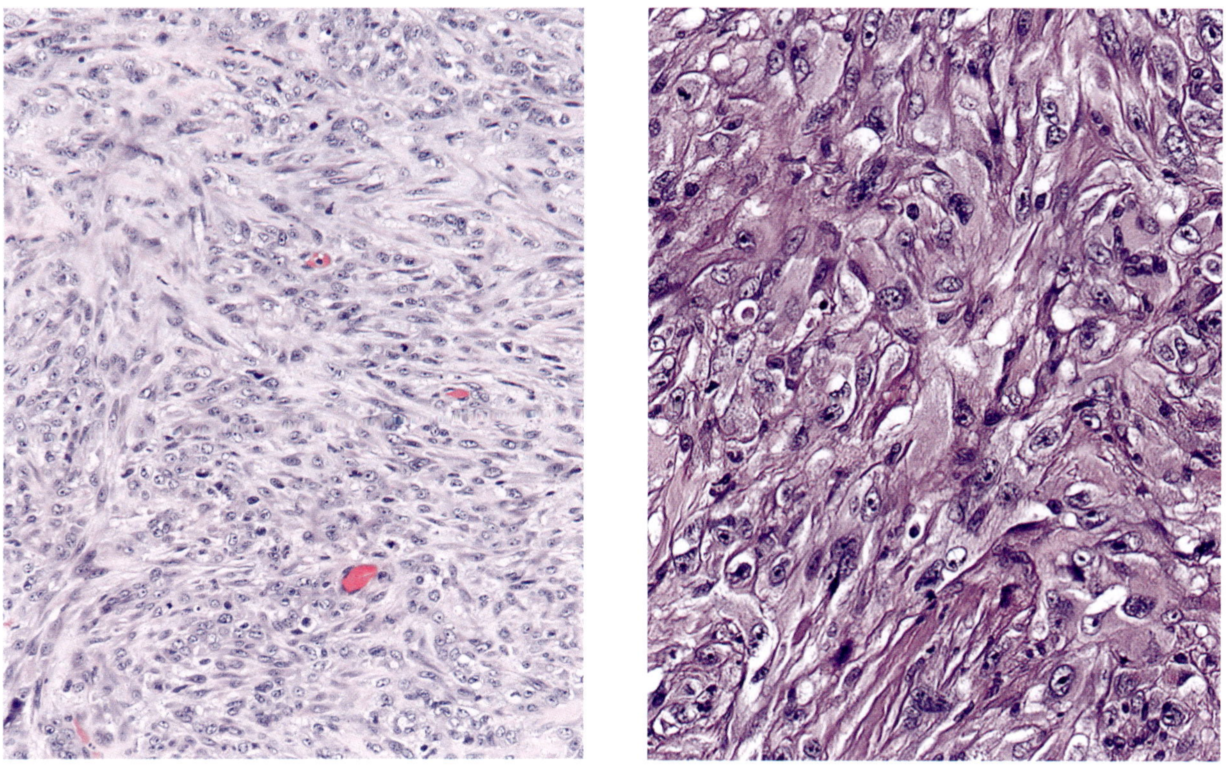

Figure 3-47

EPITHELIOID MESOTHELIOMA: TRANSITIONAL

Left: The tumor cells are somewhat spindled but have more cytoplasm, giving an appearance that is between epithelioid and sarcomatoid.

Right: High-power view shows ovoid cells morphologically between overt epithelioid and sarcomatoid. (Courtesy of Dr. F. Galateau-Salle, Lyon, France)

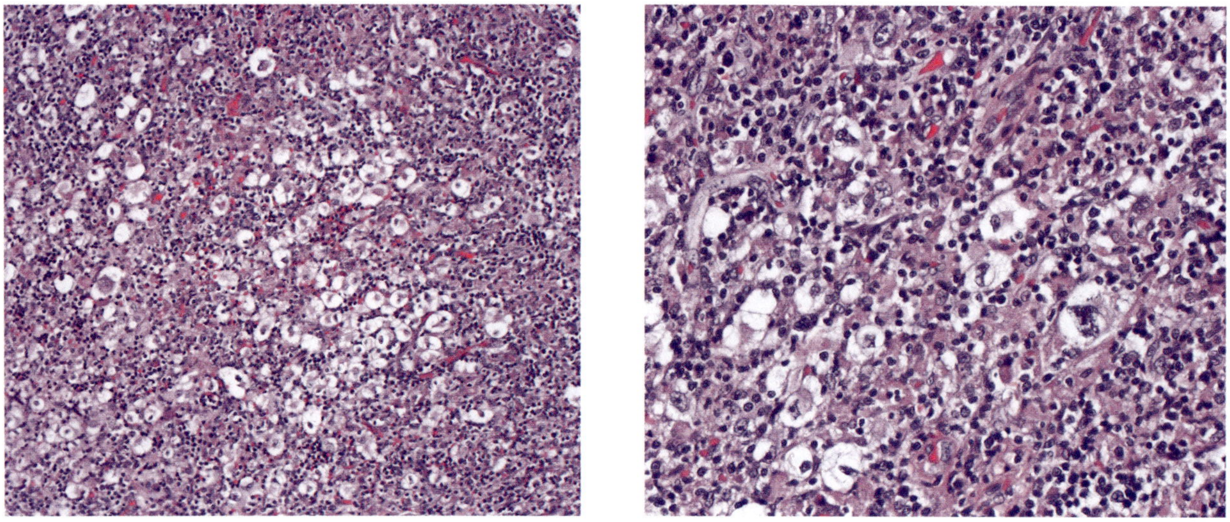

Figure 3-48

EPITHELIOID MESOTHELIOMA: LYMPHOHISTIOCYTOID

Left: At low power, a mixture of large pale cells and lymphocytes is seen.
Right: Higher-power view shows single large tumor cells with abundant pale cytoplasm and atypical nuclei with prominent nucleoli among numerous lymphocytes, raising the differential diagnosis of lymphoma.

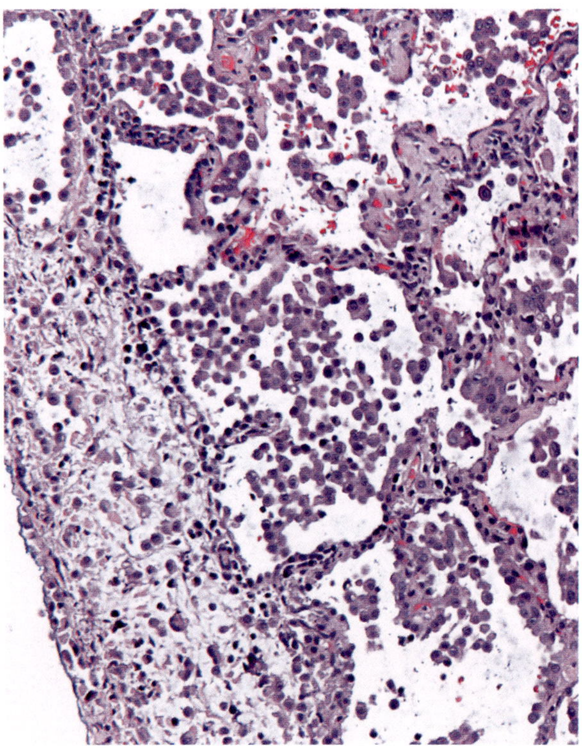

Figure 3-49

EPITHELIOID MESOTHELIOMA

The tumor is infiltrating the lung with a desquamative interstitial pneumonia (DIP)-like pattern.

neoplastic cells and mimicking non-neoplastic inflammatory processes, non-Hodgkin lymphoma, Hodgkin lymphoma, or lymphoepithelioma-like carcinoma. The tumor cells are large, with abundant pale cytoplasm, pale nuclei, and prominent nucleoli, resembling histiocytes (fig. 3-48). The mitotic rate is variable. The tumor cells stain for mesothelioma-associated markers. As with any other mesothelioma, this may also be mixed with other patterns (169).

Stromal Variations in Mesothelioma. In addition to architectural and cytologic variations, the background stroma has also been used as a feature to propose subtypes of mesothelioma. The *myxoid variant* has been discussed above in relation to a microcystic architecture, which often has a rich myxoid stroma. These tumors have a more indolent behavior than other epithelioid mesotheliomas when more than 50 percent of the tumor shows a myxoid background (see figs. 3-21, 3-33) (155).

A *desquamative interstitial pneumonia (DIP)-like pattern* of growth may be seen when mesotheliomas infiltrate the lung, a feature similar to so-called spread through airspaces (STAS) (fig. 3-49). This should not be mistaken for either interstitial lung disease or nonsmall cell carcinoma (97,170).

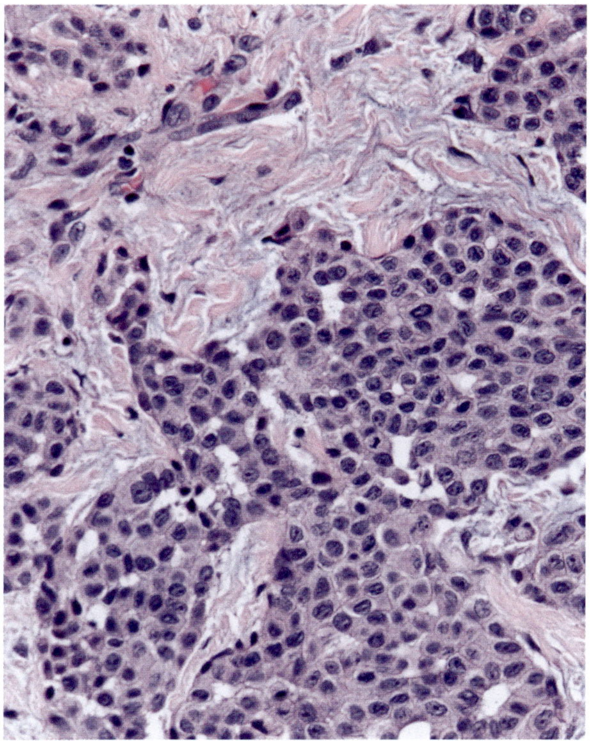

Figure 3-50

EPITHELIOID MESOTHELIOMA: NUCLEAR ATYPIA SCORE 1

The tumor nuclei are relatively small and uniform, and have inconspicuous nucleoli.

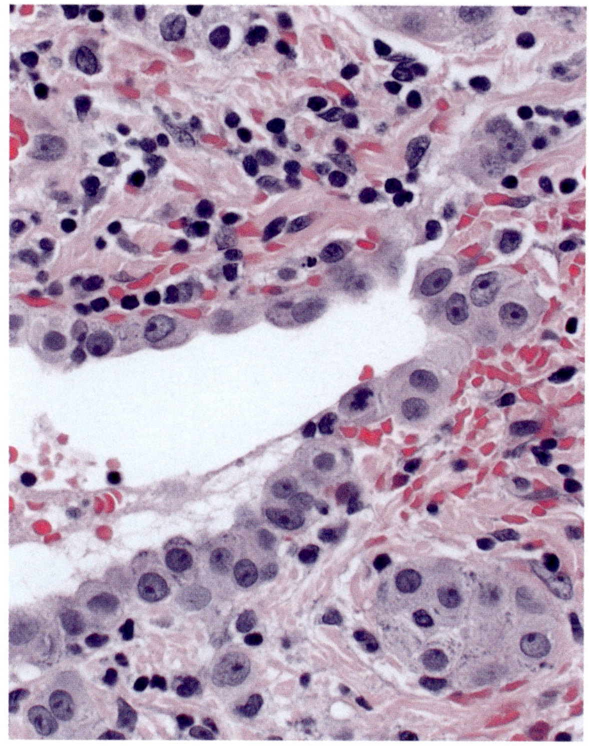

Figure 3-51

EPITHELIOID MESOTHELIOMA: NUCLEAR ATYPIA SCORE 2

The tumor nuclei are of intermediate size, are more irregular, and have prominent nucleoli. A mitotic figure is seen.

Grading of Epithelioid Mesothelioma. It is well established that tumors with epithelioid histology have a favorable prognosis compared to sarcomatoid and biphasic types; however, within the epithelioid group, there is a large variation in survival. The terms well, moderately, and poorly differentiated are not recommended for use in this setting since they are not well defined and can be confused with well-differentiated papillary mesothelial tumors.

For epithelioid tumors, a 3-tier grading system using nuclear atypia and mitotic count was shown to be an independent prognostic predictor of overall survival (171). This has been validated in studies that also showed that the presence of necrosis further stratified overall survival (122,172). A grading system has been proposed that covers both epithelioid and sarcomatoid variants (173), although the current consensus view is that grading should be limited to epithelioid variants (99).

Nuclear atypia is assessed in the area with the highest degree of atypia, and assigned a score of 1, 2, or 3 based on the degree of atypia. Uniform small nuclei with inconspicuous nucleoli are mildly atypical, and score 1 (fig. 3-50). Nuclei intermediate in size with variability in shape and variably prominent nucleoli are moderately atypical, and score 2 (fig. 3-51). Bizarre, enlarged nuclei, multinucleation, and macronuclei in more than 5 percent of tumor cells are severely atypical, and score 3 (fig. 3-52). The nuclear atypia may vary over different fields; the most atypical areas are scored (fig. 3-53).

Mitoses are counted in the areas with the highest mitotic activity as an average of mitotic figures per 10 high-power fields (HPF). Tumors are given three groups based on the mitotic count: score 1 is 0-1/10HPF, score 2 is 2-4/10HPF, and score 3 is 5 or more /10HPF. Adding the atypia and mitotic score gives three nuclear grades: nuclear grade I is a score of 2-3, nuclear grade II is a score of 4-5, and nuclear grade III is a score of 6.

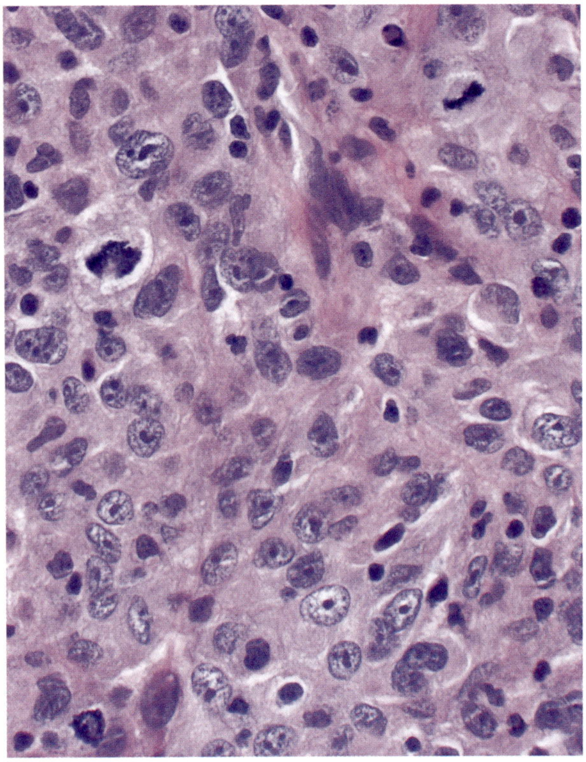

Figure 3-52

EPITHELIOID MESOTHELIOMA: NUCLEAR ATYPIA SCORE 3

The tumor nuclei are more atypical than those seen in grade 2, with markedly enlarged nucleoli, but are not as pleomorphic as is seen in the pleomorphic pattern of epithelioid mesothelioma. Mitotic figures are present.

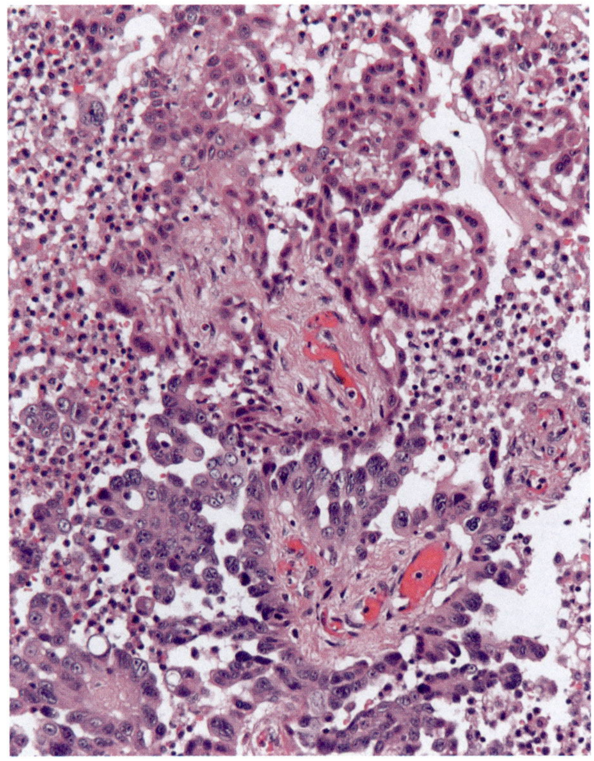

Figure 3-53

EPITHELIOID MESOTHELIOMA: NUCLEAR VARIATION

This papillary mesothelioma shows a nuclear score of 1 (top) and a nuclear score of 2 (bottom); the higher score should be used to calculate the nuclear grade.

Necrosis is recorded as either present or absent (fig. 3-54). This is then divided into low (nuclear grade I and nuclear grade II without necrosis) and high (nuclear grade II with necrosis and nuclear grade III) grades. This grading system should be applied to biopsies and resection specimens to determine prognosis (99). A composite score consisting of growth pattern and 2-tier nuclear grade did not improve prognostication compared with nuclear grade alone (166).

Sarcomatoid Mesothelioma

Sarcomatoid mesothelioma is characterized by a proliferation of neoplastic spindle cells. The cells are typically arranged in fascicles or in a haphazard pattern. Cellularity is highly variable, ranging from densely packed tumor cells to sparsely infiltrated, dense hyaline fibrosis in the desmoplastic variant. This variability is frequently seen in individual tumors (140,140a). Sarcomatoid mesotheliomas are highly aggressive tumors and overall survival is short despite a highly variable morphologic appearance, nuclear atypia, and mitotic rate.

The tumor cells vary from plump spindle to thin long cells with scant cytoplasm (fig. 3-55). Nuclei are oval to elongated, hyperchromatic, or large and irregular (fig. 3-56). Mitoses are frequent (fig. 3-57) except in the desmoplastic variant described below. Necrosis is common (fig. 3-58). Some tumors have many atypical giant cells with large multilobulated, bizarre, hyperchromatic nuclei with a high mitotic count and atypical mitoses, resembling pleomorphic carcinoma (fig. 3-59). Osteoclast-like giant cells may occasionally be present (fig. 3-60).

Variants of Sarcomatoid Mesothelioma. Similar to epithelioid mesotheliomas, recognition of variants of sarcomatoid mesothelioma helps the

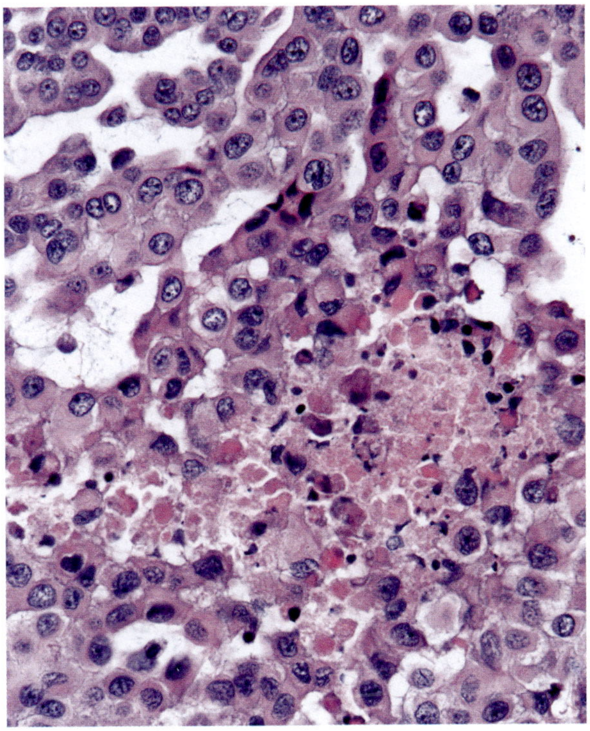

Figure 3-54

EPITHELIOID MESOTHELIOMA: NECROSIS

There is tumor necrosis in this nuclear grade 2 mesothelioma.

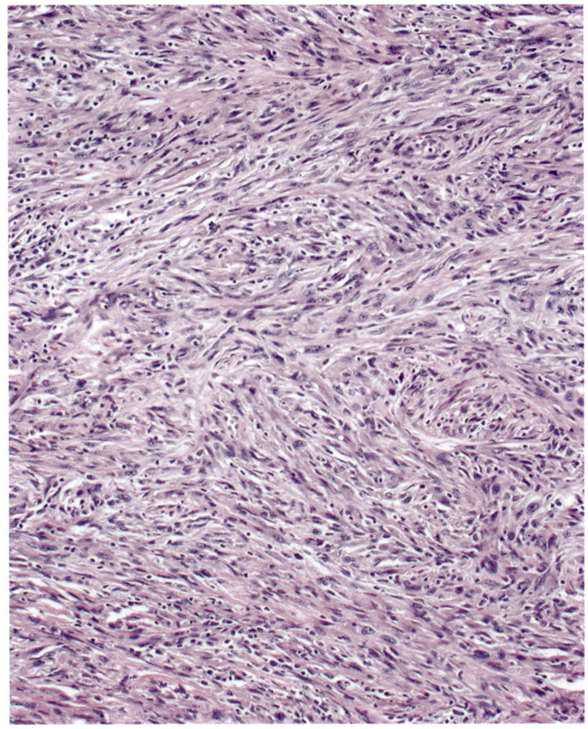

Figure 3-55

SARCOMATOID MESOTHELIOMA

The tumor is composed of spindled cells arranged in fascicles, with plump ovoid nuclei and prominent nucleoli. Many tumor infiltrating lymphocytes are present.

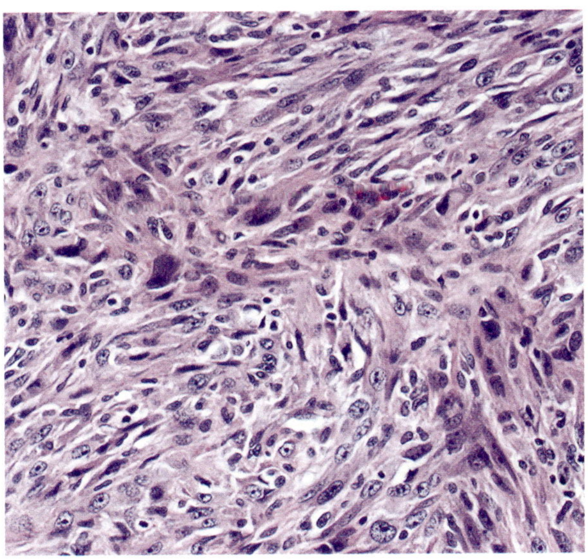

Figure 3-56

SARCOMATOID MESOTHELIOMA

There is varying nuclear pleomorphism with prominent nucleoli.

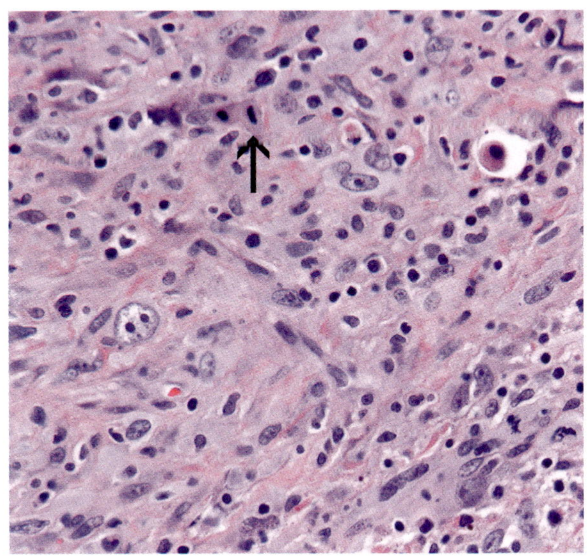

Figure 3-57

SARCOMATOID MESOTHELIOMA

High power shows mitotic figures in sarcomatoid mesothelioma (arrow).

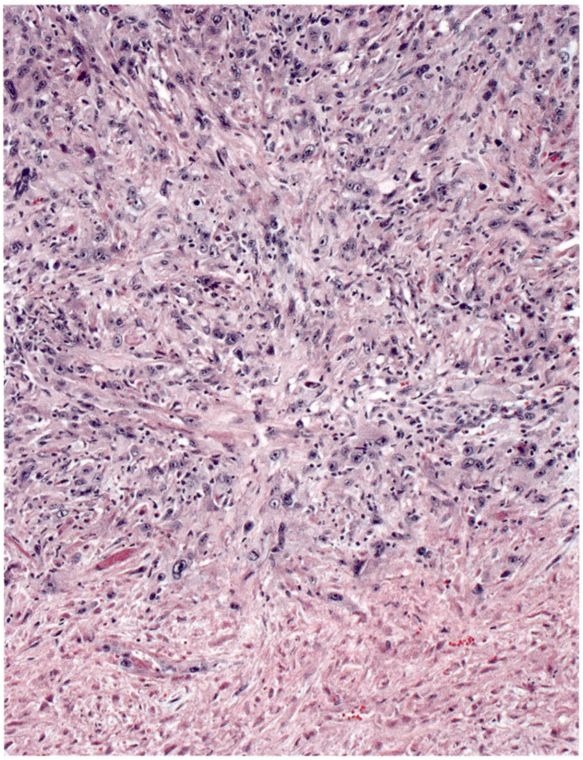

Figure 3-58

SARCOMATOID MESOTHELIOMA

An area of necrosis is present at the bottom.

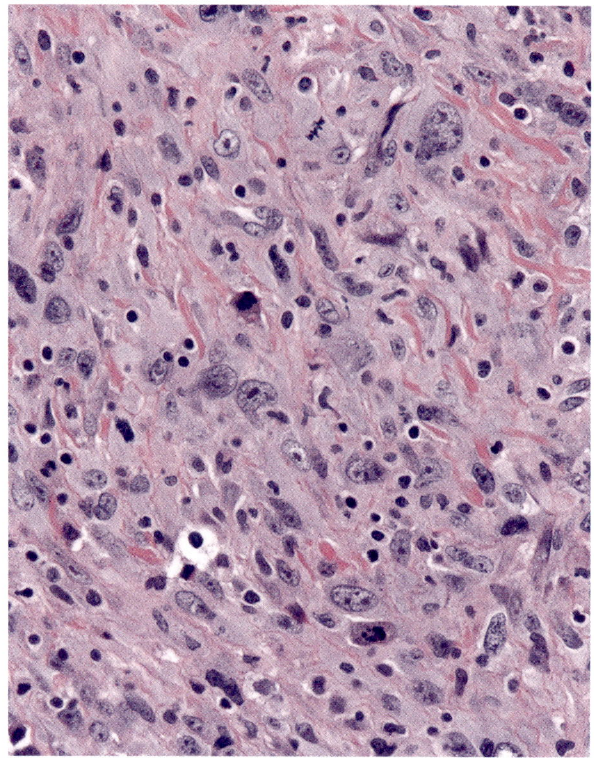

Figure 3-59

SARCOMATOID MESOTHELIOMA

This sarcomatoid mesothelioma has focal pleomorphic and transitional cytologic features, which may cause diagnostic problems in relation to other pleomorphic malignancies.

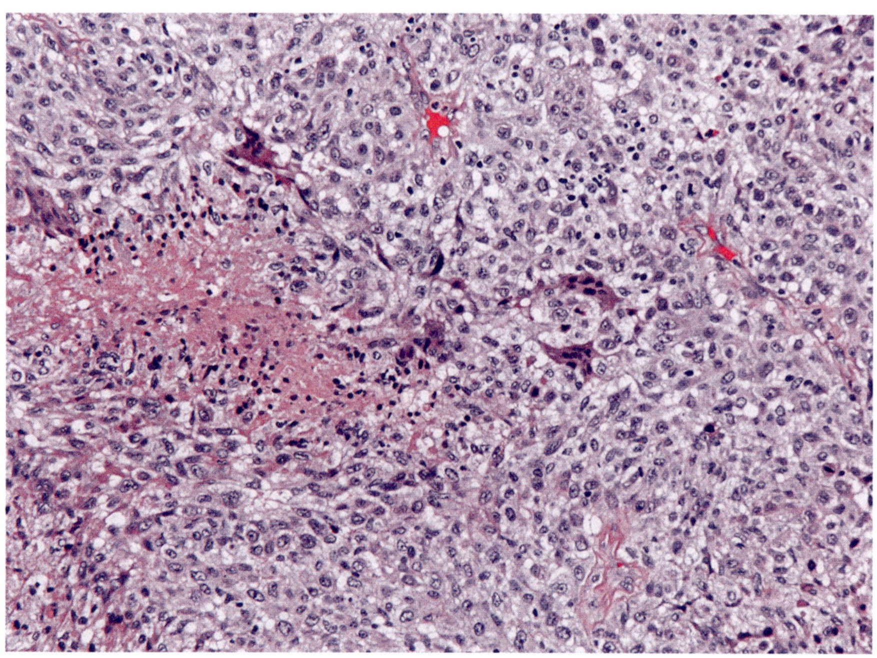

Figure 3-60

SARCOMATOID MESOTHELIOMA

Osteoclast-like benign multinucleated cells are present in this sarcomatoid mesothelioma which also shows areas of necrosis. In this field, there is more cytoplasmic overlap between sarcomatoid and epithelioid morphologies (transitional cytologic features) in the tumor cells.

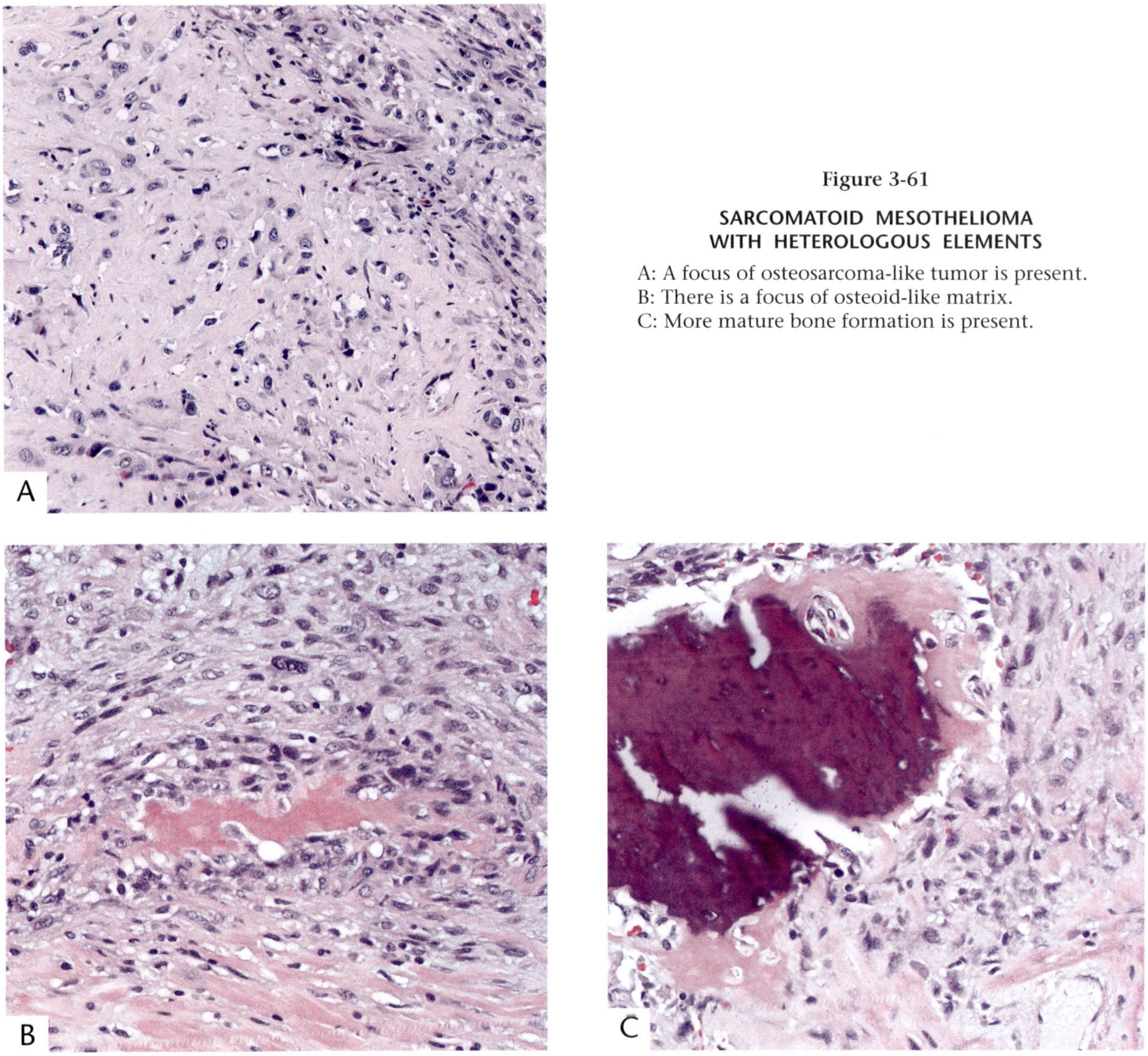

Figure 3-61

SARCOMATOID MESOTHELIOMA WITH HETEROLOGOUS ELEMENTS

A: A focus of osteosarcoma-like tumor is present.
B: There is a focus of osteoid-like matrix.
C: More mature bone formation is present.

pathologist make a diagnosis of mesothelioma when faced with complex histologic features.

Mesothelioma with Heterologous Elements. Mesothelioma with heterologous elements is the term used to describe sarcomatoid mesotheliomas with rhabdomyosarcomatous, osteosarcomatous (fig. 3-61), or chondrosarcomatous components. About 2 percent of sarcomatoid mesotheliomas show osteosarcomatous or chondrosarcomatous differentiation (174). Of the heterologous elements, osteosarcoma is most common, followed by a mixture of osteosarcoma and chondrosarcoma, rhabdomyosarcoma only, and chondrosarcoma only (175). Muscle markers may be required to confirm rhabdomyosarcomatous areas. When there is a diffuse pleural tumor with heterologous elements, even if the keratin stain is negative, by convention the diagnosis of mesothelioma is preferred over osteosarcoma, chondrosarcoma, or rhabdomyosarcoma (140,175). Rare cases with angiosarcomatous heterologous elements have also been reported (176).

Desmoplastic Mesothelioma. This is characterized by areas constituting at least 50 percent of the tumor of atypical spindle cells arranged haphazardly (so-called patternless pattern) within a dense hyalinized, fibrous stroma (fig. 3-62) (140), although tumor may have lesser amounts

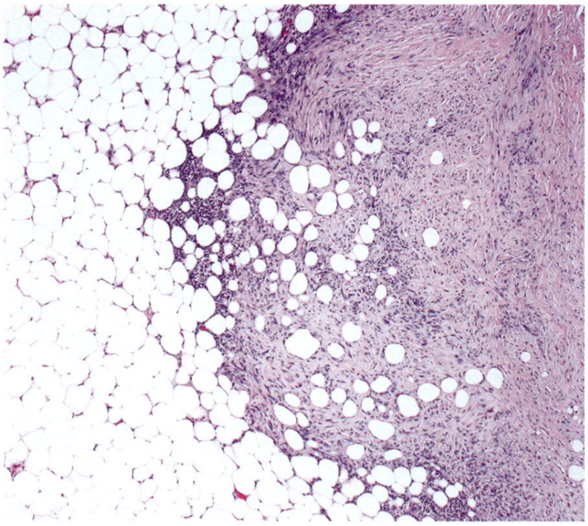

Figure 3-62

SARCOMATOID MESOTHELIOMA: DESMOPLASTIC

At low power, a paucicellular fibrous-appearing tumor is seen, which infiltrates into fat, evidence supporting that it is malignant.

of desmoplastic features (174). In one study of 326 sarcomatoid mesotheliomas, 44 percent were conventional sarcomatoid mesotheliomas, 34 percent had desmoplastic foci, and 21 percent were classified as true desmoplastic variants, although there was no standardization of sample types (174).

The tumor is paucicellular and nuclear atypia is mild, making the diagnosis difficult in small biopsies (fig. 3-63). In larger biopsies and resections, areas of conventional sarcomatoid mesothelioma are often present, which is confirmatory for diagnosis of malignancy (fig. 3-64). This highlights the need for thorough sampling of densely fibrotic tumors in order to identify overtly malignant areas that may be sparsely present. When only desmoplastic areas are seen, the best indication of malignancy is finding invasion into underlying fat or lung parenchyma (discussed below) (fig. 3-62). Identifying areas of bland "collagen necrosis," i.e., without any nuclear debris or inflammation, is also helpful in making a diagnosis (fig. 3-65).

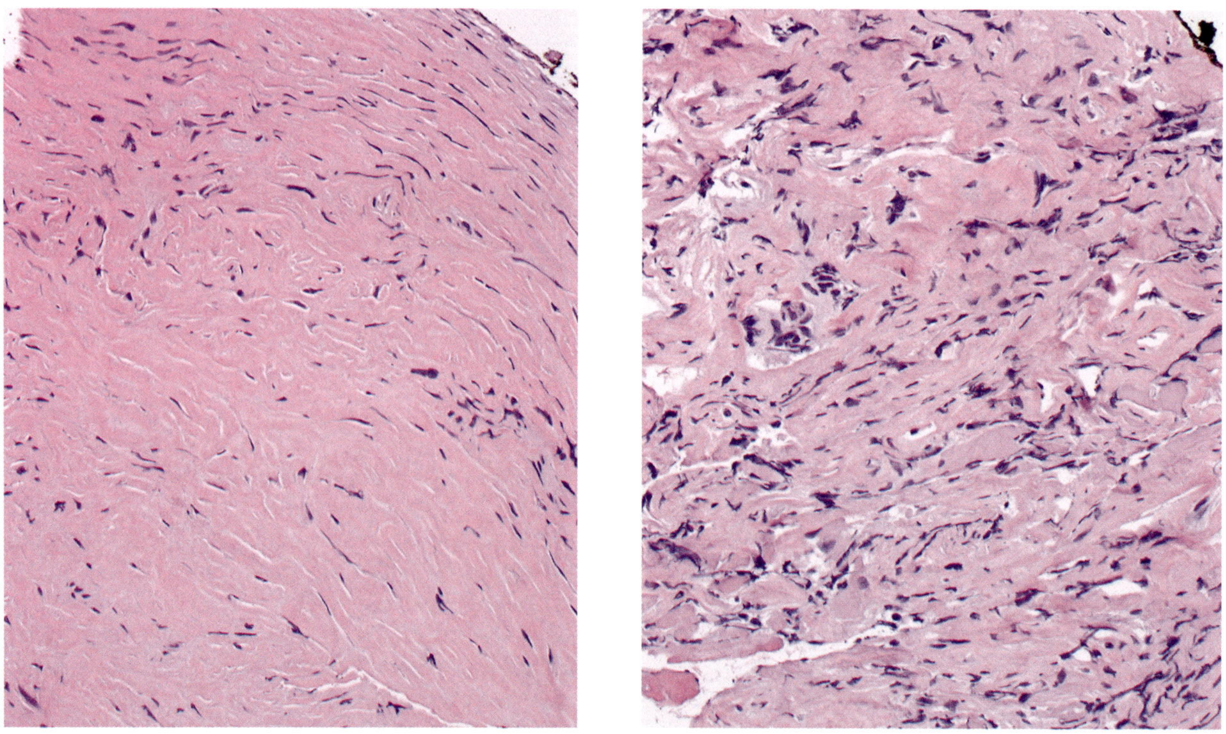

Figure 3-63

SARCOMATOID MESOTHELIOMA: DESMOPLASTIC

Left: Core needle biopsy of pleura shows a bland desmoplastic tumor. There was homozygous loss of p16, without which a diagnosis of malignancy would not be possible.

Right: A different area of the same biopsy shows more atypia and hyperchromasia, which led to testing for the *CDKN2A* deletion.

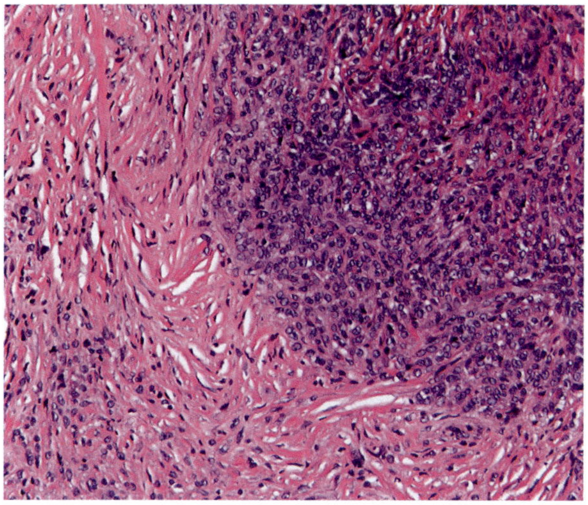

Figure 3-64

SARCOMATOID MESOTHELIOMA: DESMOPLASTIC COMPONENT

There are areas of high as well as low cellularity in this surgical biopsy specimen.

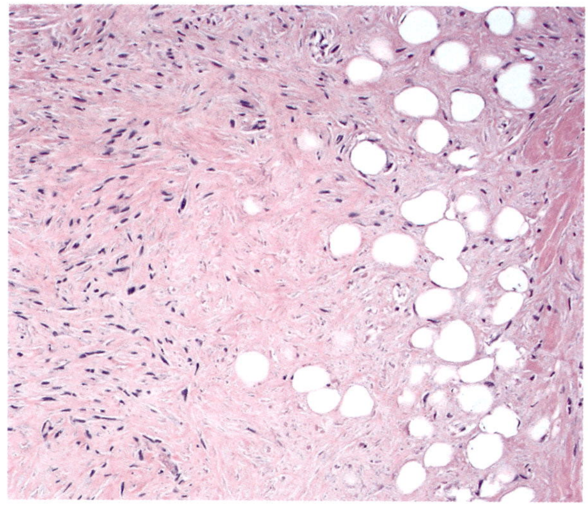

Figure 3-65

SARCOMATOID MESOTHELIOMA: DESMOPLASTIC

Areas of bland "collagen necrosis," as well as infiltration into fat, are present.

Biphasic Mesothelioma

Biphasic mesothelioma has both epithelioid and sarcomatoid components, either one of which comprises at least 10 percent of the tumor in definitive resections (fig. 3-66) (140a). The 10 percent cutoff is arbitrary, however, with different prognostically relevant cutoff points reported (125,165). In smaller samples, tumors are classified as biphasic when any sarcomatoid area is present. Various histologic patterns of sarcomatoid and epithelioid subtypes can be present in any combination, thus the morphology will be as described above. Heterologous elements may be present (fig. 3-67). Pleomorphism and necrosis are common (fig. 3-68).

One study of 144 patients undergoing pleurectomy and decortication showed that those with biphasic mesotheliomas with more than 50 percent epithelioid component had a longer overall survival period than those with less than 50 percent (11.8 versus 6.62 months; $p<0.001$) (125). In a further paper, a sarcomatoid component of less than 80 percent correlated with better survival (165).

Good interobserver agreement has been reported in identifying the extent of the sarcomatoid component in biphasic mesotheliomas (143). Reporting the amount of sarcomatoid and

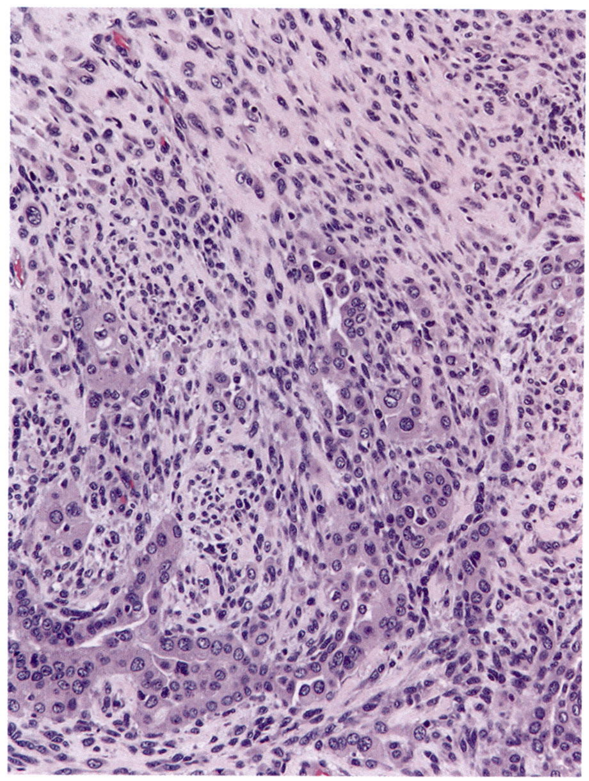

Figure 3-66

BIPHASIC MESOTHELIOMA

The tumor has both epithelioid and sarcomatous components.

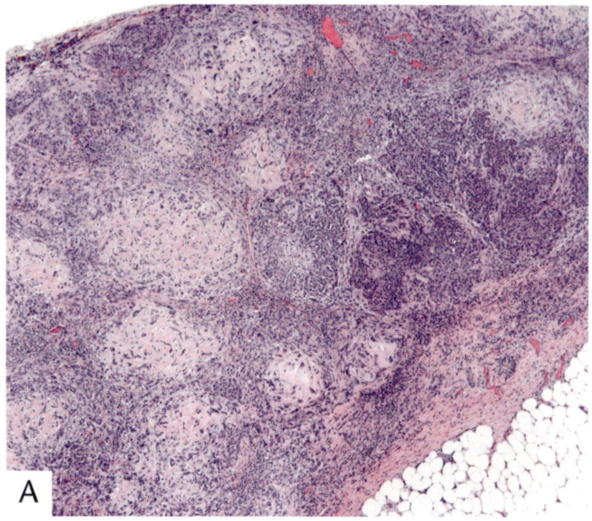

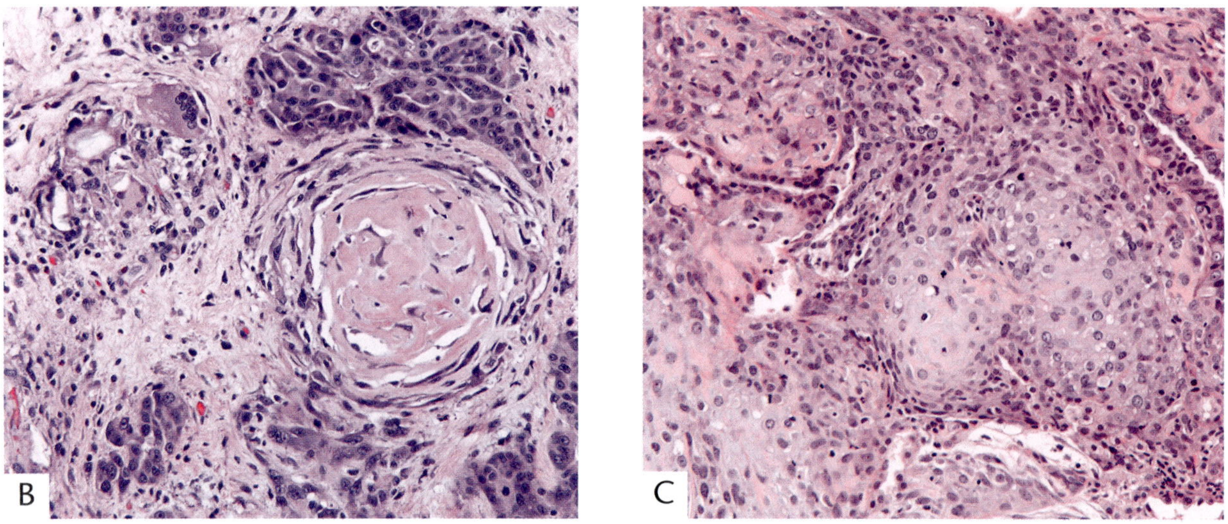

Figure 3-67

BIPHASIC MESOTHELIOMA

A: Nodules of osteoid-like matrix are present in this biphasic mesothelioma.

B: Higher-power view shows a biphasic malignancy with a nodule of osteoid. Foreign body giant cells with talc are also present in this patient who had prior talc pleurodesis.

C: Chondrosarcoma is present in another biphasic tumor.

epithelioid components is therefore recommended, because of its potential impact on prognosis and the therapeutic management of the patient (140a), although the point of greatest prognostic relevance has yet to be determined. In biphasic mesotheliomas, the extent of epithelioid and sarcomatoid elements needs to be documented in a standardized fashion prospectively, so that the significance of the subtypes, in terms of prognosis and management, can be refined.

Care needs to be taken not to interpret benign stroma as malignant (fig. 3-69) because it can be difficult to determine whether the spindle cell proliferation is malignant or a reaction to the invasive epithelioid mesothelioma. Immunostaining for BAP1 helps diagnose biphasic mesothelioma if it shows loss of staining in the spindle cell nuclei (i.e., concordant loss in both components) (177,178). Also, malignant spindle cells tend to show more intense cytokeratin staining.

IMMUNOHISTOCHEMICAL FINDINGS

The availability of highly sensitive and specific antibodies to mesothelial antigens and other tumor antigens has made immunohistochemical staining the mainstay for confirming the diagnosis of mesothelioma (179). Histochemical stains such as mucicarmine, periodic acid–Schiff (PAS) (with and without diastase), and Alcian blue (with and without hyaluronidase) were helpful in the past but are no longer

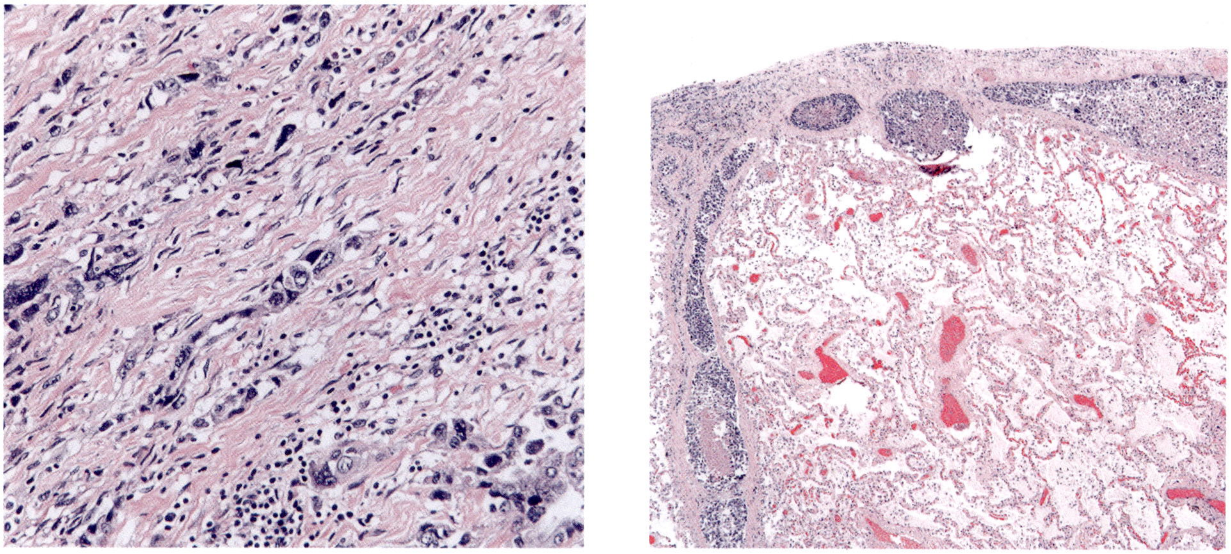

Figure 3-68

BIPHASIC MESOTHELIOMA

Left: Marked nuclear pleomorphism is present in both the sarcomatoid and epithelioid components.
Right: There is extensive lymphangitic spread of the epithelioid component, which also shows necrosis.

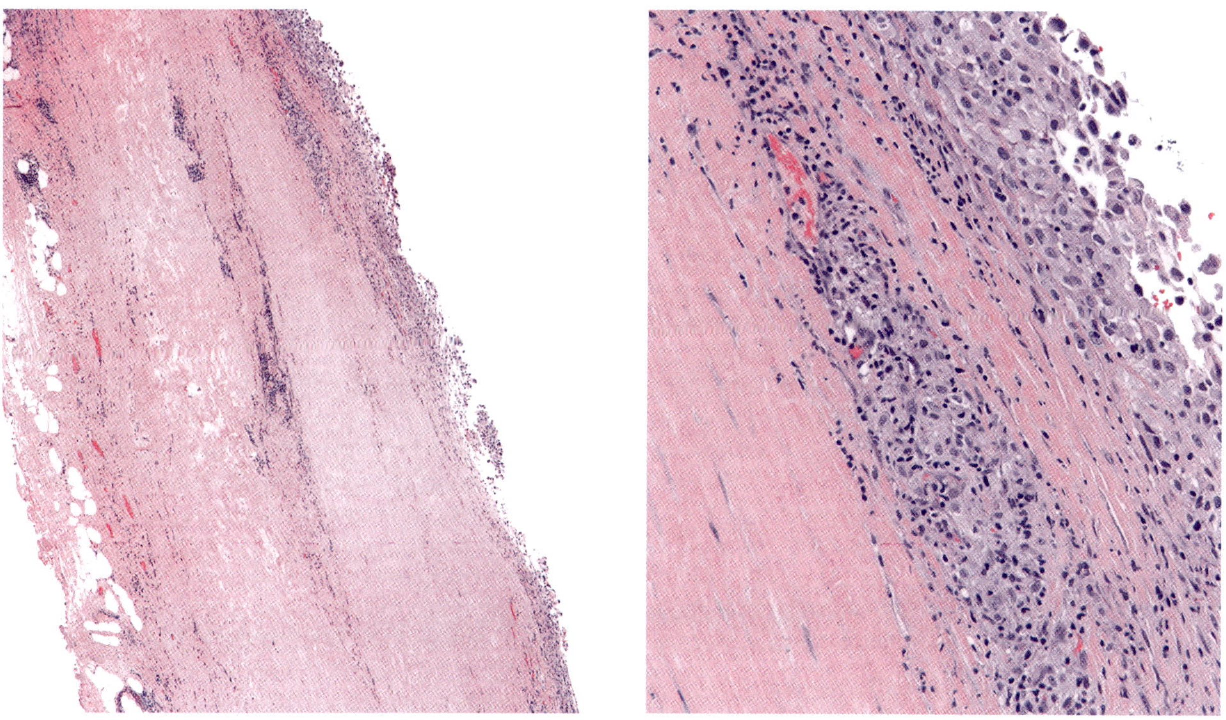

Figure 3-69

EPITHELIOID MESOTHELIOMA AND PLAQUE

Left: Low power shows the tumor growing over and into a plaque, giving the appearance of a biphasic tumor.
Right: A higher-power view shows epithelioid mesothelioma growing over and into a plaque. Note the parallel arrangement of collagen fibers of the plaque.

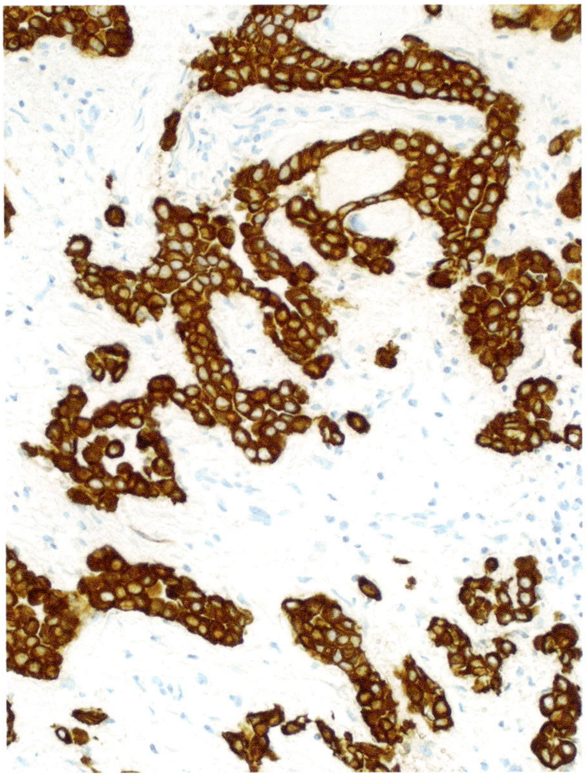

Figure 3-70

EPITHELIOID MESOTHELIOMA

All tumor cells show strong positivity with cytokeratin AE1/AE3.

used routinely. Similarly, the role of electron microscopic examination has become restricted because immunohistochemistry is faster and more widely available (180).

Since no antibody is 100 percent diagnostic, a panel of stains is usually applied, the selection of which depends on the differential diagnosis, as further discussed below, as well as on availability and experience of the laboratory. It is not possible to describe and discuss all the aspects of all the antibodies since new ones continue to be developed; thus pathologists need to keep a continuous eye on recently published literature.

Epithelioid Mesothelioma

Virtually all epithelioid mesotheliomas are positive for pancytokeratin (fig. 3-70) and most are strongly and diffusely positive for calretinin, WT-1, CK5/6, and podoplanin (D2-40), which are the main mesothelial markers (fig. 3-71). If the morphology is typical, it is recommended that staining for two mesothelial markers and two carcinoma markers be done to confirm a diagnosis of epithelioid mesothelioma (180).

In a few cases, no markers are positive, which can be caused by a variety of reasons, such as poor fixation. Selection of a different block and repeating the staining process may be helpful (180).

Sarcomatoid Mesothelioma

Most sarcomatoid mesotheliomas are much less immunoreactive than epithelioid mesothelioma. However, 90 to 95 percent are positive, even if focally, for broad spectrum cytokeratin antibodies (fig. 3-72) (174) . Calretinin is seen in only about 30 percent of cases (fig. 3-73) while other mesothelial markers (CK5/6 and WT-1) are often focal or negative (figs. 3-74, 3-75). Podoplanin (D2-40) is more likely to be positive (fig. 3-76) (174,175). Recently, GATA3 has been proposed as a useful marker in distinguishing sarcomatoid mesotheliomas from other sarcomatoid tumors (181).

CYTOLOGIC FINDINGS

Cytologic preparations may be useful in the diagnosis of mesothelioma. A common specimen type for this evaluation is effusion exfoliative cytology. Pleural, peritoneal, and pericardial fluid samples are common, and most often benign, as a result of physiologic abnormalities or inflammatory or infectious processes. Therefore, the distinction of reactive from malignant cells, including malignant mesothelial cells, is a frequent issue.

While many effusions are transudative (i.e., low in protein and generally the result of congestive heart failure), others are exudative (higher in protein and cellularity). While many exudative effusions are infectious, others are the result of malignancy (182). Most pleural malignancies, however, are secondary to carcinomatous or lymphomatous involvement rather than a mesothelial malignancy; the majority of malignant effusions are either metastatic lung or breast carcinomas (183).

Pleural fluid is examined by cytology smear, cytospin preparations, liquid-based enrichment such as Thin Prep, or cell blocks. Cell blocks are useful for immunohistochemistry as well as molecular testing.

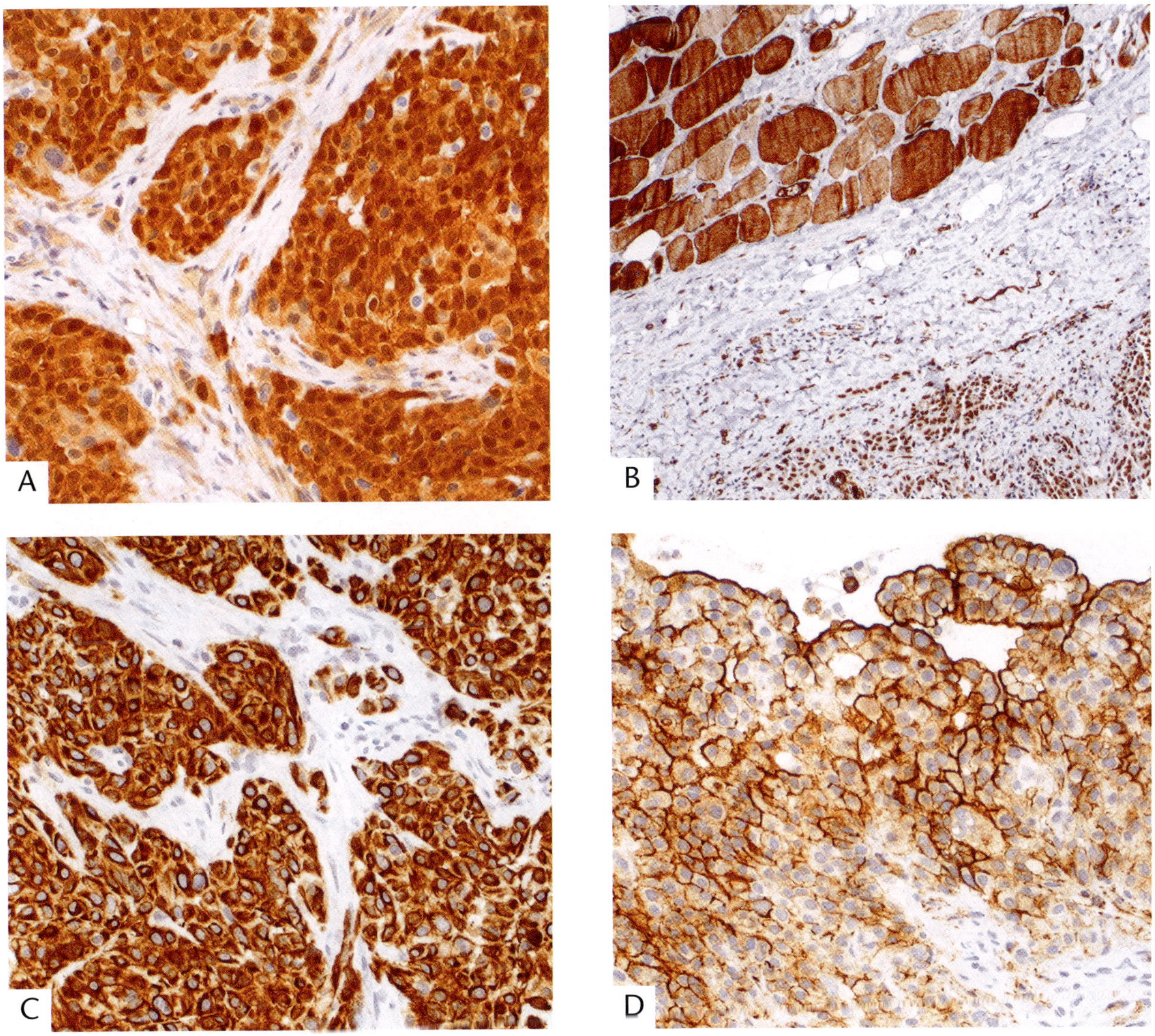

Figure 3-71

EPITHELIOID MESOTHELIOMA

A: There is strong diffuse staining with antibody to calretinin. Most tumor cells have the typical pattern of staining of stronger nuclear and weaker cytoplasmic staining.

B: There is strong nuclear staining with antibody to WT-1 in the tumor (bottom right). The stromal cells of the pleura stain in a similar fashion. Lymphatics (which are compressed and appear as linear structures) and skeletal muscle show strong cytoplasmic staining.

C: There is strong diffuse cytoplasmic staining with antibody to CK5/6.

D: There is strong membranous staining with antibody to podoplanin (D2-40) in the majority of tumor cells.

In patients with mesothelioma, pleural effusion cytology is positive in about 60 percent of cases, and the majority of positive cases can be obtained from one sample of at least 50 mL. Serial samples improve yield, but with diminishing returns (184). Effusion cytology can be diagnostically challenging for a variety of reasons. This includes obscuring blood and inflammation, as well as an admixture with reactive mesothelial cells (185).

The concordance of diagnosis among observers is higher for malignant diagnoses than benign ones (186). Simultaneous use of different preparations improves diagnostic outcomes,

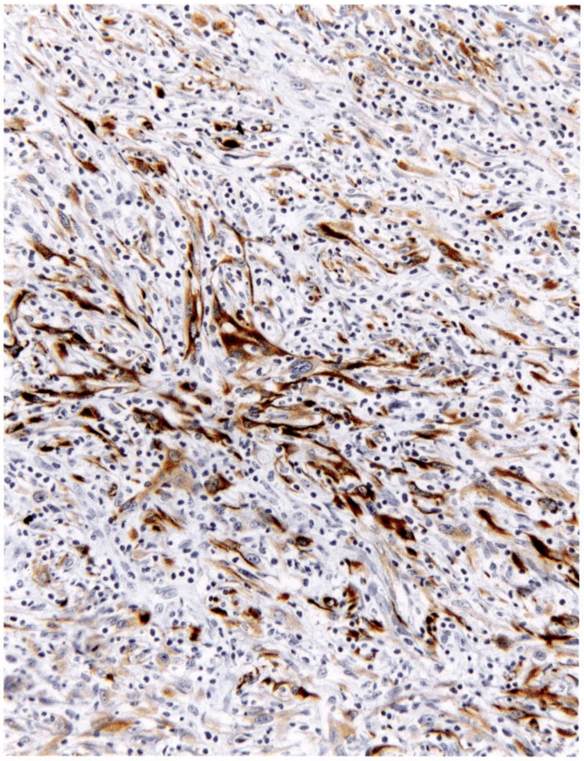

Figure 3-72

SARCOMATOID MESOTHELIOMA

Many of the tumor cells show strong cytoplasmic positivity with cytokeratin AE1/AE3. Numerous tumor-infiltrating lymphocytes, which are mainly T cells, are seen.

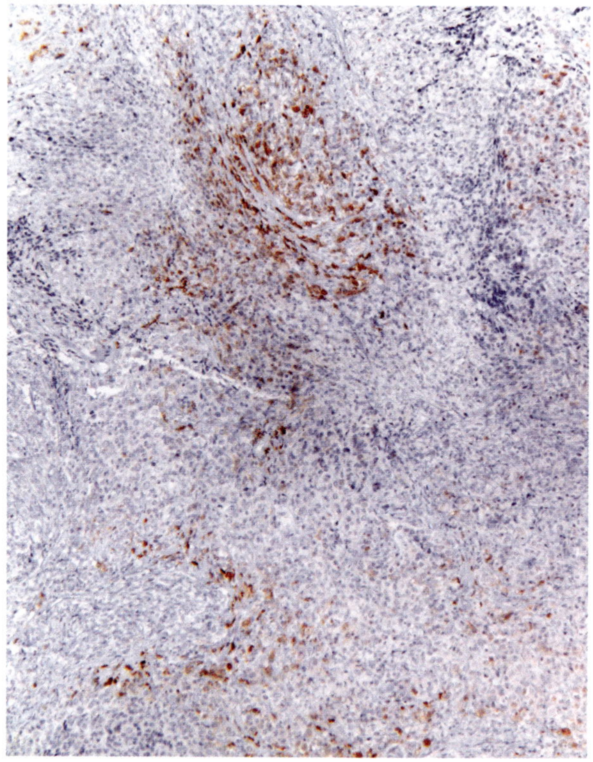

Figure 3-73

SARCOMATOID MESOTHELIOMA

Calretinin stains only some of the tumor cells.

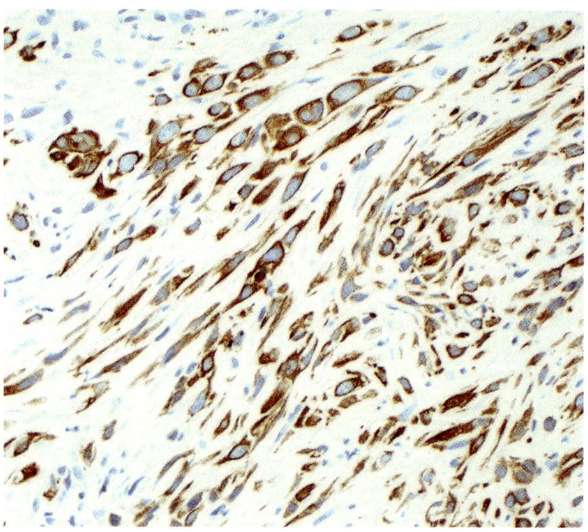

Figure 3-74

SARCOMATOID MESOTHELIOMA

Focal staining for CK5/6 is present in this tumor.

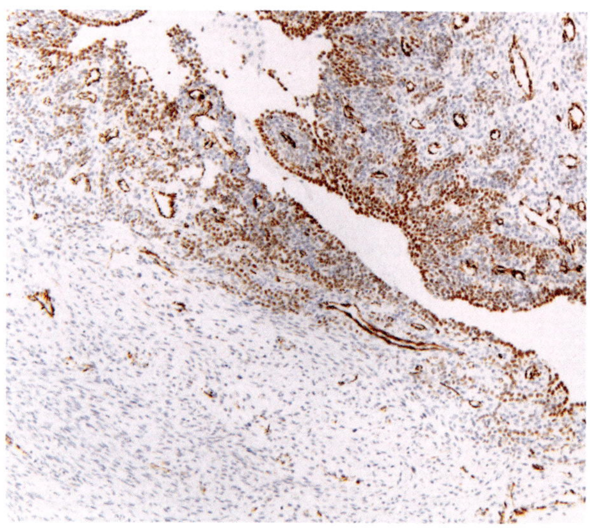

Figure 3-75

BIPHASIC MESOTHELIOMA

WT-1 stains only the epithelioid component while the sarcomatoid component is negative.

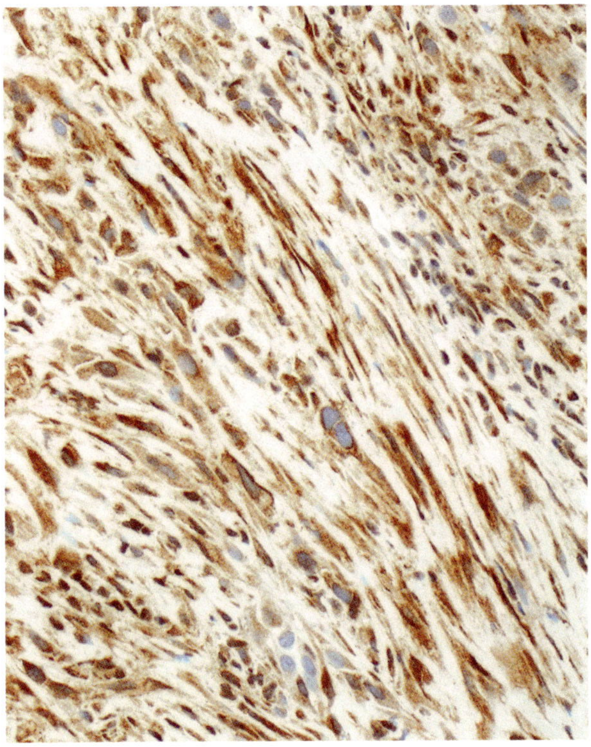

Figure 3-76

SARCOMATOID MESOTHELIOMA

Antibody to podoplanin (D2-40) strongly stains this sarcomatoid mesothelioma. The staining is more granular and cytoplasmic than seen in epithelioid mesothelioma.

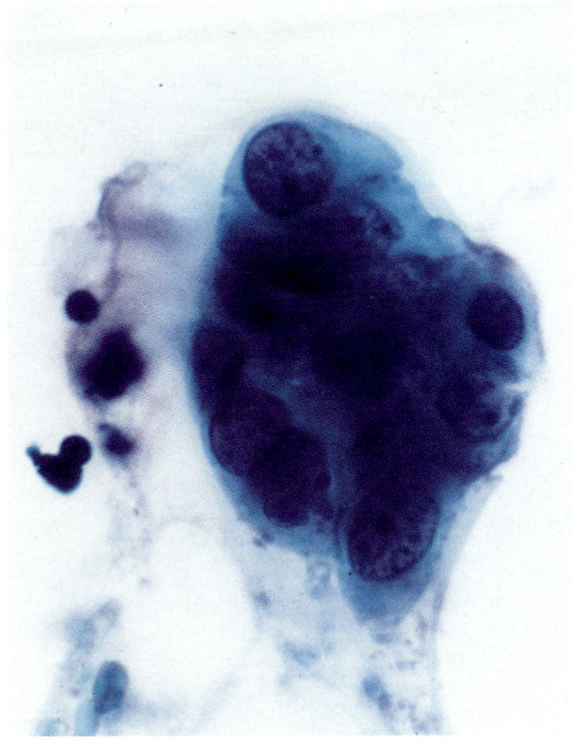

Figure 3-77

ADENOCARCINOMA IN PLEURAL EFFUSION

While the nuclear features of malignancy are seen at the periphery, evaluation of the entire cluster requires changing the focal plane in a 3-D structure.

and the addition of clinical information and immunohistochemistry may help in the subclassification of malignancy.

Cytologic Features of Malignancy

Cancer cells in effusions occur in nests, sheets, three-dimensional clusters, and single cells. The nests, and especially the three-dimensional structures, suggest malignancy (fig. 3-77), since reactive mesothelial cells are more often single or in flat sheets (fig. 3-78). As previously noted, most malignant effusions result from secondary involvement of the pleura and are often adenocarcinoma. These can additionally form acini or papillary arrangements, and psammoma bodies may be seen. While cellular samples with morular formations are diagnostically helpful, challenging cases include those with paucicellular samples or with relatively bland single cells, such as metastatic lobular breast carcinoma. In addition to adenocarcinoma, other secondary malignancies, such as small cell carcinoma (fig. 3-79), malignant melanoma (fig. 3-80), sarcomas, and lymphomas, are encountered. In such cases, adjunctive use of immunohistochemistry can help convert an atypical or suspicious impression into a definitive positive diagnosis.

A classic cellular feature of malignancy is cytomegaly with a high nuclear to cytoplasmic ratio (fig. 3-81, left). Chromatin may be coarse with hyperchromasia, and macronucleoli or multiple nucleoli can be seen. Atypical mitoses, when present, add to the certainty of diagnosis (fig. 3-81, right).

Reactive Atypia in Mesothelial Cells

One of the challenges of exfoliative cytology of effusions is the overlap of reactive atypia of mesothelial cells with malignancy. While a uniform background of single or sheet-like mesothelial cells is typically benign or reactive (fig. 3-78), mesothelial cells can exhibit nuclear

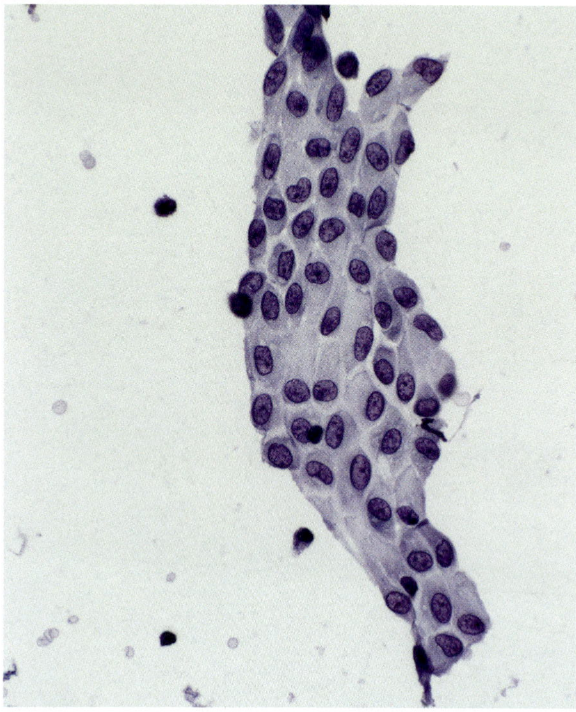

Figure 3-78

REACTIVE MESOTHELIAL CELLS IN PLEURAL EFFUSION

The flat sheet is in focus throughout, and the cells are relatively uniform with a pavemented appearance.

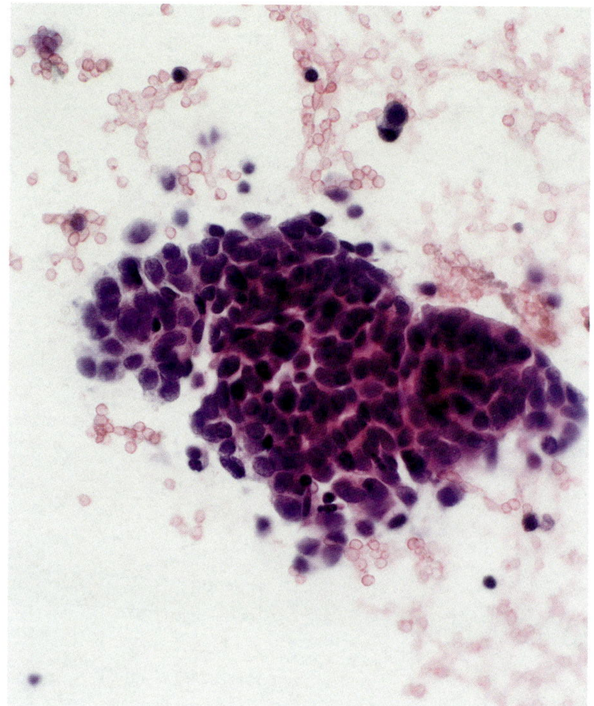

Figure 3-79

SMALL CELL CARCINOMA IN EFFUSION

The cell block shows uniform chromatin, nuclear molding, and scant cytoplasm, supportive diagnostic features.

enlargement, multinucleation (fig. 3-82), irregular nuclei, and nucleoli (fig. 3-83). Although round to oval in most cells, nuclear membrane irregularity can be seen. Cytoplasmic vacuoles may mimic intracellular mucin, although these vacuoles are larger than what is usually seen in signet ring cells (fig. 3-84). In some instances, reactive effusions are very cellular, even with clusters of cells (fig. 3-85).

Mesothelial cells have distinct cytomorphologic features. The cells have abundant cytoplasm, and as previously mentioned, can have an intracytoplasmic vacuole. The periphery of the cytoplasm can appear denser, and the surface of the cell can have a fuzzy or blurry outer rim, the light microscopic visualization of the long microvillous processes of these cells. When mesothelial cells are adjacent to each other, their surfaces do not directly touch, creating a space or "window" (fig. 3-86).

There are two components to consider when effusion cytology is evaluated: the assessment of lineage and the assessment of malignancy. Mesothelial lineage is confirmed by a combination of morphology, and if needed, immunohistochemistry for calretinin, WT-1, and CK5/6 (fig. 3-87). Epithelial lineage is supported using immunohistochemistry, often as part of a panel, for MOC31, B72.3, carcinoembryonic antigen (CEA), and claudin 4 (fig. 3-88). If the lineage is epithelial and morphologically malignant, additional markers help determine organ of origin, such as thyroid transcription factor 1 (TTF1), PAX8, and CDX2 (fig. 3-89) (180,187,188). One caveat of this approach is the low but consistent rate of PAX8 immunoreactivity in mesothelioma, especially of peritoneal origin. Also, GATA3 should not be used as a single marker for breast or urothelial carcinoma since it is immunoreactive in a subset of mesothelial cells and mesotheliomas (fig. 3-90) (181).

Once determined as mesothelial lineage, more challenging is the determination of malignancy in mesothelial cells. Statistically, a reactive mesothelial proliferation is more likely than mesothelioma. Therefore, the evidence in favor of malignancy needs to be strong. In the absence of evidence of

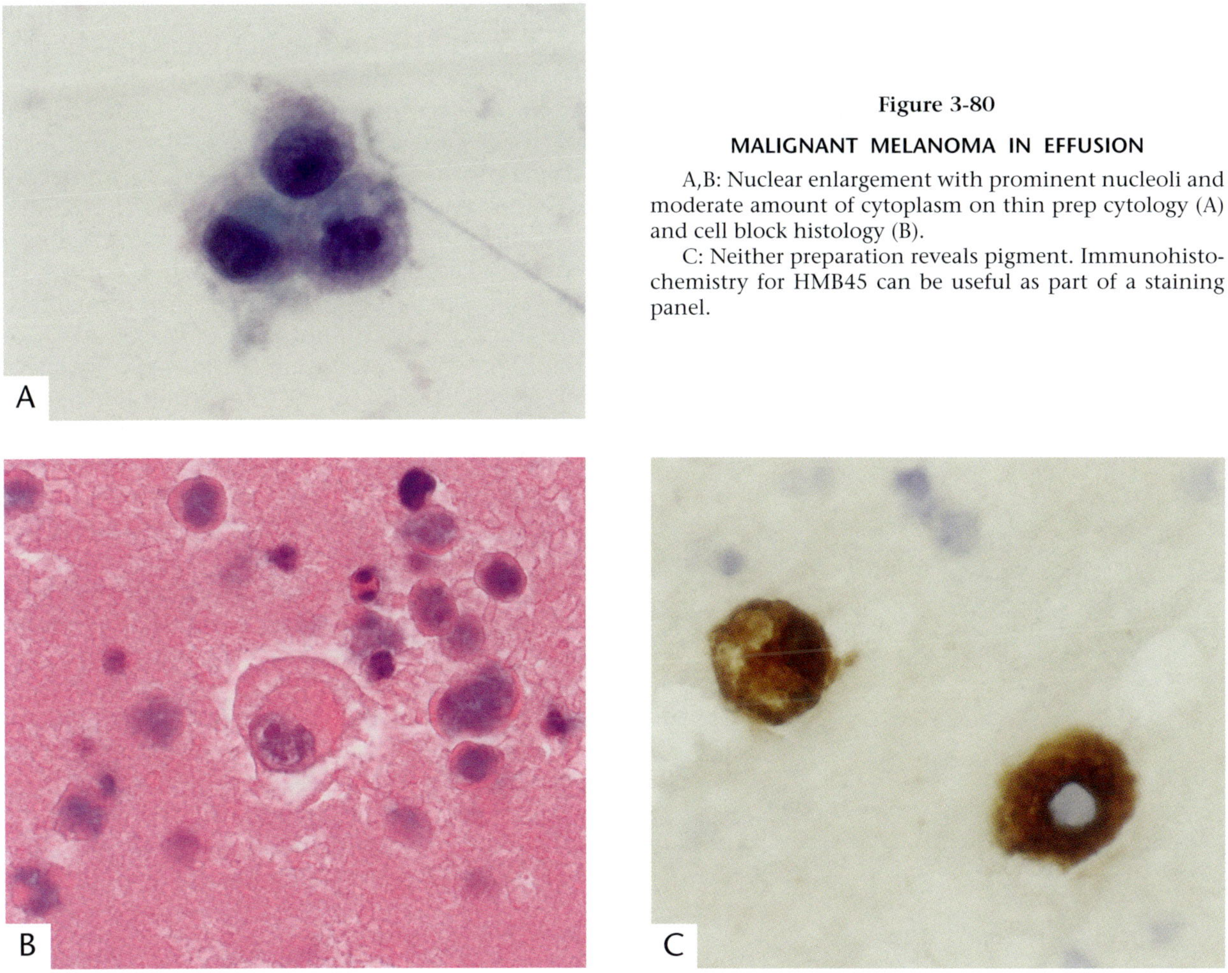

Figure 3-80

MALIGNANT MELANOMA IN EFFUSION

A,B: Nuclear enlargement with prominent nucleoli and moderate amount of cytoplasm on thin prep cytology (A) and cell block histology (B).

C: Neither preparation reveals pigment. Immunohistochemistry for HMB45 can be useful as part of a staining panel.

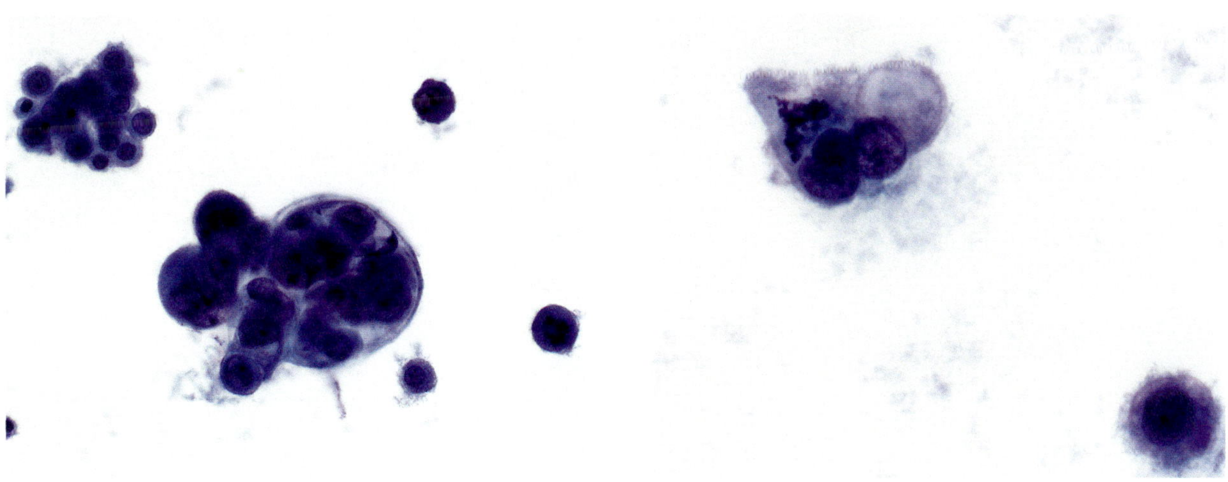

Figure 3-81

FEATURES OF MALIGNANCY

Left: Nucleomegaly and high nuclear to cytoplasmic ratio (compare to background cells). Right: Mitotic activity and atypical mitoses are malignant features.

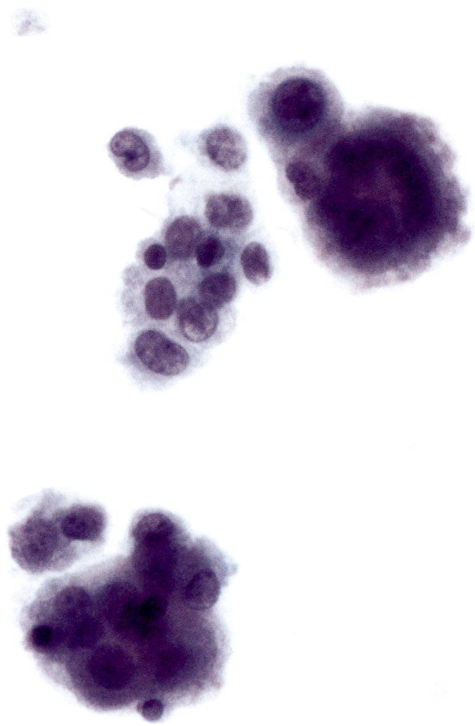

Figure 3-82

REACTIVE MESOTHELIAL ATYPIA

Nuclear enlargement and multinucleation are seen in reactive mesothelial cells.

tissue invasion, cytologic diagnosis of mesothelioma must be approached with caution, especially in the primary diagnosis of the disease.

In cases of epithelioid mesothelioma, malignant cells are larger than the background reactive mesothelial cells, resulting in recognizable differences within the smear, thin prep slide, or cell block slide (fig. 3-91). However, mesothelioma cells are often bland when compared to carcinomas (189–191). Mesothelioma cells are more likely to appear in clusters, acini, or papillary structures when compared to reactive mesothelial cells (fig. 3-92A–C), and can also have psammoma bodies (fig. 3-92D). High cellularity is also a clue, especially with cellular fragments (fig. 3-93). Cell membrane blebs and cytoplasmic vacuoles are also seen. Invasion, the single most reliable parameter in the diagnosis of malignancy, is not a feature generally obtained by cytology, but cell block samples can contain tissue fragments suggestive of infiltration (fig. 3-94). In diagnostic cases, all the criteria should be met, so that the specificity can be high for malignancy, but the resulting sensitivity may be relatively low (192,193).

The diagnosis of sarcomatoid mesothelioma may be especially difficult in cytology (192). Sarcomatoid mesotheliomas are less likely to

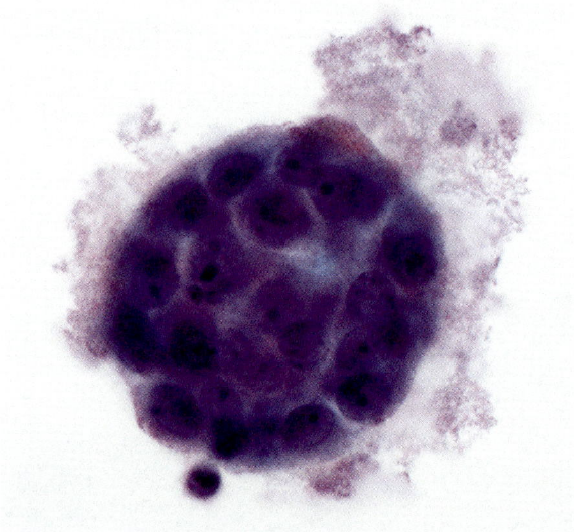

Figure 3-83

REACTIVE MESOTHELIAL ATYPIA

Reactive mesothelial cell clusters have nuclei showing nucleoli and coarse chromatin, an overlapping feature with malignancy.

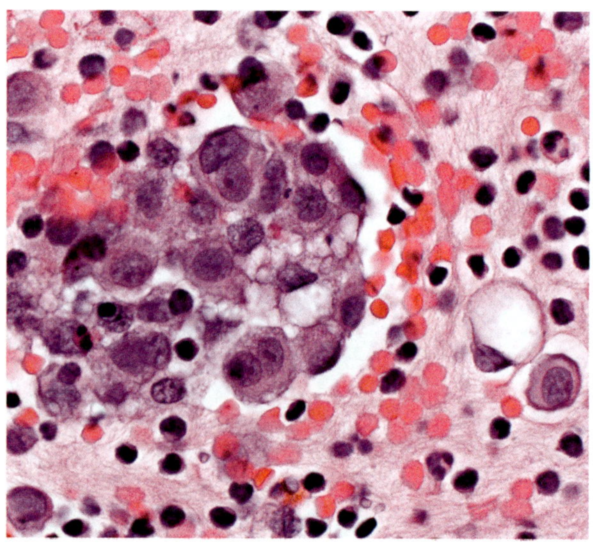

Figure 3-84

REACTIVE MESOTHELIAL ATYPIA

Mesothelial cells can contain vacuoles but these do not contain mucin. Some vacuoles dwarf the nucleus.

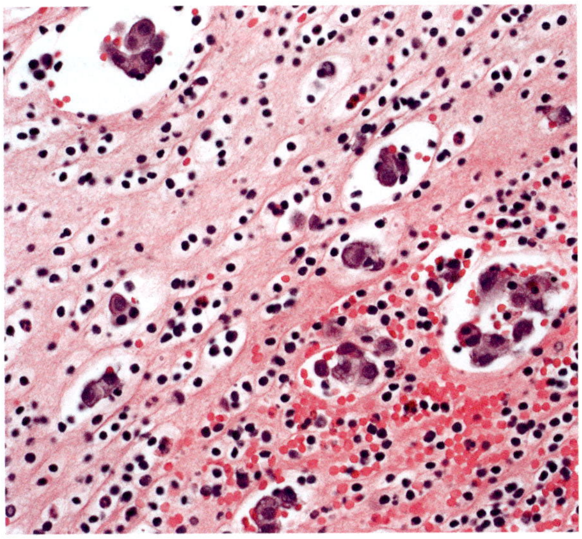

Figure 3-85

REACTIVE EFFUSION

Clusters and cellularity can be seen in reactive effusions.

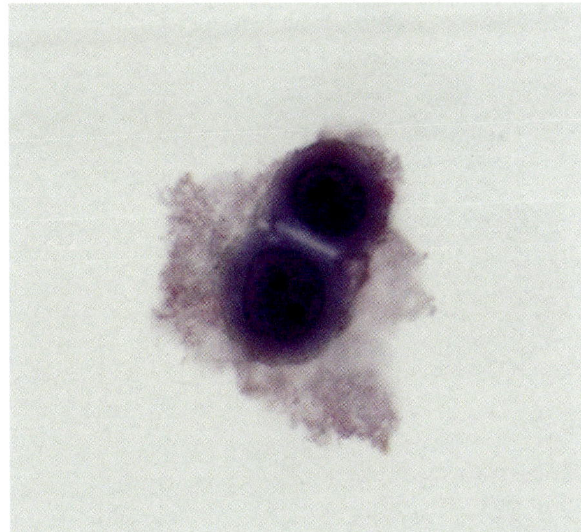

Figure 3-86

MESOTHELIAL CELL: CYTOMORPHOLOGY

An intercellular space, or "window," and a blurry outer rim of mesothelial cells are present.

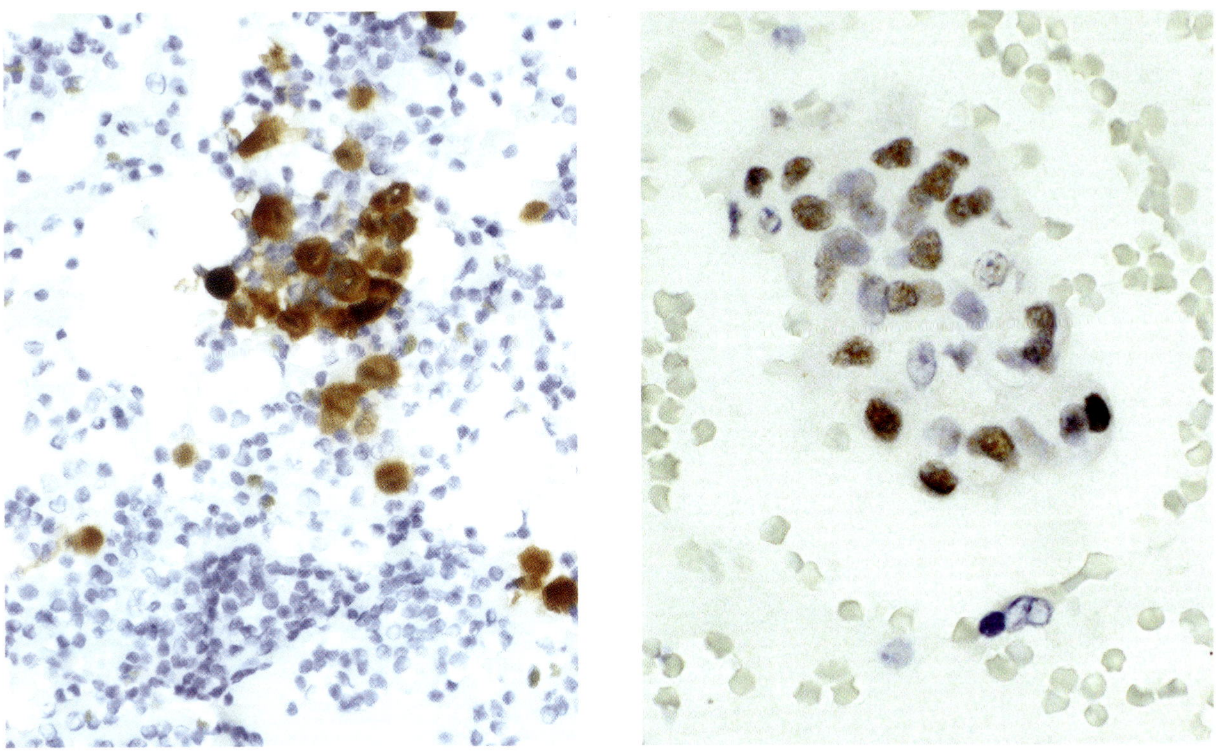

Figure 3-87

MESOTHELIAL CELL: IMMUNOHISTOCHEMISTRY

Left: Calretinin immunoreactivity should be nuclear but is often accompanied by cytoplasmic staining. Right: WT-1 immunoreactivity is nuclear in mesothelial cells.

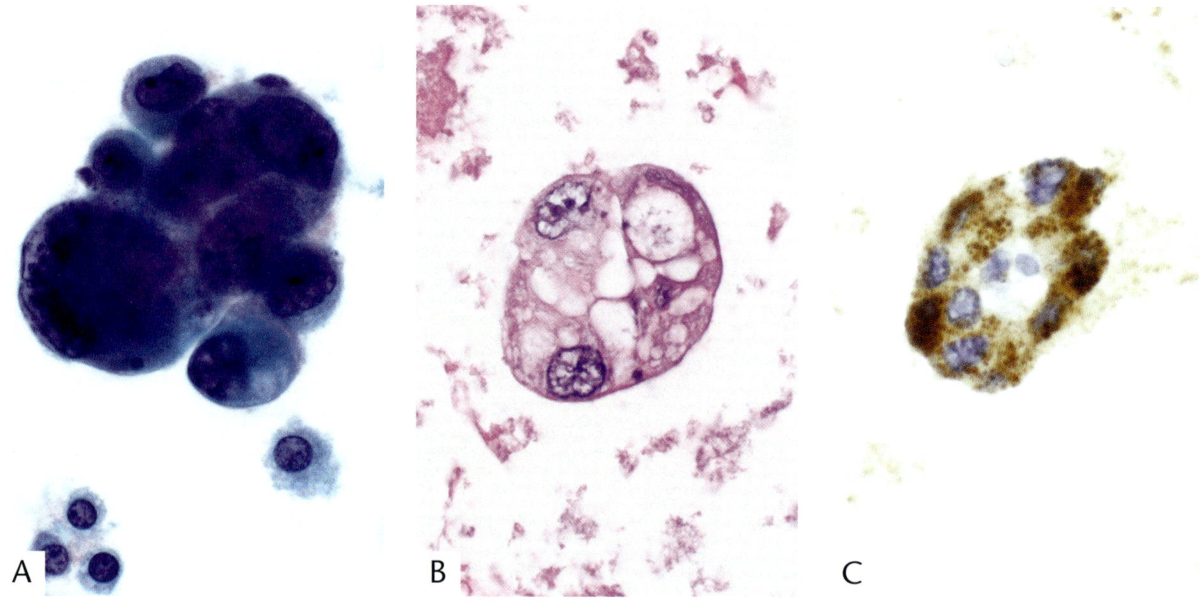

Figure 3-88

ADENOCARCINOMA: CYTOMORPHOLOGY

A: Markedly enlarged and irregular nuclei, with 3-D structure, on liquid-based preparation.
B: Cell block shows cytoplasmic vacuolization.
C: Immunoreactivity for B72.3 as part of a panel supports the diagnosis of adenocarcinoma over mesothelioma.

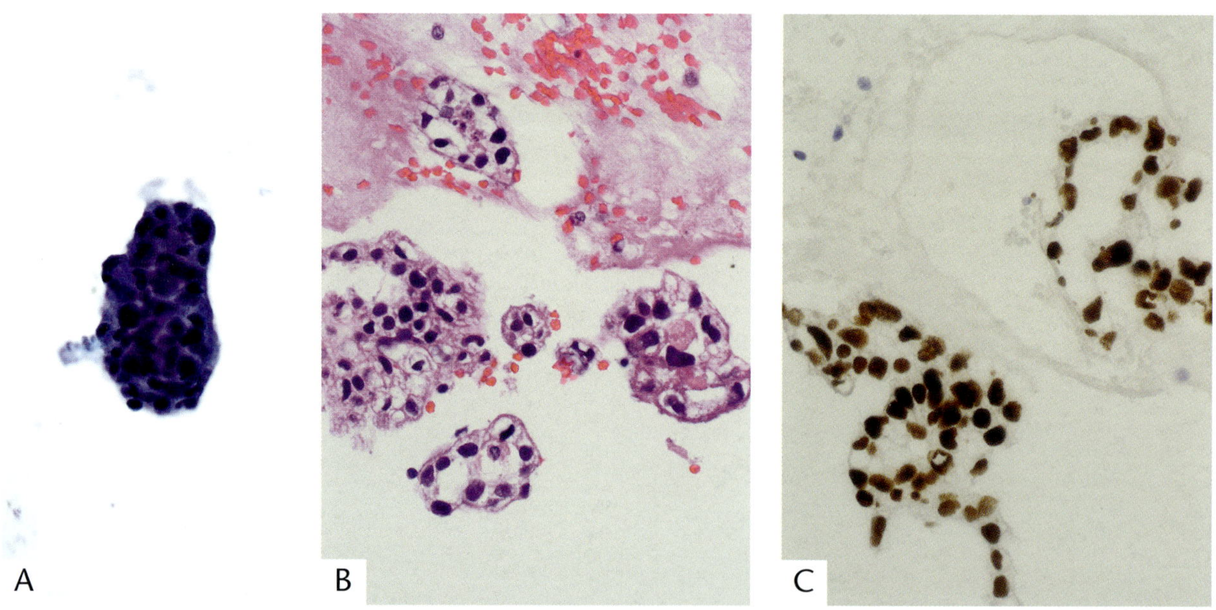

Figure 3-89

ADENOCARCINOMA: CYTOMORPHOLOGY

A: Cytomorphology shows cohesive clusters and dense darkly stained chromatin on a Papanicolaou-stained liquid-based preparation.
B: Cell block morphology suggests gland formation and cytoplasmic vacuolization.
C: Nuclear immunoreactivity for CDX2 supports a gastrointestinal origin for the tumor.

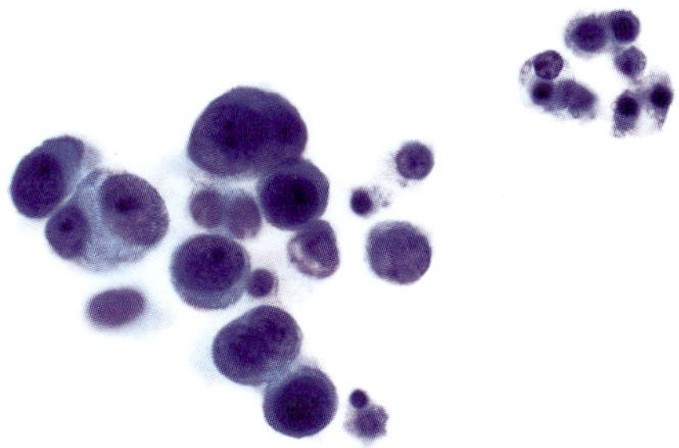

A

Figure 3-90

ADENOCARCINOMA: CYTOMORPHOLOGY

A: Cytomorphology with dyscohesion, nuclear enlargement, relatively smooth nuclear contour, and small nucleoli.

B: The cell block shows similar features of relatively bland and uniform cells.

C: Nuclear immunoreactivity for GATA3 supports a breast origin. However, a panel approach is needed, as mesothelial cells also can be immunoreactive for GATA3.

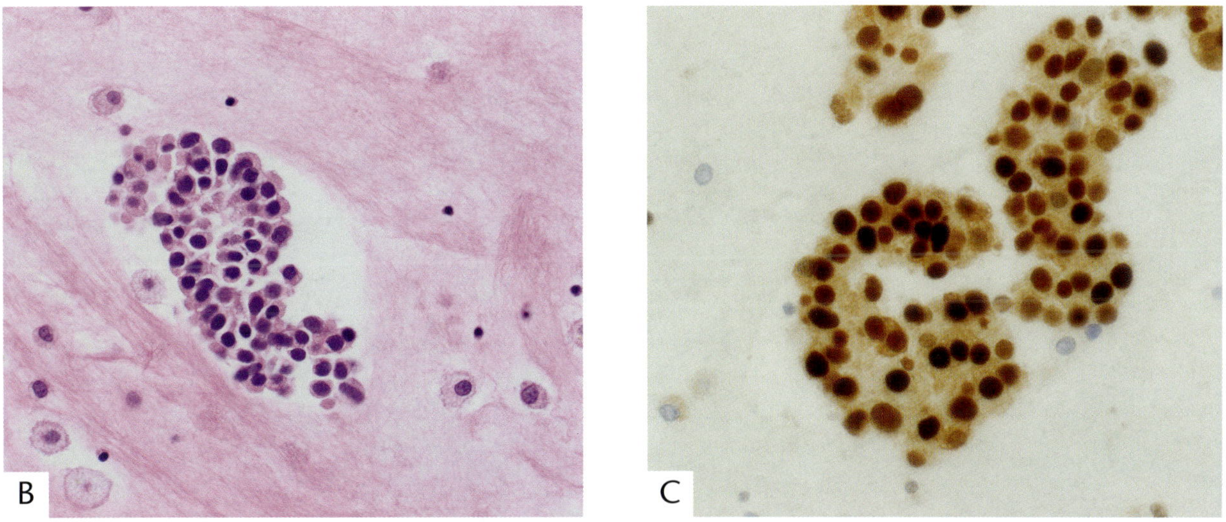

B

C

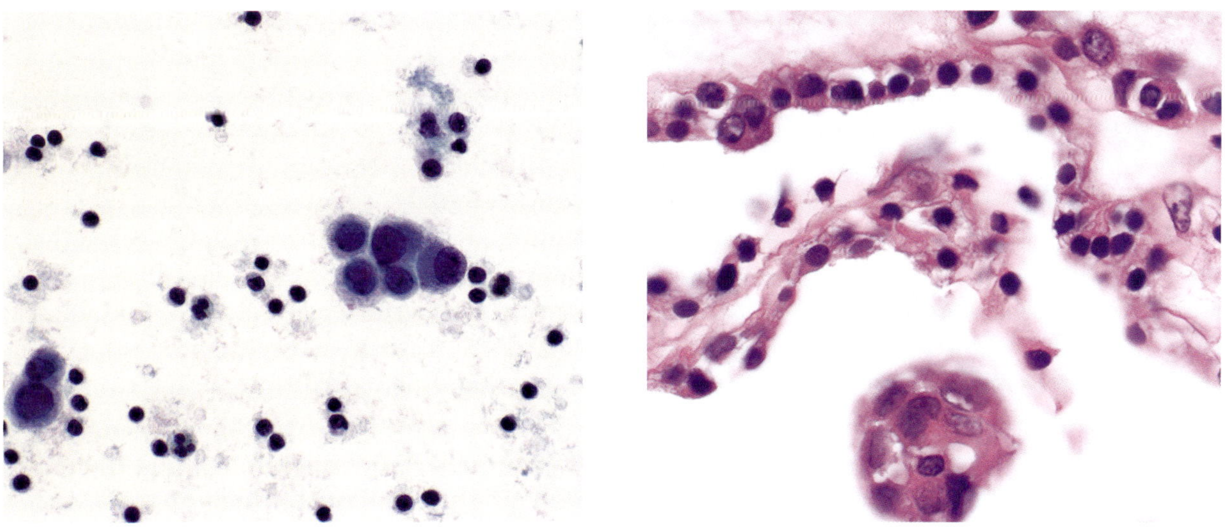

Figure 3-91

EPITHELIOID MESOTHELIOMA: NUCLEAR ENLARGEMENT

On liquid-based Papanicolaou stain (left) or cell block H&E stain (right), comparison of the malignant population to background benign mesothelial cells may be a first clue to malignancy.

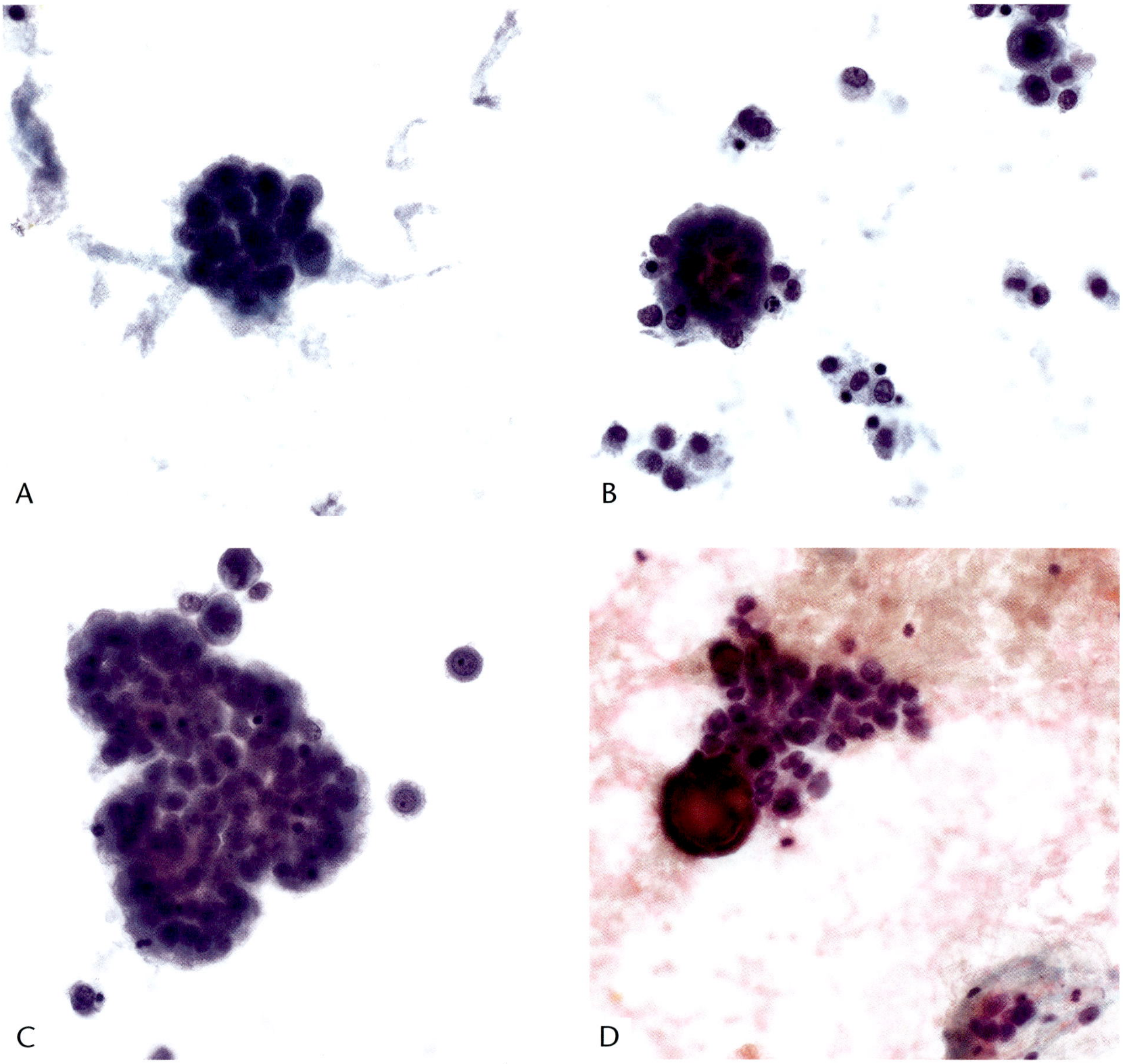

Figure 3-92

EPITHELIOID MESOTHELIOMA: CYTOLOGY

Cell clusters (A), acini (B), papillary architecture (C), and psammoma bodies (D) are seen. While these can be seen in mesothelioma, individually these features overlap with reactive mesothelial cells and carcinoma.

shed cells into fluid. By fine needle aspiration, the differential diagnosis of these tumors includes sarcomatoid carcinoma and sarcoma, requiring immunohistochemistry to resolve, and a diagnosis may remain challenging even after immunohistochemistry.

There is increasing recognition of markers that may distinguish reactive mesothelial cells from mesothelioma. Currently, a combination of BAP1 (negative nuclear immunoreactivity) and p16 fluorescence in situ hybridization (FISH) testing (99,194,195) showing homozygous deletion may be helpful. However, the sensitivity of this approach varies by histologic type and also by site (196). Immunohistochemistry for MTAP may also be of value (197,198).

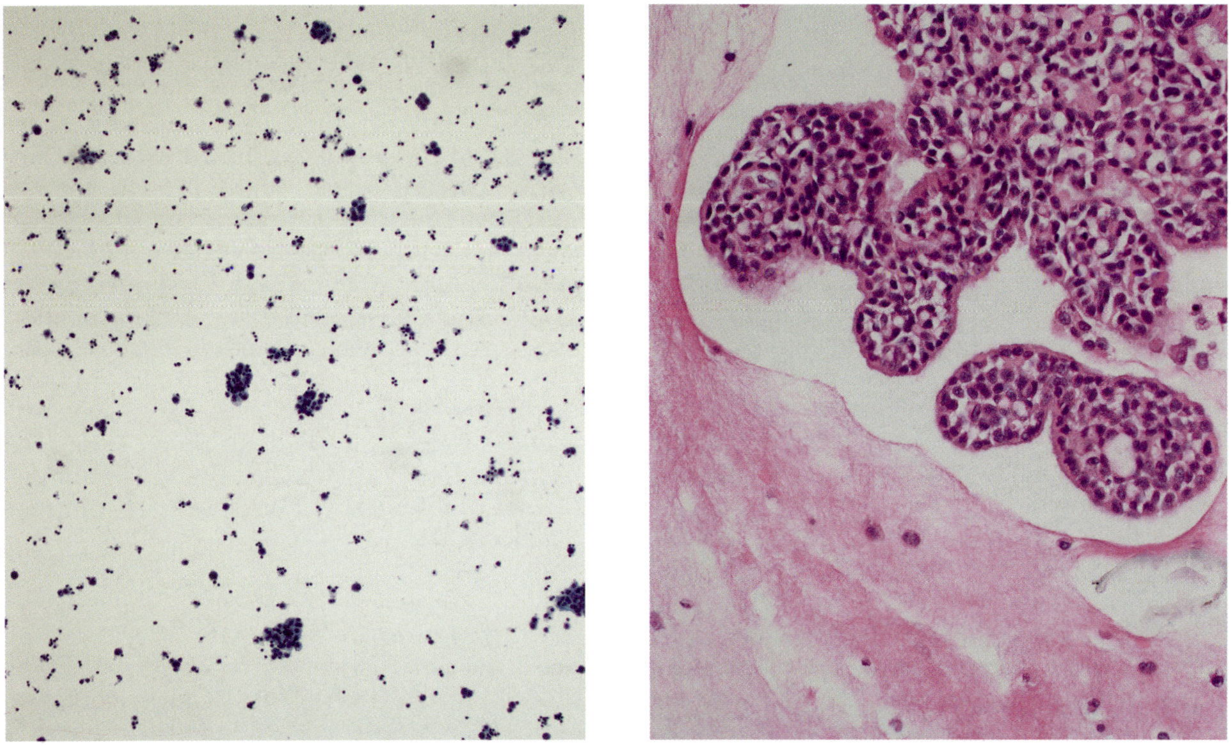

Figure 3-93

EPITHELIOID MESOTHELIOMA: CELLULARITY

Left: A Papanicolaou-stained smear highlights the high cellularity. Right: Large fragments on cell block also suggest neoplasm.

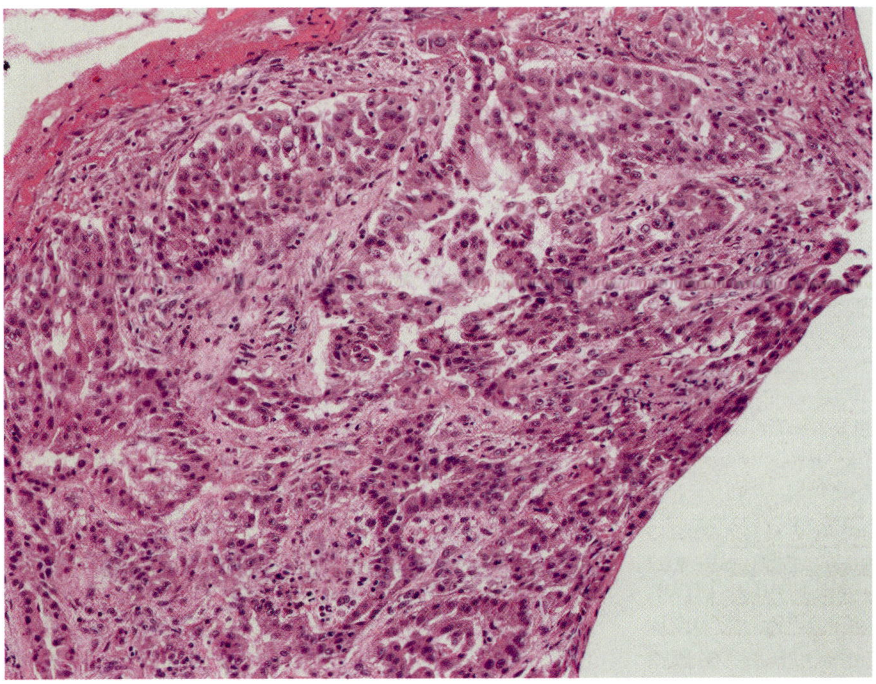

Figure 3-94

EPITHELIOID MESOTHELIOMA: TISSUE FRAGMENT

While invasion is not generally assessed in cytology, fine needle aspirate samples can show expansile mesothelial nests in stroma suggestive of infiltration.

MOLECULAR GENETIC FINDINGS

Biomarkers in Mesothelioma

Many serologic biomarkers have been proposed for the purpose of early detection or disease monitoring in mesothelioma. These have largely focused on mesothelin and related molecules such as osteopontin, and most recently the high mobility group box 1 (HMGB1), fibulin-3, and thioredoxin. Mesothelin is a cell surface protein that is seen on mesothelial cells as well as on other cell types. Splice variants such as soluble mesothelin-related peptide (SMRP) shed extracellular portions of the protein, such as megakaryocyte-potentiating factor (MPF), have also been studied.

Mesothelin is measured in serum or pleural fluid. It has been proposed as a prognostic marker (199) and as a means to monitor treatment effect (200) and recurrence. Its diagnostic utility has also been analyzed as an adjunct to imaging modalities and cytology (201,202). Its limitation is its variable sensitivity among histologic subtypes, with epithelioid tumors associated with much higher measurable levels than biphasic or sarcomatoid variants. Measurements of mesothelin may also have therapeutic implications in relation to amatuximab, a potential therapeutic antibody targeting mesothelin (203). Chimeric antigen receptor T-cell therapy has been explored in mesothelioma, using mesothelin as a target antigen (204).

SMRP may represent the best disease-monitoring biomarker in current use. It can be detected in pleural fluid and serum (205). In asbestos-exposed individuals, its increase may precede the development of mesothelioma (206). SMRP levels may decrease after debulking, so serial measurements may correlate with recurrence (207). MPF was developed as a biomarker in the hope that its biology (shed into the circulation) would improve sensitivity, but overall it has not shown better performance than SMRP or mesothelin (202,207).

Osteopontin is a cell surface glycoprotein that is involved in cell-matrix interaction. In contrast to mesothelin, its elevation is seen across mesothelioma histologic subtypes. Its limitation is its increase in other diseases and tumor types. The higher level in serum is associated with the development of mesothelioma (208,209) and may be prognostic (202,210). As an assay, osteopontin measurement has been complex and its usefulness limited by these challenges.

New biomarkers have also been investigated. Acetylated HMGB1 is associated with cytoplasmic and secreted levels of this protein. Interest in HMGB1 rests in its increase in asbestos-exposed individuals and its further increase in mesothelioma patients (211). Fibulin-3 has been proposed to have diagnostic and prognostic use in patients with mesothelioma of pleura (206,211–214). Thioredoxin 1 is also a potential biomarker that may be useful in screening asbestos-exposed individuals (206,215). Calretinin has been proposed as a blood biomarker (216). Combinations of the above may improve diagnostic accuracy (217).

Molecular Markers

Examination of molecular alterations in mesothelioma using the Catalog of Somatic Mutations in Cancer (COSMIC) revealed alterations in cyclin-dependent kinase inhibitor 2A (*CDKN2A*), BRCA1-associated protein 1 (*BAP1*), and neurofibromin 2 (*NF2*) genes in a significant proportion of tumors, with less frequent alterations in tumor protein 53 (*TP53*), and telomerase reverse transcriptase (*TERT*) (218). Rarer alterations have also been recently reported. A further study on integrative molecular characterization of mesotheliomas defined an additional novel genomic subtype with *TP53* and *SETDB1* mutations and extensive loss of heterozygosity (219). It also identified strong expression of the immune-checkpoint gene, V-domain Ig-containing suppressor of T-cell activation (*VISTA*), in epithelioid subtypes; *VISTA* gene expression is also assessable using immunohistochemistry (220).

CDKN2A is a critical tumor suppressor gene that controls the cell cycle in the retinoblastoma pathway. Loss of its protein product, p16, allows unregulated cellular entry into the cell cycle, causing cell division. Mechanisms of loss include homozygous deletion of its chromosomal locus at 9p21 (the most common event), as well as point mutations or indels. The high frequency of homozygous deletion in pleural mesothelioma and the retention of the *CDKN2A* locus in reactive conditions has made this molecular alteration a valuable diagnostic tool. Several studies (145,221–223) have shown an absence of homozygous deletions in reactive conditions, and deletions rates from 60 to 85 percent in pleural mesotheliomas. Peritoneal mesotheliomas have lower rates of homozygous deletion of 25 to 50 percent, so that it is less sensitive in this disease (145,223,224).

P16 loss may also be mediated by epigenetic mechanisms (225), and loss of cell cycle regulation in the retinoblastoma pathway may result from other alterations. Loss of p16 expression, detected by gene expression profiling, has prognostic significance in both pleural (226) and peritoneal disease (227), although immunohistochemistry for p16 does not correlate well as a surrogate for FISH. Loss of methylthioadenosine phosphorylase (MTAP) expression using immunohistochemistry, the *MTAP* gene being in close proximity to the *CDKN2A* gene, has recently been shown to be a highly specific marker of p16 deletion (151,153,228,229).

BAP1 is an important element in pathways associated with chromatin remodeling and double-stranded DNA repair; it partners with TP53, transcriptional repressor protein YY1, and E2F transcription factor 1 (E2F1) in various transcriptional regulatory pathways (230). Loss of *BAP1* function through deletion at its chromosomal locus 3p and/or mutation occurs in a substantial subset (55 to 60 percent) of pleural (231–233) and peritoneal (87,234,235) mesotheliomas. Loss of BAP1 by immunohistochemistry was not seen in reactive proliferations (149,232,236–238) but can be demonstrated in 60 to 80 percent of epithelioid pleural mesotheliomas, with lower rates in biphasic and sarcomatoid disease. Peritoneal disease shows a similar rate of loss immunohistochemically (235,239). Interestingly, in one series of six reactive biopsies showing BAP1 loss in mesothelial cells, patient follow-up showed the subsequent development of mesothelioma (236). Use of BAP1 and CDKN2A loss, either via FISH or using immunohistochemistry for MTAP as a surrogate, can be used as a dual test to improve diagnostic utility (145).

The *NF2* gene, with its protein product merlin, is an important link between cellular growth and intercellular adhesion. Loss of merlin function encourages cell proliferation and survival despite cellular attachment. In mesothelioma, *NF2* loss is mediated by heterozygous deletion of 22q as well as point mutations and small indels. Loss of *NF2* has not been investigated as extensively as *CDKN2A* and *BAP1* in the diagnostic arena (240), although immunohistochemistry does not seem to be an effective diagnostic modality using *NF2* pathway elements.

Mutations in the telomerase reverse transcriptase (*TERT*) gene are also important. TERT is a polymerase whose function is to maintain telomeric ends. TERT activity is decreased in somatic cells, and as a result, normal cells show progressive telomere shortening. Malignant cells show dysregulation of TERT, and in mesothelioma, mutations in the *TERT* promoter mediate this dysregulation (241).

Tumor protein 53 (*TP53*) is a critical tumor suppressor gene whose role is to respond to cellular stress by mediating pathways of repair or cell death. Its loss is central to tumorigenesis in many tumor types, but the mechanisms of functional loss are diverse. In mesothelioma, the loss of 9p21 associated with *CDKN2A* loss results in simultaneous loss of *p16* and *p14* and loss of *p14* causes loss of *p53* function (242). Rare mutations in *p53* have also been reported in both pleural and peritoneal mesotheliomas (219,235).

Comprehensive genomic profiling of mesothelioma has revealed, in addition to mutations in *BAP1, TP53,* and *NF2,* mutations in *SETD2, DDX3X, ULK2, RYR2, SETDB1,* and *DDX51.* Relevant pathways of activation include the Hippo, MTOR, histone methylation, helicase, and p53 pathways (243). In addition to mutations in *NF2,* mutations in *LATS2,* also in the Hippo pathway, have been reported (244). Receptor tyrosine kinase pathway activation in EGFR, VEGFR, IGFR, and MET (145) occurs in mesothelioma (245), but activating mutations in these receptors are rare.

Immune checkpoint biomarkers are also starting to be used in diagnostic as well as research practice. Programmed death-ligand 1 (PD-L1) is positive in about 20 percent of mesotheliomas (246), although sarcomatoid subtypes have a higher rate of immunoreactivity (247). Treatment using immune checkpoint inhibitors remains under study. In one study, mesothelioma patients whose tumors showed PDL1 expression in 1 percent or more of tumor cells were treated with the PD1 inhibitor pembrolizumab and showed a response rate of 20

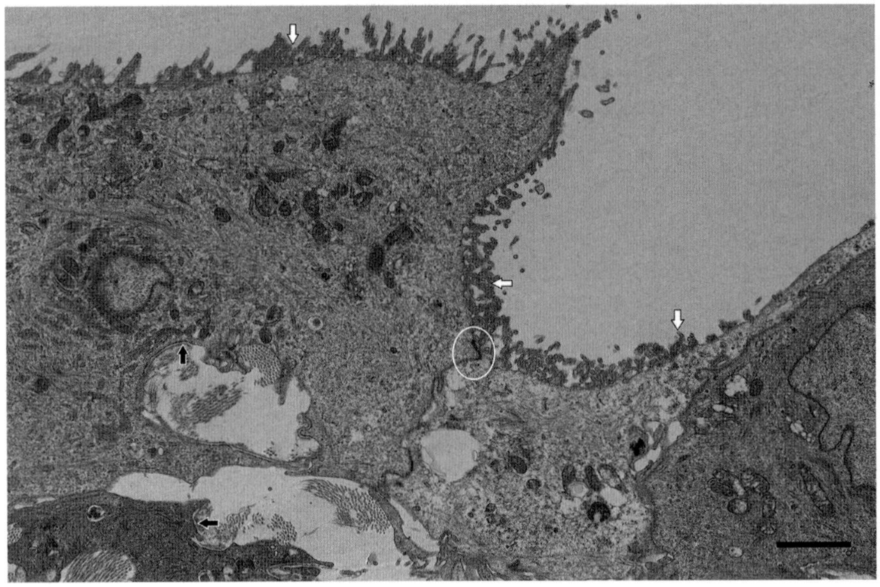

Figure 3-95

EPITHELIOID MESOTHELIOMA: ULTRASTRUCTURE

An electron micrograph shows long, slender microvilli (white arrows), desmosomes (circle), and basal lamina (black arrows). (Scale bar = 2 μm) (Courtesy of Dr. A. Rogers, London, UK)

percent (248). A study of the CTLA4 inhibitor tremelimumab showed possible response to this therapy in some patients (249), but no impact on overall survival (250). VISTA is a further option for assessing immune checkpoint status in mesotheliomas (219,220).

While peritoneal mesothelioma has alterations similar to those of pleural disease, copy number alterations in *CDKN2A* and *NF2* occur at a lower rate and *BAP1* losses at a similar rate as in pleural disease. Overall, chromosomal copy gains are seen more often than losses in peritoneal disease, especially in nonasbestos-related cases (234). Translocations in the *ALK* gene are reported in rare but well-characterized peritoneal mesotheliomas (251). In young adults, rare examples of *EWSR1* and *FUS* translocations in mesothelioma without *BAP1* mutations have been reported (252). There are no published data yet on molecular findings in mesotheliomas arising in pericardium and tunica vaginalis.

ULTRASTRUCTURAL FINDINGS

Although ultrastructural analysis is rarely used in routine diagnosis in current practice, a brief review of the diagnostic features is still warranted, since ultrastructural features may be useful in the diagnosis of tumors of the serosal membranes. The most striking feature of epithelioid mesothelioma is long, often branching, surface microvilli that are devoid of a glycocalyx (fig. 3-95) (253–257). The microvilli often have a length to diameter ratio that exceeds 10 to 1 (258,259). The morphologic features of the microvilli are seen particularly well by scanning electron microscopy (260). In contrast, most adenocarcinomas have short and stubby microvilli covered with a glycocalyx and are often associated with glycocalyceal bodies (fig. 3-96). These contrasting features are not absolute, and there is often overlap (258). Rootlets at the base of the microvilli favor adenocarcinoma while microvilli abutting extracellular collagen is more common in mesothelioma.

Other ultrastructural features commonly seen in epithelioid mesotheliomas include perinuclear tonofibrillar bundles, cytoplasmic glycogen, basal lamina, and long desmosomes (256). Mesotheliomas lack lamellar bodies, club cell granules, mucus granules, and dense core neurosecretory granules. The findings of pinocytotic vesicles and Weibel-Palade bodies are suggestive of epithelioid hemangioendothelioma, while the identification of pre-melanosomes favors melanoma.

There is no single ultrastructural feature that is the hallmark of mesothelioma, rather there is a constellation of findings that may be diagnostically useful. Even classic mesotheliomas show considerable variation in ultrastructural features (255). The absence of any or all of these features does not necessarily exclude mesothelioma, and ultrastructure, like all ancillary investigations, requires integration with clinical, imaging, histologic, and immunohistochemical data.

Generally, samples require specific fixation for electron microscopy, but some features, such as long surface microvilli and perinuclear tonofilaments, are seen in samples retrieved from paraffin-embedded blocks (261).

Electron microscopy is less useful in sarcomatoid mesothelioma, where the cells usually resemble fibrosarcoma and other fibroblastic tumors (fig. 3-97, left). Epithelioid differentiation, however, is identified in some cases (fig. 3-97, right), manifesting as occasional microvilli, incomplete basal lamina, intercellular junctions, or tonofibrils. The ultrastructural features of other sarcomatoid tumors such as angiosarcoma, schwannoma, leiomyosarcoma, rhabdomyosarcoma, or spindle cell melanoma are useful in distinguishing these tumors from sarcomatoid mesothelioma (256).

DIFFERENTIAL DIAGNOSIS

Benign Versus Malignant Mesothelial Proliferations

Differentiating benign and malignant mesothelial proliferations is typically most difficult in small biopsies. Once the lesion has been recognized as mesothelial, the diagnostic approach depends on whether it is predominantly epithelioid or spindled.

Reactive Mesothelial Hyperplasia Versus Epithelioid Mesothelioma. Reactive mesothelial proliferations may show high cellularity (fig. 3-98), numerous mitotic figures, cytologic atypia, formation of papillary groups (fig. 3-99), and entrapment of mesothelial cells within fibrosis that mimics invasion (fig. 3-100) (180). The morphologic features favoring a benign process are lack of stromal invasion, surface cellularity, simple papillae, loose sheets of cells without stroma (fig. 3-98), absence of necrosis, and uniform growth (which can be highlighted by a cytokeratin stain). Features that favor mesothelioma include stromal invasion, dense cellularity including cells surrounded by stroma, complex papillae, bulky tumor (fig. 3-101), multifocal necrosis, and expansile nodules with disorganized growth (fig. 3-102). Occasionally, reactive mesothelial cells are present within lymphatics and should be interpreted in context (fig. 3-103).

When additional studies are needed to determine the nature of a mesothelial proliferation, it is useful to stain immunohistochemically for BAP1, which is lost in about 60 percent of epithelioid mesotheliomas (fig. 3-104). If BAP1 nuclear expression is diffusely lost, the diagnosis of epithelioid mesothelioma is strongly supported by this finding (145). Loss of staining for MTAP has recently been described as being 100 percent specific and 76.5 percent sensitive when combined with BAP1 staining (229). For MTAP, although both nuclear and cytoplasmic staining is often lost, it is the loss of cytoplasmic staining that is seen in malignancy (fig. 3-105) (150,229). If the expression of both BAP1 (fig. 3-106) and MTAP is retained, FISH studies for loss of *CDKN2A/p16* are helpful since homozygous deletion is only present in malignancies (discussed above), although this test is not 100 percent sensitive. Enhancer of Zeste Homolog

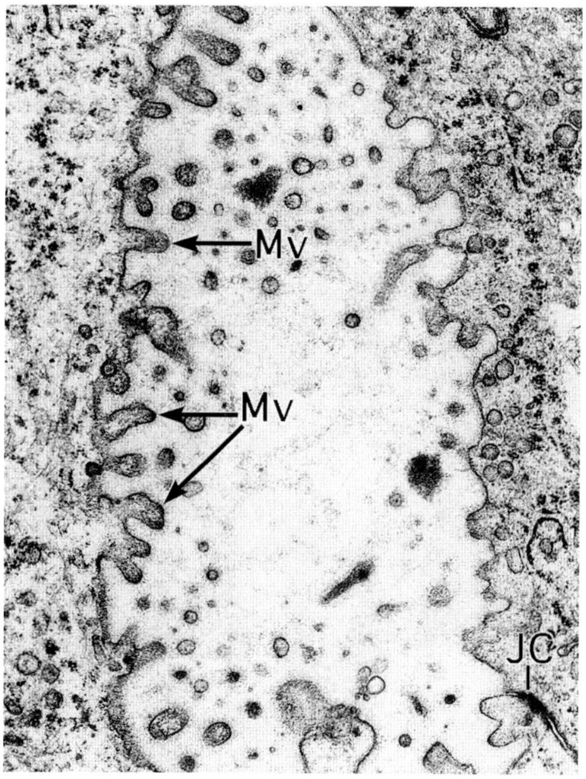

Figure 3-96

METASTATIC ADENOCARCINOMA: ULTRASTRUCTURE

In comparison to the epithelioid mesothelioma, surface microvilli (Mv) are blunt and a junctional complex (JC) is observed. Intermediate filaments and tonofibrils are not seen. (Courtesy of Dr. S. Hammar, Bremerton, WA)

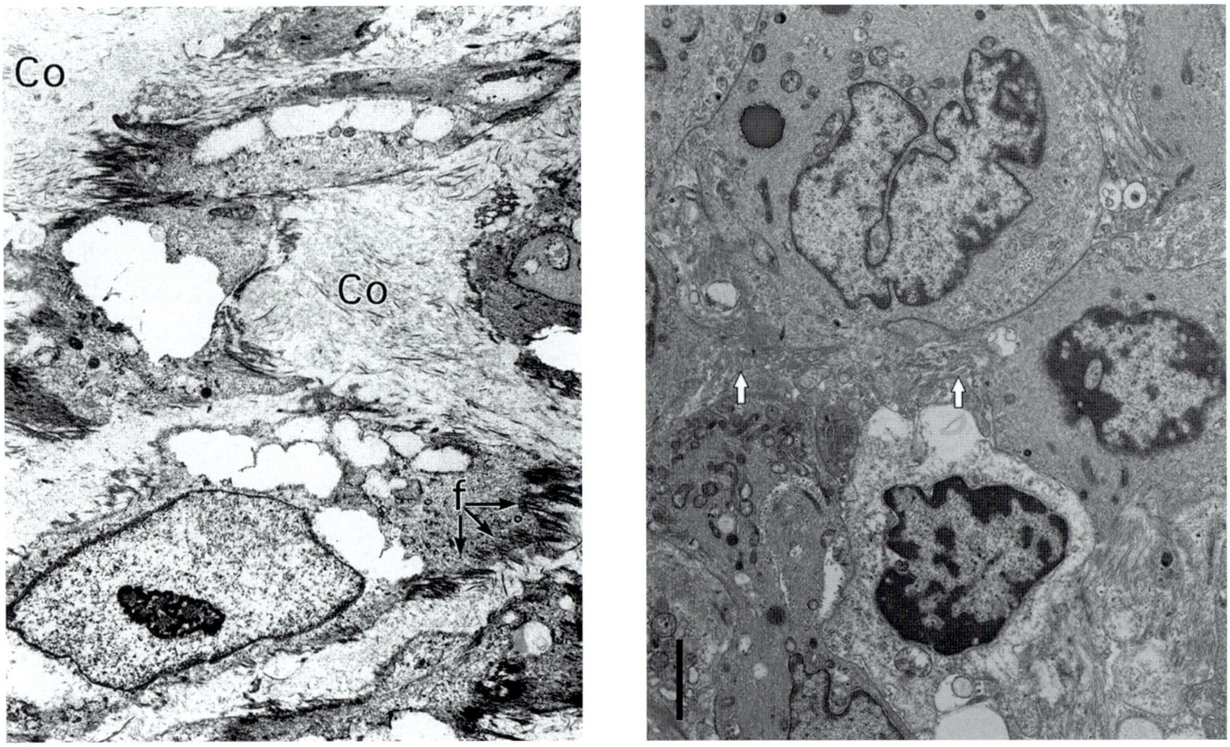

Figure 3-97

SARCOMATOID MESOTHELIOMA: ULTRASTRUCTURE

Left: Spindle cells with cytoplasmic filaments (f) and abundant extracellular collagen (Co) are seen. (Fig. 17 from Roggli VL, Kolbeck J, Sanfilippo, Shelbourne JD. Pathology of human mesothelioma: etiologic and diagnostic considerations. Pathol Annu 1987;22(Pt 2):110.)

Right: Focal evidence of mesothelial ultrastructural morphology is seen in the form of scattered long microvilli (white arrows indicate long microvilli). (Scale bar = 1 µm.) (Courtesy of Dr. A. Rogers, London, UK)

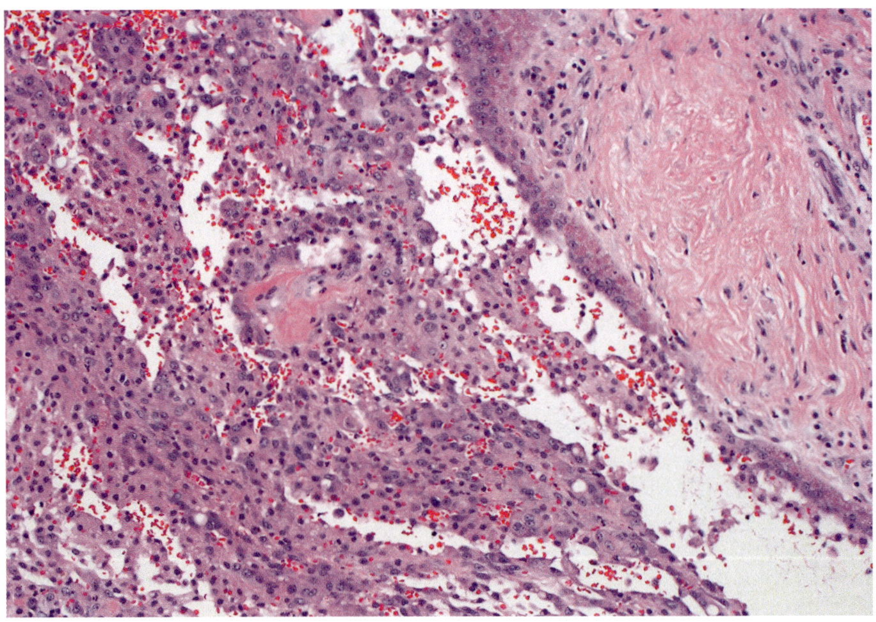

Figure 3-98

MESOTHELIAL HYPERPLASIA

Marked mesothelial hyperplasia is present in this patient undergoing surgery for aortic dissection. The mesothelial cells form solid sheets without stroma.

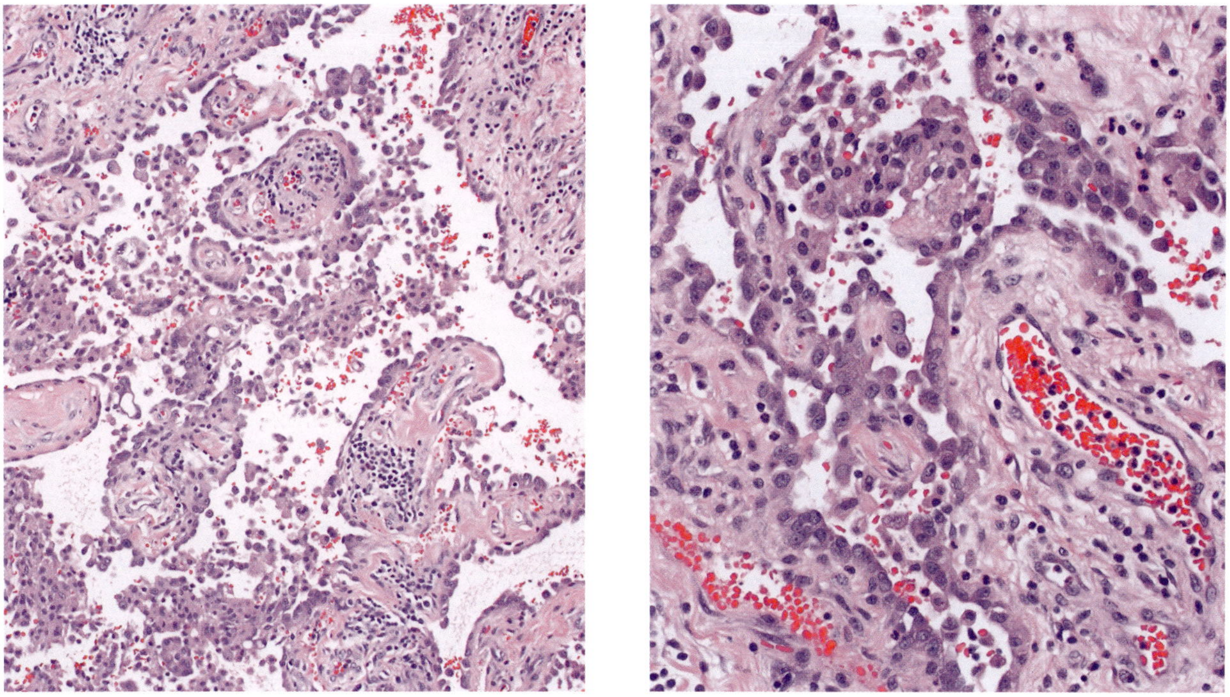

Figure 3-99

MESOTHELIAL HYPERPLASIA

Left: A different area from the same patient as in fig. 3-98 shows papillary formations.
Right: Higher power shows papillae lined by a single layer of mesothelial cells with prominent nucleoli.

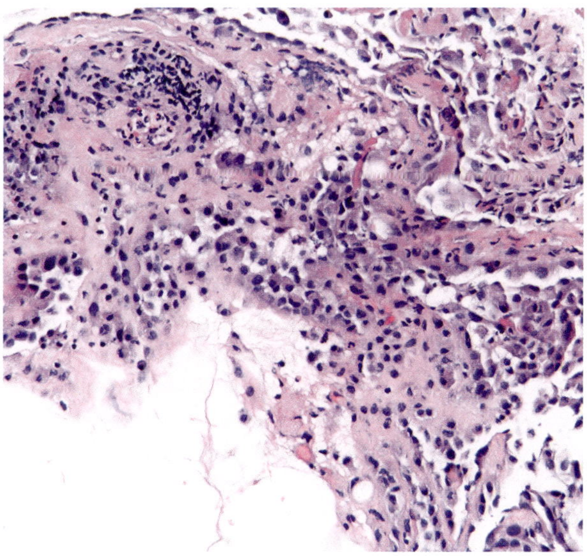

Figure 3-100

MESOTHELIAL HYPERPLASIA

Mesothelial hyperplasia in a 65-year-old patient with pleural effusion and lung cancer. There are clusters of mesothelial cells within fibrous tissue, mimicking mesothelioma.

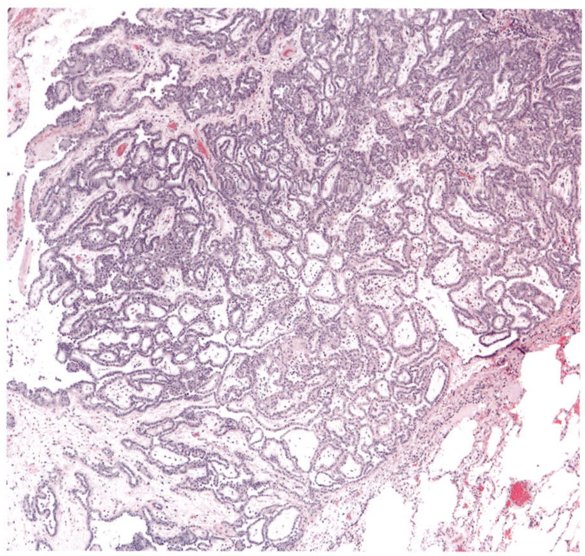

Figure 3-101

EPITHELIOID MESOTHELIOMA

There is a mesothelial proliferation on the visceral pleura that does not invade the attached lung. The amount of proliferation alone would favor the diagnosis of mesothelioma.

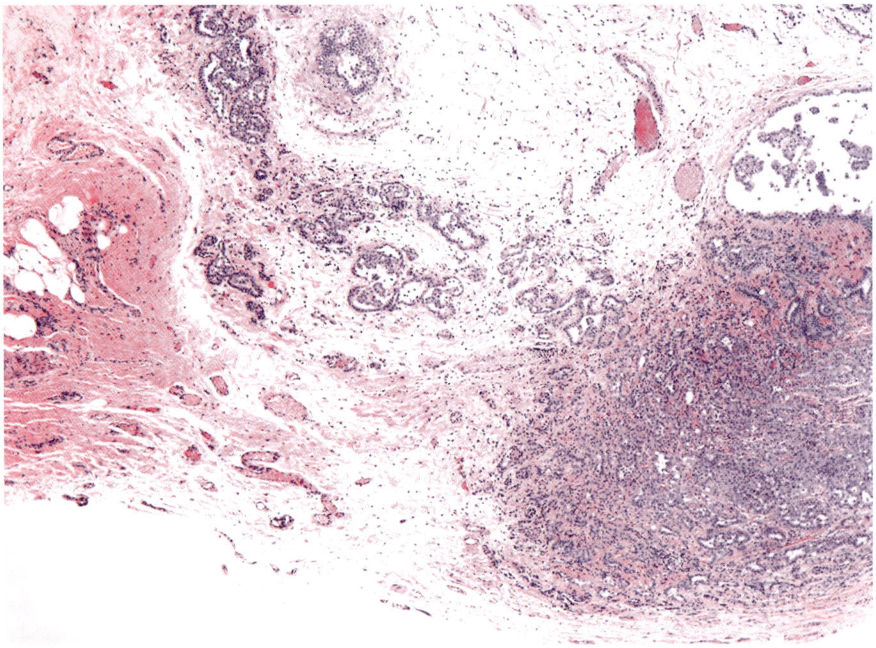

Figure 3-102

EPITHELIOID MESOTHELIOMA

Although there is no invasion of the fat seen at the middle left, the nodular proliferation favors the diagnosis of mesothelioma.

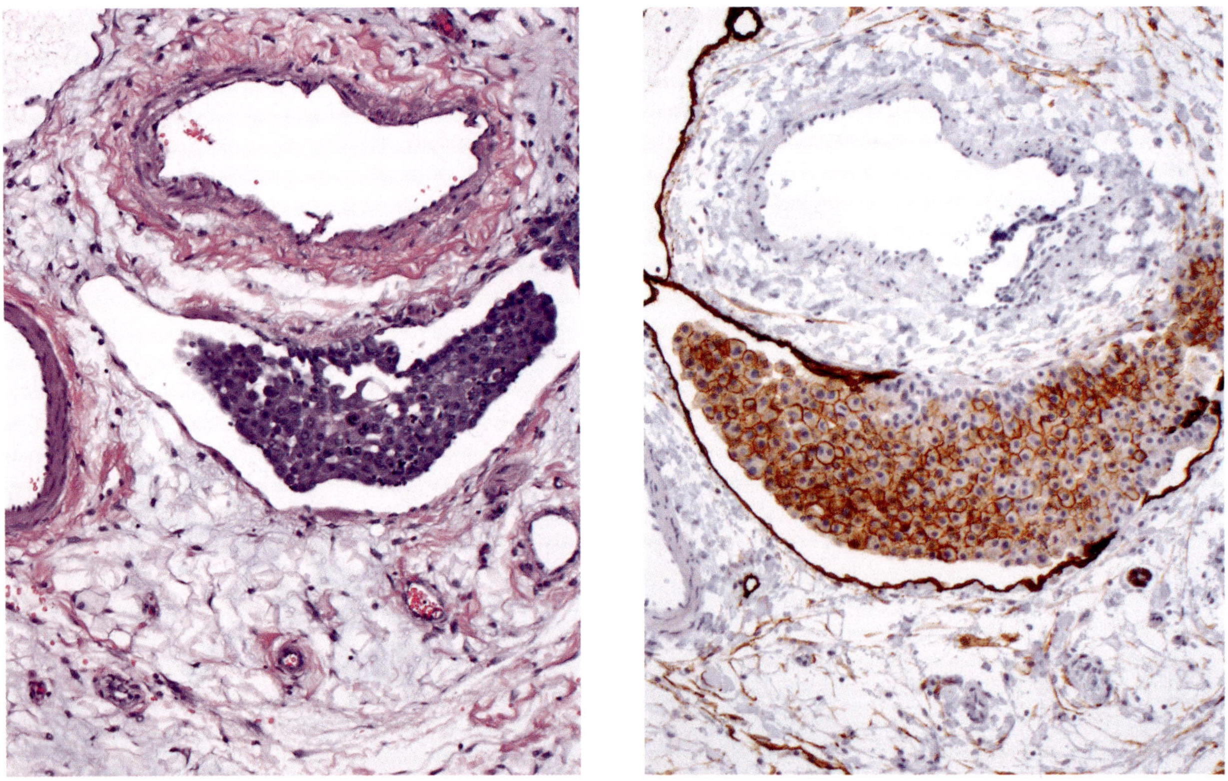

Figure 3-103

REACTIVE MESOTHELIAL CELLS

Left: Pericardial resection for chronic pericarditis shows a cluster of reactive mesothelial cells within the lymphatic space. Right: D2-40 stain confirms the mesothelial nature of the intralymphatic cells.

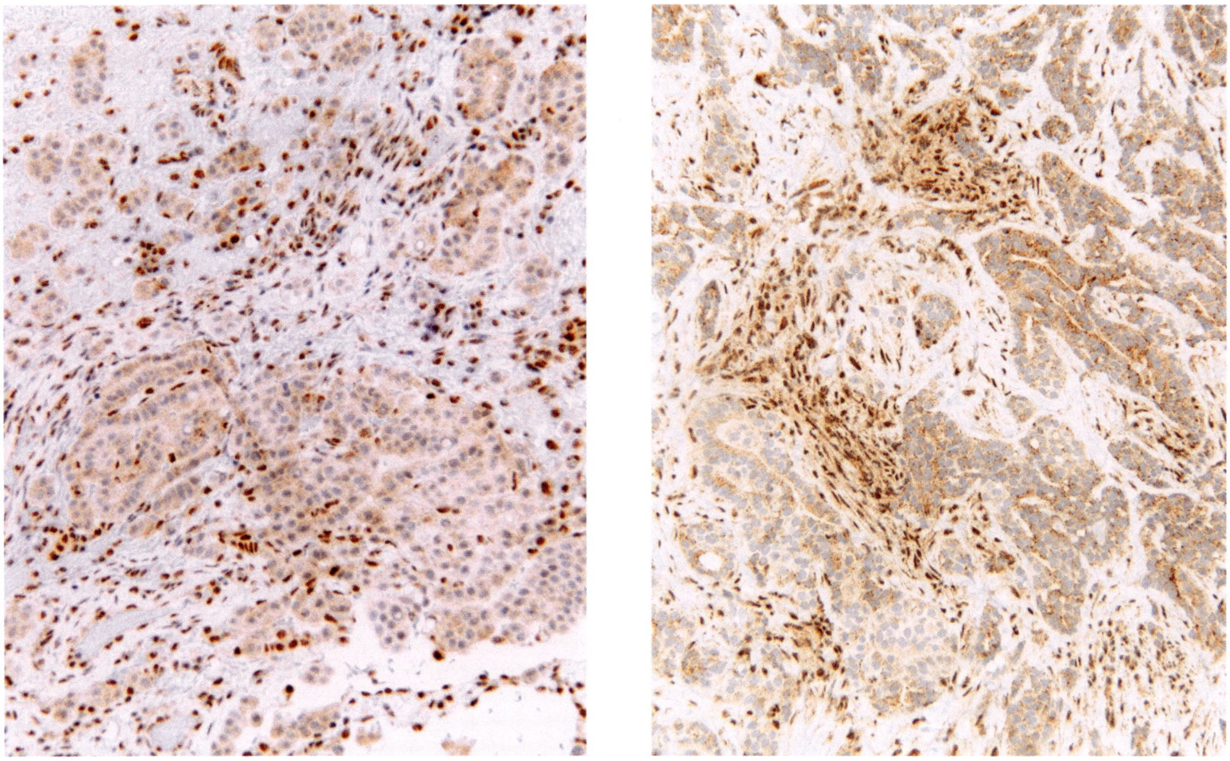

Figure 3-104

EPITHELIOID MESOTHELIOMA: BAP1 STAIN

Left: There is loss of nuclear staining of BAP1 in the mesothelial cells. The strong positivity in inflammatory cells and stromal nuclei serves as internal control.

Right: The tumor cells have lost expression of BAP1 protein in the nuclei which is diagnostic for malignancy; however, there is cytoplasmic granular staining which should not be interpreted as retained BAP1.

2 Protein (EZH2) is an additional marker that may have value in this setting (262,263), as well as expression of cyclin D1 (264).

Fibrous Pleurisy Versus Desmoplastic Variant of Sarcomatoid Mesothelioma. The morphologic features of benign spindle cell proliferations include relatively parallel growth of spindle cells, absence of stromal invasion, only surface necrosis where it is often associated with acute inflammation, uniform thickness, hypercellularity at the surface with maturation and decreased deep cellularity (zonation), and perpendicularly oriented vessels (fig. 3-107). Nuclear pleomorphism may also be seen in organizing pleuritis (fig. 3-108) and occasional mitotic figures are seen (fig. 3-107B), but there are no atypical mitoses.

Features that favor desmoplastic mesothelioma include a prominent storiform pattern, fat or other tissue invasion which can be highlighted with a keratin stain (fig. 3-109), bland necrosis of paucicellular collagenized tissue (fig. 3-110), disorganized growth with uneven thickness (fig. 3-111), expansile nodules, abrupt change in cellularity (fig. 3-112), lack of zonation, and a few haphazardly arranged vessels (fig. 3-113) (180). Fat-like spaces are seen in some cases of reactive fibrosis that may be due to pulling of tissue during biopsy, the "fake fat phenomenon" (fig. 3-114) (265). Care must be taken not to confuse this with invasion into adipose tissue since the keratin stain is positive in reactive submesothelial cells (fig. 3-114). S-100 protein, laminin, and collagen IV stain fat cells but not the artefactual spaces (fig. 3-115) (180).

Immunohistochemistry for BAP1 shows loss of nuclear staining in only about 10 percent of sarcomatoid mesotheliomas. FISH to look for *CDKN2A/p16* loss, or immunohistochemistry to look for MTAP loss as a surrogate marker, may be more helpful as this molecular alteration is

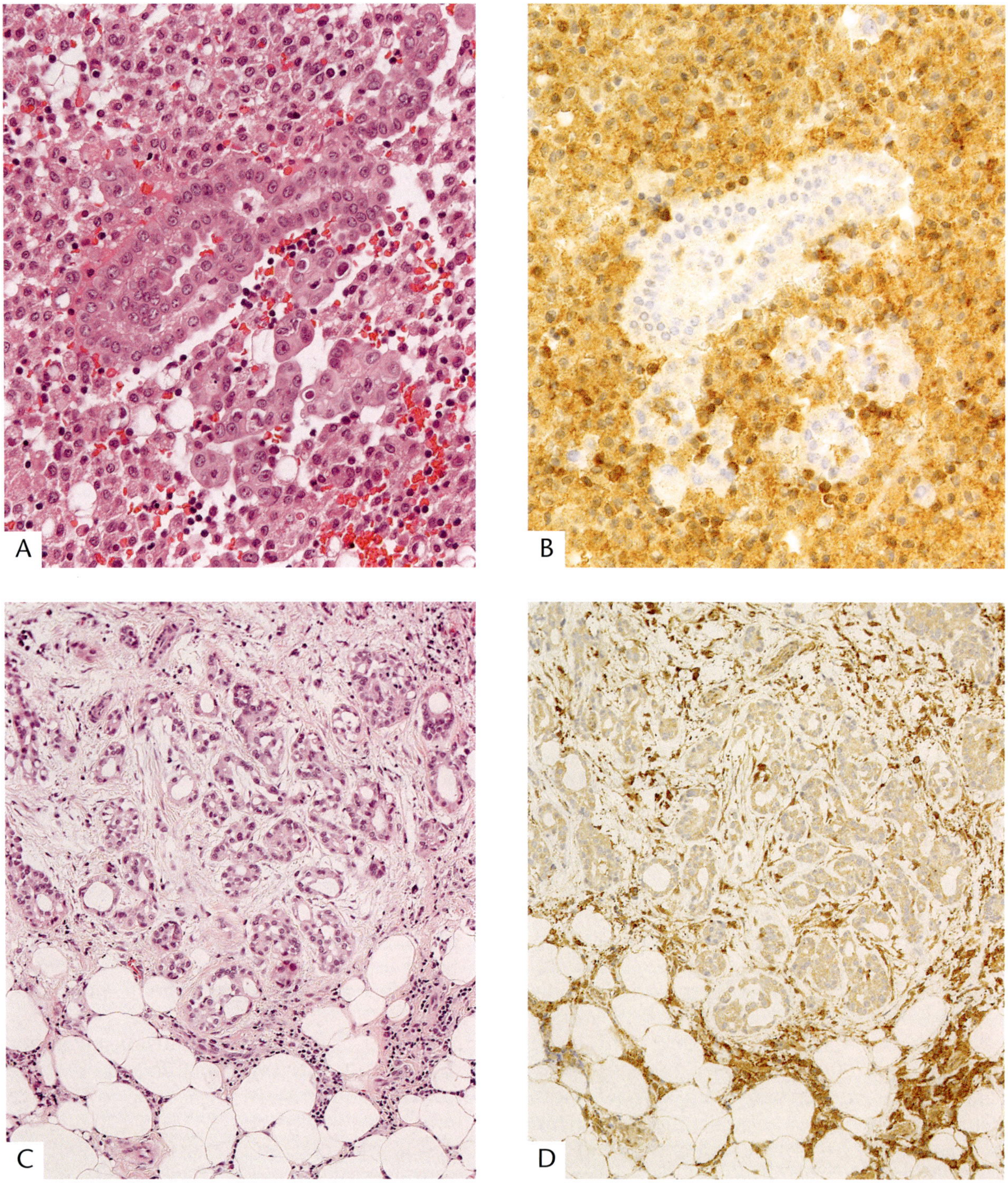

Figure 3-105

EPITHELIOID MESOTHELIOMA: MTAP STAIN

A,B: In a cell block stained with H&E (A), MTAP expression is lost in the mesothelioma cells, while it is retained in histiocytes and inflammatory cells (B).

C,D: MTAP expression is lost in mesothelioma cells, while it is retained in lymphocytes and fat cells. (Courtesy of Prof. K. Nabeshima, Fukuoka, Japan)

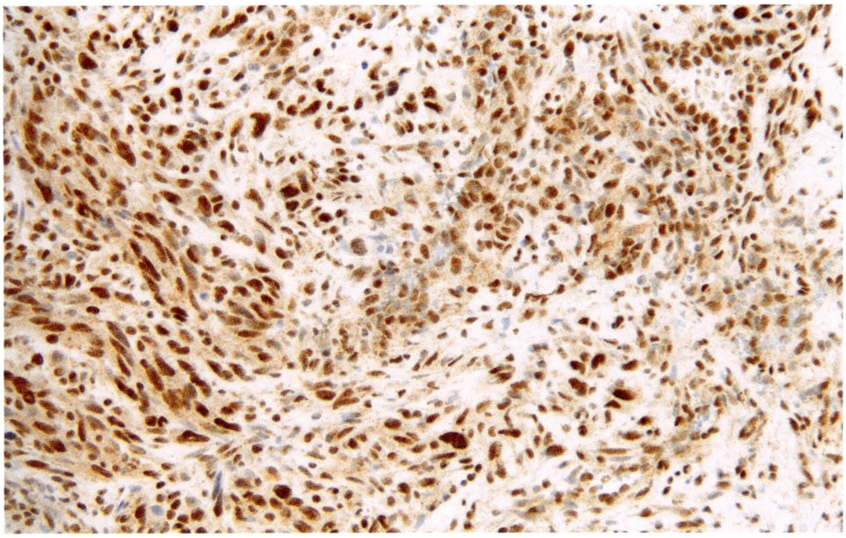

Figure 3-106

BIPHASIC MESOTHELIOMA: BAP1 STAIN

BAP1 is retained in the nuclei of both the epithelioid and sarcomatous components of this tumor, which does not help in differentiating benign from malignant.

Figure 3-107

ORGANIZING PLEURITIS

A: Under the surface fibrin there is granulation tissue with capillaries perpendicular to the surface (zonation), characteristic of organizing pleuritis.

B: Deeper to the granulation tissue, there is more mature collagen within which foci of fibrin are still present. There is a mitotic figure in the center (likely to be an endothelial cell).

C: In the deepest layer, mature collagen fibers are present which are arranged in an organized parallel manner. Several small blood vessels are also present.

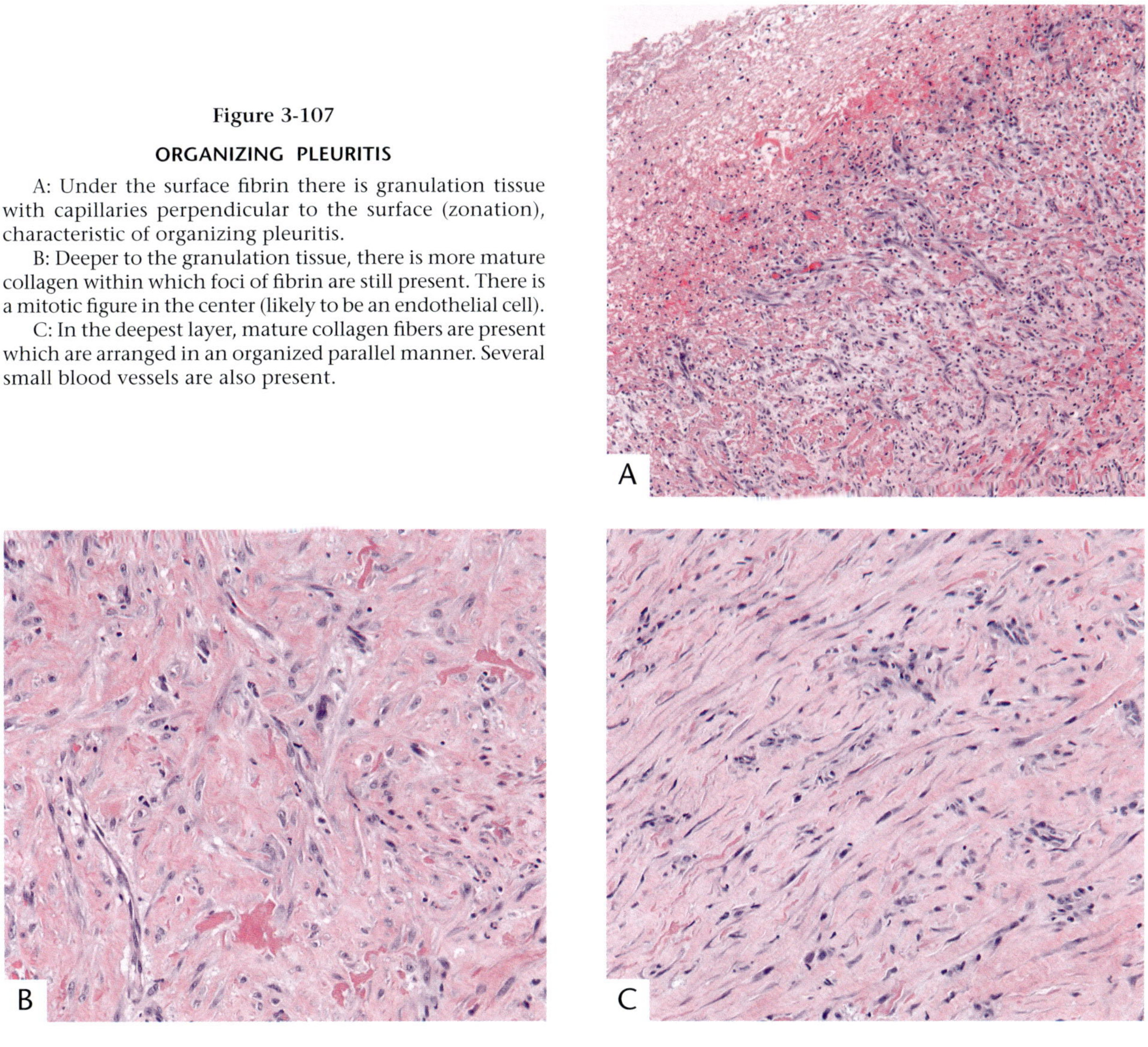

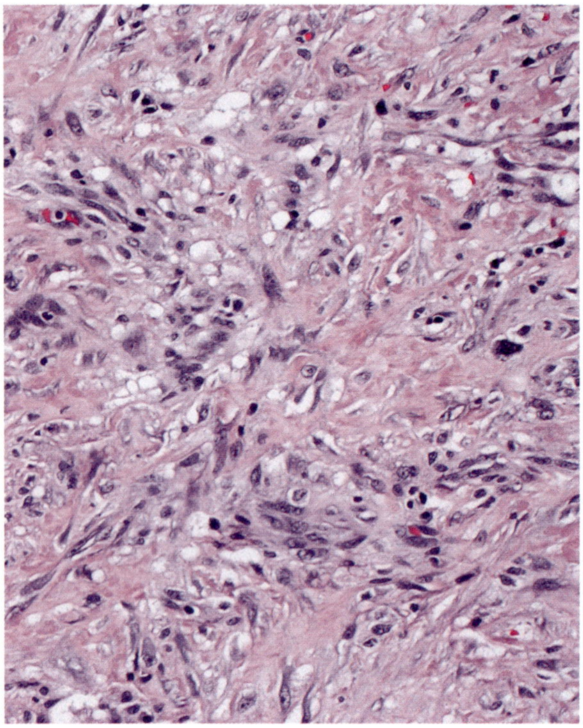

Figure 3-108

ORGANIZING PLEURITIS

Nuclear atypia is seen in this pleural peel from a 56-year-old with empyema.

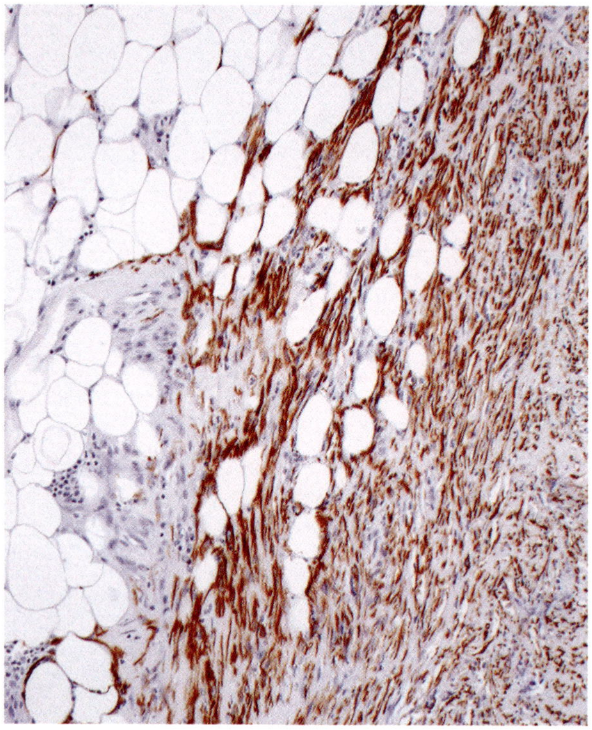

Figure 3-109

DESMOPLASTIC MESOTHELIOMA

Keratin stain (CAM5.2) highlights sarcomatoid mesothelioma invading adipose tissue.

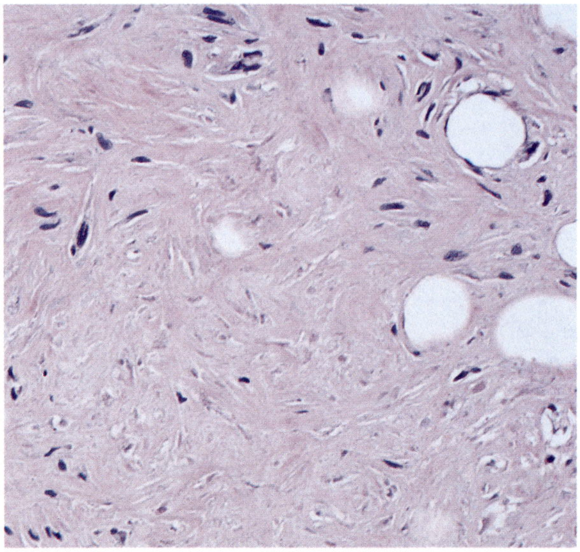

Figure 3-110

DESMOPLASTIC MESOTHELIOMA

There is focal necrosis of paucicellular and paucivascular proliferation.

commoner in sarcomatoid mesotheliomas (fig. 3-116) (178). Broader molecular profiling tools may also be helpful (266,267). Strong PD-L1 expression helps distinguish sarcomatoid mesothelioma from reactive fibrous pleuritis (268).

Epithelioid Mesothelioma Versus Metastatic Carcinoma

With the wide availability of many highly sensitive and specific immunohistochemical markers, used in the appropriate clinical setting, this differential diagnosis has become much easier. It can be problematic if there is inadequate tissue or there is extensive crush artefact. The most common problem is differentiating lung carcinomas, especially peripheral ones, which can grow along the pleura (so-called pseudomesotheliomatous carcinoma) (fig. 3-117). Markers that are currently most useful in distinguishing lung adenocarcinoma (fig. 3-118) and squamous cell carcinoma (fig. 3-119) from epithelioid mesothelioma are shown in Tables 3-4 and 3-5, respectively (180,269).

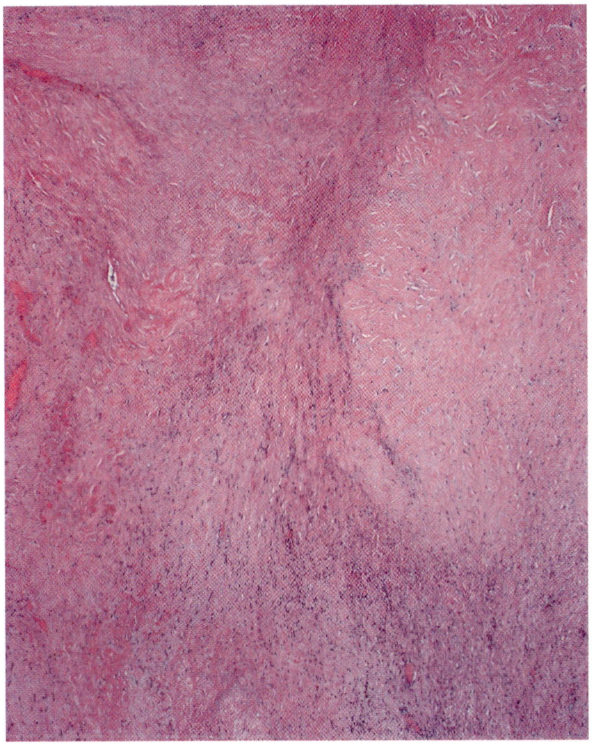

Figure 3-111

DESMOPLASTIC MESOTHELIOMA

At low power, haphazard growth of paucicellular spindle cells is seen.

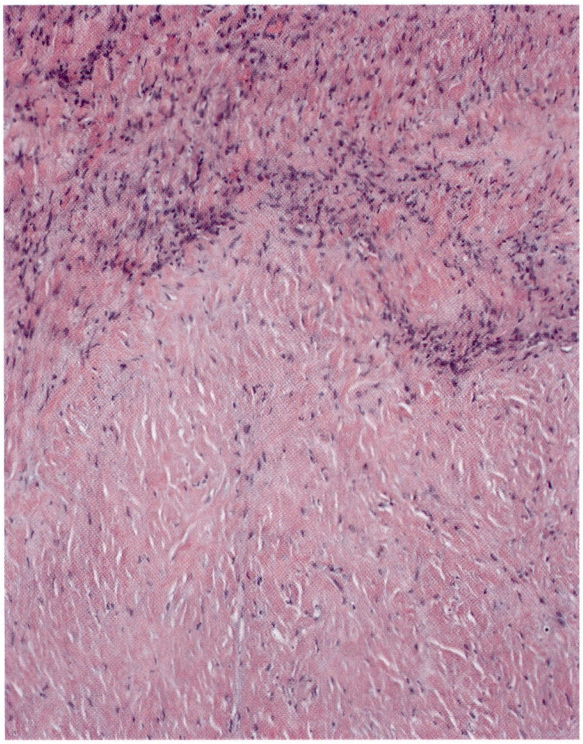

Figure 3-112

DESMOPLASTIC MESOTHELIOMA

There is abrupt change in cellularity, a feature favoring the diagnosis of malignancy.

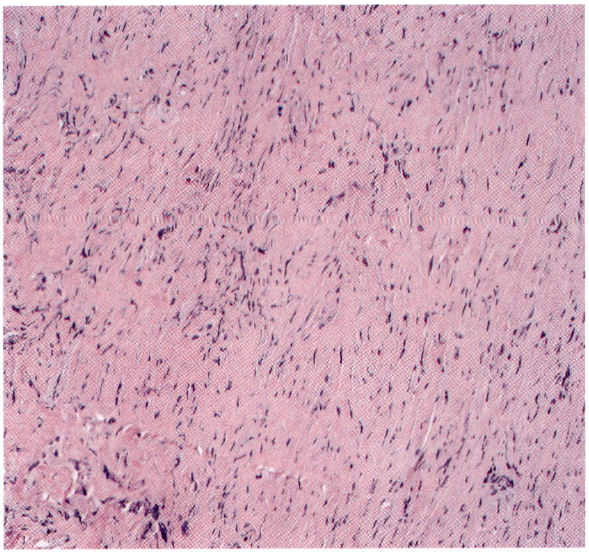

Figure 3-113

DESMOPLASTIC MESOTHELIOMA

The spindle cells are haphazardly arranged. Note the paucity of blood vessels.

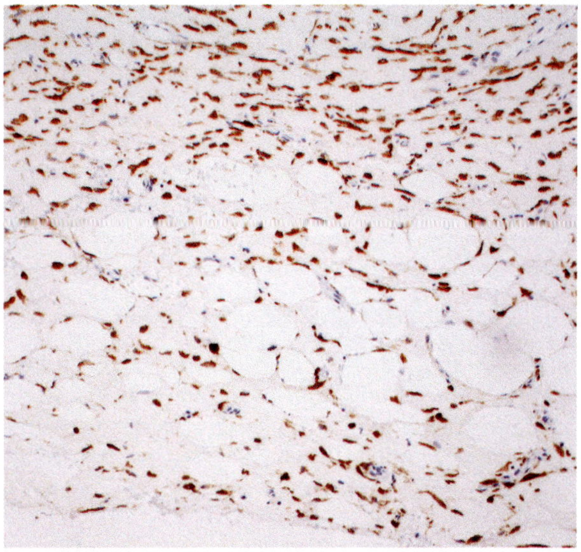

Figure 3-114

FAKE FAT PHENOMENON

Keratin highlights benign fibrous tissue that appears to be invading fat; however, the spaces are artifactual.

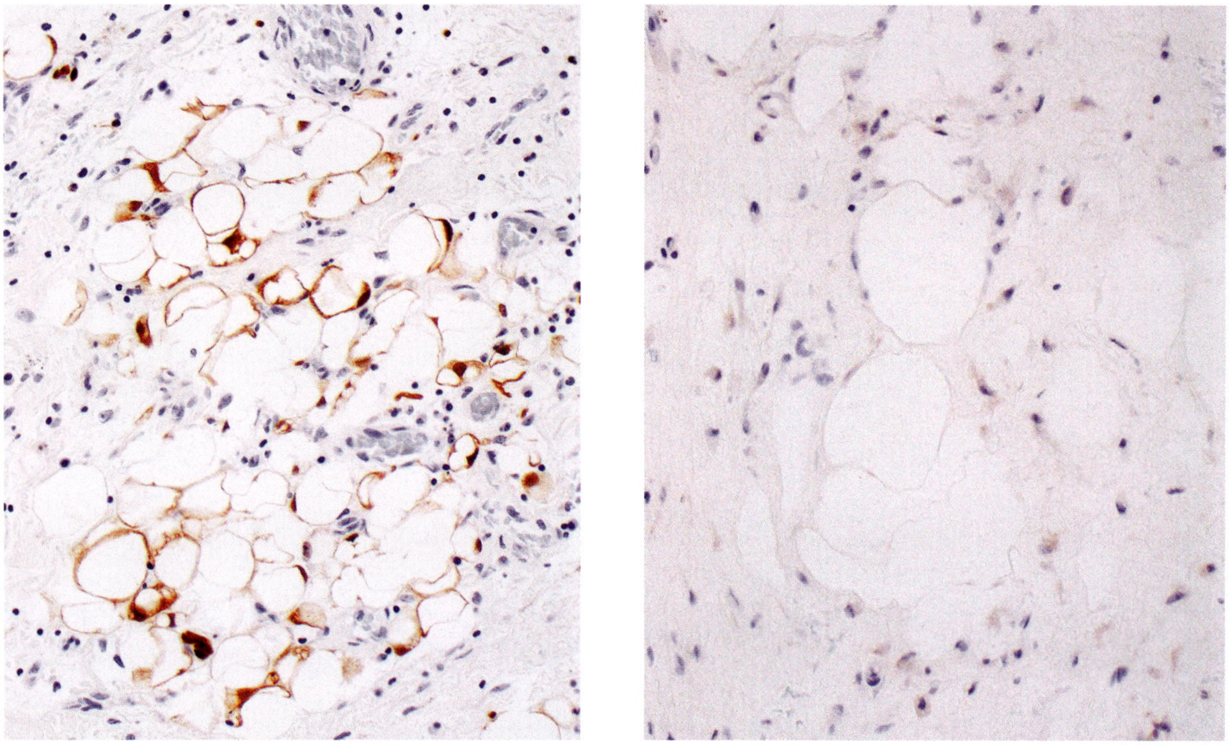

Figure 3-115

REAL FAT AND FAKE FAT PHENOMENON

Left: Adipose tissue is highlighted by a S-100 protein stain.
Right: S-100 protein is negative in fake fat.

Breast carcinoma markers include estrogen and progesterone receptor, gross cystic disease fluid protein-15, and mammaglobin (although the last two are not very sensitive). GATA3 is frequently positive in breast carcinomas but can also stain up to half of epithelioid mesotheliomas (and most sarcomatoid mesotheliomas) (181,270,271). Markers to distinguish renal cell carcinomas (272) are shown in Table 3-6.

Sarcomatoid Mesothelioma Versus Other Spindle Cell Malignancies

The differential diagnosis of a keratin-positive sarcomatoid malignancy of the pleura includes sarcomatoid carcinoma of the lung, synovial sarcoma, angiosarcoma, and other sarcomatoid tumors such as renal cell carcinoma. Positivity for claudin 4, TTF1, Napsin A, or p40 (or p63 if p40 is not available) strongly supports the diagnosis of sarcomatoid carcinoma of the lung. A recent study has demonstrated that GATA3 is positive in most sarcomatoid mesotheliomas, including the desmoplastic subtype (fig. 3-120), and is negative in sarcomatoid lung carcinomas (181). It should be noted, however, that GATA3 variably stains other sarcomatoid carcinomas such as from the larynx and kidney.

Synovial sarcoma can easily be distinguished based on the definitive finding of t(X;18) (fig. 3-121) (273). Solitary fibrous tumors typically mark with STAT6 (fig. 3-122) and CD34, and are usually keratin negative. Other specific markers should be used according to the differential diagnosis (liposarcoma, myogenic sarcomas, neurogenic sarcomas).

Other Differential Diagnoses

If a pleural tumor is keratin negative after using multiple keratin stains such as AE1/AE3, CAM5.2, and CK5/6, other possible differential diagnoses should be considered and a screening panel selected accordingly. This could include markers such as CD45, CD3, CD20, and CD30 for large cell lymphoma; S-100 protein, HMB45,

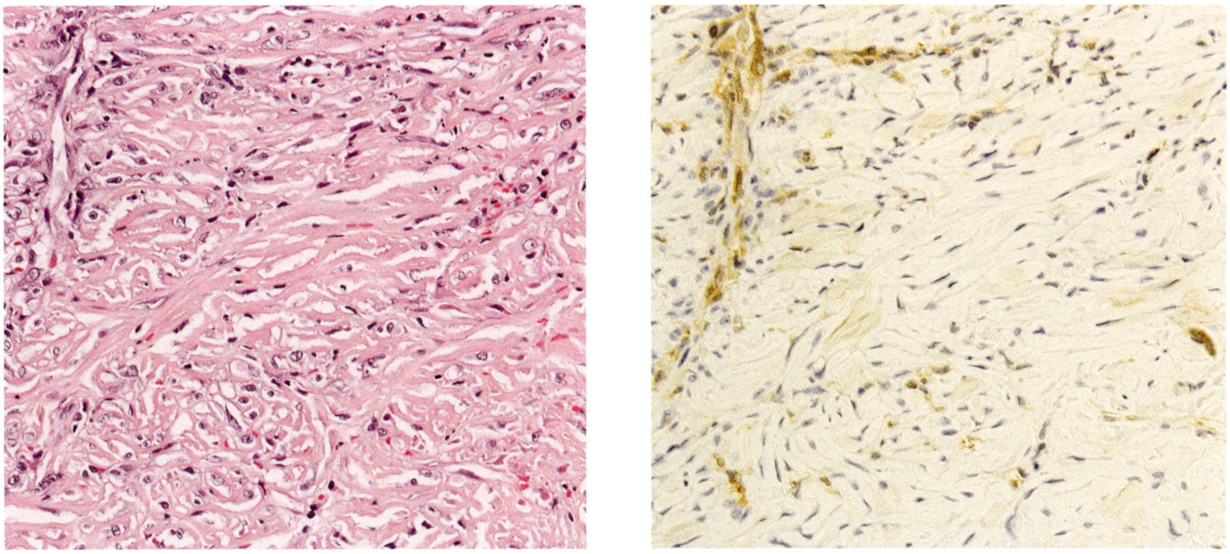

Figure 3-116

DESMOPLASTIC MESOTHLIOMA

In this desmoplastic mesothelioma (left), the endothelial cells (right), as internal positive control, retain MTAP expression, while MTAP expression is lost in mesothelioma cells. (Courtesy of Prof. K. Nabeshima, Fukuoka, Japan)

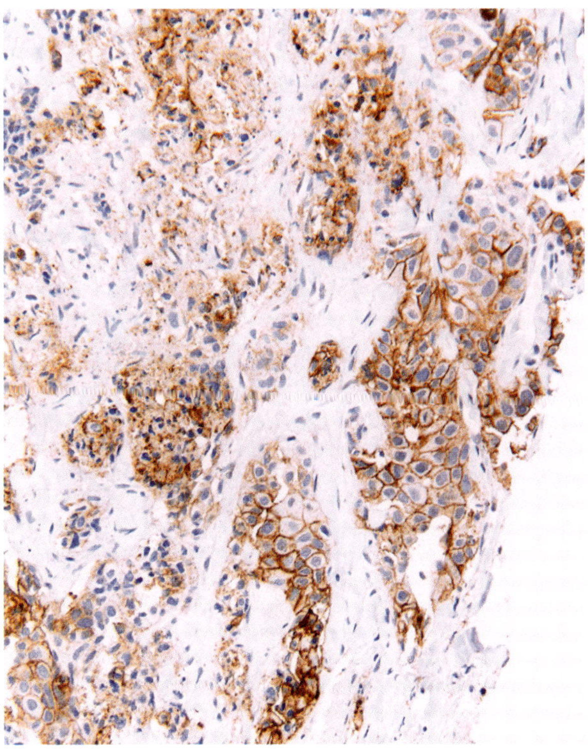

Figure 3-117

PSEUDOMESOTHELIOMATOUS CARCINOMA

Claudin-4 shows strong membranous staining of the carcinoma cells growing along the pleura.

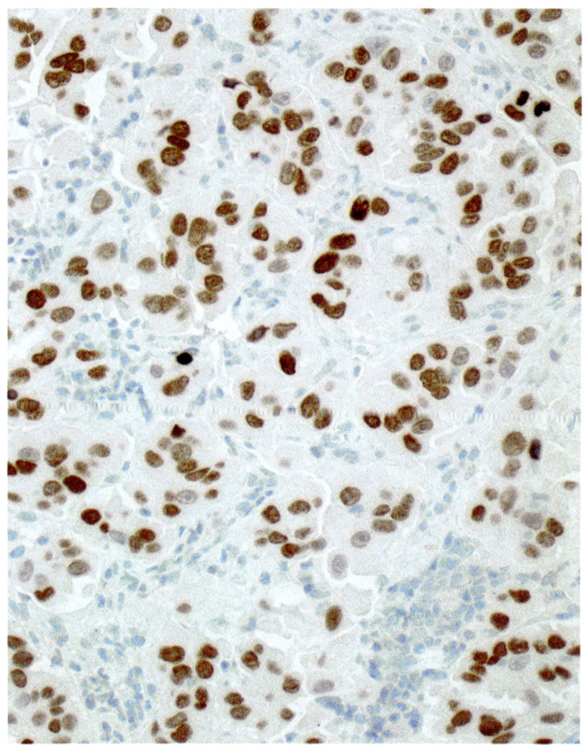

Figure 3-118

LUNG ADENOCARCINOMA

TTF-1 shows strong nuclear staining in a lung adenocarcinoma.

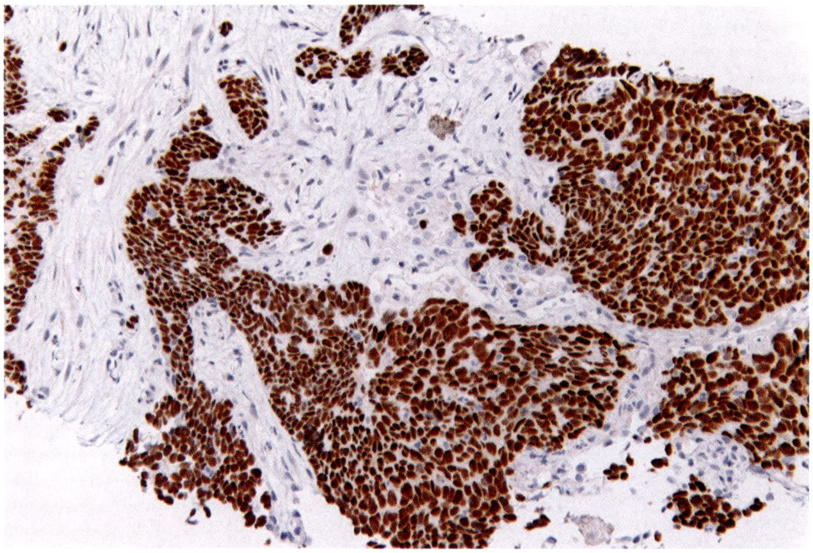

Figure 3-119

SQUAMOUS CELL CARCINOMA

p40 shows strong nuclear staining in this squamous cell carcinoma.

Table 3-4

IMMUNOHISTOCHEMICAL MARKERS USED IN THE DIFFERENTIAL DIAGNOSIS OF PLEURAL MESOTHELIOMA VERSUS LUNG ADENOCARCINOMA INVOLVING THE PLEURA[a]

Marker	Current Value/Comments
Epithelioid mesothelioma (positive mesothelioma markers)	
Calretinin	Very useful. Is demonstrated in nearly all epithelioid mesotheliomas when antibodies to human recombinant calretinin are used. The staining is often strong and diffuse and is both nuclear and cytoplasmic; 5% to 10% of lung adenocarcinomas are positive, but the staining is usually focal
Cytokeratin 5 or 5/6	Very useful. Expressed in 75% to 95% of mesotheliomas; 2% to 20% of lung adenocarcinomas can be focally positive
WT-1[b]	Very useful. About 70% to 95% of mesotheliomas show nuclear positivity. Lung adenocarcinomas are negative
Podoplanin (D2-40)	Very useful. About 90% to 100% of mesotheliomas show positivity along the cell membranes; ≤15% of lung adenocarcinomas are focally positive
Lung adenocarcinoma (possible carcinoma markers)	
Claudin 4	Very useful. Nearly all lung adenocarcinomas are positive. Immunoreaction is often strong and diffuse and occurs along the cell membrane in a continuous or punctate pattern. Mesotheliomas are negative
MOC31	Very useful. About 95% to 100% of lung adenocarcinomas are positive; 2% to 10% of mesotheliomas show focal staining
CEA	Very useful. About 80% to 100% of lung adenocarcinomas are positive; <5% of mesotheliomas are focally positive
B72.3	Very useful. About 75% to 85% of lung adenocarcinomas are positive. Very few mesotheliomas are positive
BER-EP4	Very useful. About 95% to 100% of lung adenocarcinomas are positive; <20% of mesotheliomas are locally positive
BG8 (Lewis^y)	Very useful. About 90% to 100% of lung adenocarcinomas are positive; 3% to 7% of mesotheliomas show focal reactivity
TTF-1	Very useful. About 75% to 85% of lung adenocarcinomas show nuclear positivity (most nonmucinous lung adenocarcinomas are positive). It is rarely expressed in mesotheliomas
Napsin A	Very useful. About 80% to 90% of lung adenocarcinomas show cytoplasmic staining. It is not expressed in mesotheliomas

[a]Table 3 from Husain AN, Colby TV, Ordonez NG, et al. Guidelines for pathologic diagnosis of malignant mesothelioma 2017 update of the consensus statement from the International Mesothelioma Interest Group. Arch Pathol Lab Med 2018;142:89-108.

[b]WT-1 = Wilms tumor-1; BG8 = blood group 8; CEA = carcinoembryonic antigen; TTF-1 = thyroid transcription factor-1.

Table 3-5

IMMUNOHISTOCHEMICAL MARKERS USED IN THE DIFFERENTIAL DIAGNOSIS OF PLEURAL MESOTHELIOMA VERSUS SQUAMOUS CELL CARCINOMA OF THE LUNG INVOLVING THE PLEURA[a]

Marker	Current Value/Comments
Epithelioid mesothelioma (positive mesothelioma markers)	
WT-1[b]	Very useful. Up to 95% of mesotheliomas show nuclear positivity. Lung squamous cell carcinomas are negative
Calretinin	Somewhat useful. Essentially all mesotheliomas are positive, often strongly and diffusely, with nuclear and cytoplasmic staining. About 40% of lung squamous cell carcinomas are positive, but the staining is often focal
Podoplanin (D2-40)	Not useful. About 80% to 100% of mesotheliomas are positive; 50% of lung squamous cell carcinomas also stain
Cytokeratin 5 or 5/6	Not useful. Expressed in 75% to 100% of mesotheliomas and 100% of lung squamous cell carcinomas
Lung squamous cell carcinoma (positive carcinoma markers)	
p40 or p63	Very useful. 100% of lung squamous cell carcinomas show strong and diffuse nuclear positivity for either marker. About 2.5% and 7% of mesotheliomas are focally positive for p40 and p63, respectively.
Claudin 4	Very useful. About 95% of squamous cell carcinomas are positive. Mesotheliomas are negative.
MOC31	Very useful. About 97% to 100% of lung squamous cell carcinomas are positive; 2% to 10% of mesotheliomas show focal staining
BG8 (Lewis^y)	Very useful. About 80% of lung squamous cell carcinomas are positive; 3% to 7% of mesotheliomas show focal staining
BER-EP4	Useful. Approximately 85% to 100% of lung squamous cell carcinomas are positive; ≤20% of mesotheliomas are focally positive
Cytokeratin 5 or 5/6	Not useful. Nearly all lung squamous cell carcinomas (100%) and most mesotheliomas (75% to 100%) are positive

[a]Table 4 from Husain AN, Colby TV, Ordonez NG, et al. Guidelines for pathologic diagnosis of malignant mesothelioma 2017 update of the consensus statement from the International Mesothelioma Interest Group. Arch Pathol Lab Med 2018;142:89-108.

[b]WT-1 = Wilms tumor-1; BG8 = blood group 8.

and SOX10 for melanoma; and CD31, CD34, and ERG (or FLI-1) for angiosarcoma and epithelioid hemangioendothelioma. Vascular tumors may be positive with antibodies against podoplanin (D2-40), which is not useful in this differential. Since 5 to 10 percent of sarcomatoid mesotheliomas are negative for cytokeratins, this diagnosis can still be made after thorough exclusion of other spindle cells tumors.

TREATMENT

Pleura

Surgical therapy plays a role in disease management. Specifically, pleurectomy decortication to debulk disease and talc pleurodesis to prevent fluid accumulation can be performed. In selected patients, maximal cytoreductive surgery (extended pleural decortication and rarely, extrapleural pneumonectomy) is used (121), but, as it is not possible to achieve microscopically complete resection with diffuse mesothelioma, there appears to be no role for surgery alone (274–277). Thus, it is likely that the use of surgery will continue to be part of a multimodality approach and trials will be required to assess what works best with cytoreductive surgery (277).

Pleural mesothelioma is responsive to combination chemotherapy, specifically to cisplatin and pemetrexed (278), however, these responses are not curative and of limited durability. The addition of bevacizumab may improve survival duration, although side effects also increase (279). Studies using a variety of agents, including immune checkpoint inhibitors, have not replaced platinum/pemetrexed combinations (250), although combinations of immunotherapeutic drugs are showing promise (280,281). The addition of nintedanib to combination chemotherapy showed promise in a phase 2 trial (282), but progression-free survival

Table 3-6

IMMUNOHISTOCHEMICAL MARKERS USED IN THE DIFFERENTIAL DIAGNOSIS OF PLEURAL MESOTHELIOMA VERSUS METASTATIC RENAL CELL CARCINOMAS[a]

Marker	Current Value/Comments
Epithelioid mesothelioma (positive mesothelioma markers)	
Calretinin	Very useful. Essentially all mesotheliomas are positive, and the staining is often strong and diffuse with nuclear and cytoplasmic staining; 4% to 10% of renal cell carcinomas are focally positive
Cytokeratin 5 or 5/6	Very useful. About 75% to 100% of mesotheliomas are positive. Renal cell carcinomas are negative
Podoplanin (D2-40)	Very useful. About 80% 100% of mesotheliomas show positivity along the cell membrane; renal cell carcinomas are negative
Mesothelin	Very useful. All (100%) of mesotheliomas are positive. Renal cell carcinomas are negative
WT-1[b]	Useful. Approximately 70% to 93% of mesotheliomas show nuclear positivity; 4% of renal cell carcinomas are positive
Renal cell carcinoma (possible carcinoma markers)	
PAX8	Very useful. About 85% to 100% of renal cell carcinomas are positive. Mesotheliomas are mostly negative
PAX2	Useful. About 60% to 75% of renal cell carcinomas are positive. Mesotheliomas are negative
Claudin 4	Useful. About 90% of renal cell carcinomas are positive. Mesotheliomas are negative
CD15 (Leu-M1)	Useful. About 65% of renal cell carcinomas are positive. Mesotheliomas only rarely show focal positivity. Can stain any necrotic tissue
RCC Ma	Somewhat useful. About 50% to 70% of renal cell carcinomas are positive; 28% of mesotheliomas are focally positive
Napsin A	Limited utility. About 30% of renal cell carcinomas are positive. Mesotheliomas are negative
MOC31	Limited utility. About 50% of renal cell carcinomas are positive; 2% to 10% of mesotheliomas show focal staining
BER-EP4	Not useful. About 40% of renal cell carcinomas are positive; $\leq$20% of mesotheliomas are focally positive
CD10	Not useful. About 80% of renal cell carcinomas are positive. About 50% of mesotheliomas are positive
BG8 (Lewis^{y})	Not useful. About 4% of renal cell carcinomas and 3% to 7% of mesotheliomas are positive

[a]Table 5 from Husain AN, Colby TV, Ordonez NG, et al. Guidelines for pathologic diagnosis of malignant mesothelioma 2017 update of the consensus statement from the International Mesothelioma Interest Group. Arch Pathol Lab Med 2018;142:89-108.
[b]WT-1 = Wilms tumor-1; BG8 = blood group 8.

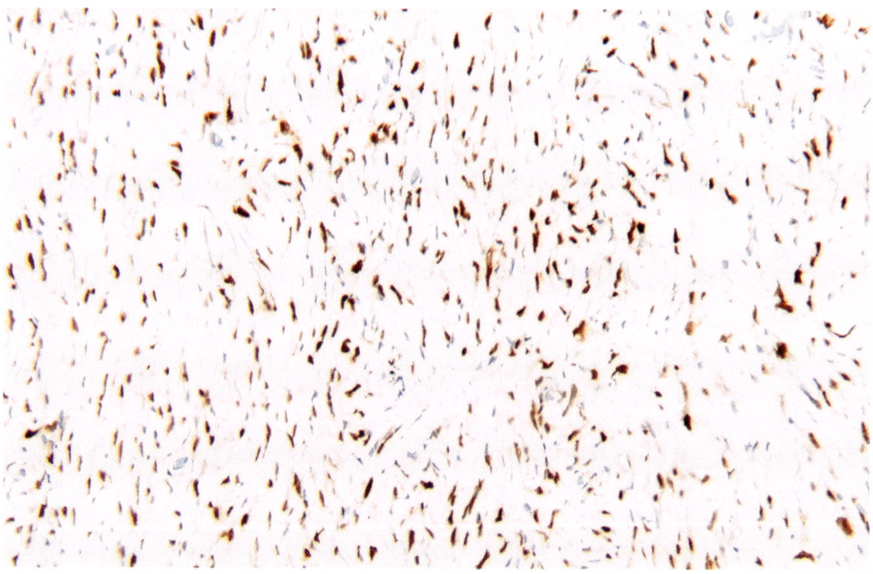

Figure 3-120

DESMOPLASTIC MESOTHLIOMA

GATA3 shows strong nuclear positivity.

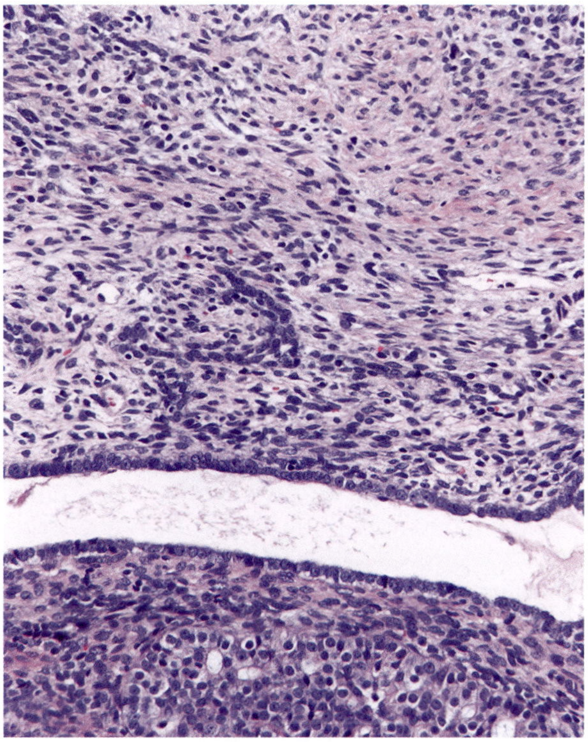

Figure 3-121

BIPHASIC MESOTHELIOMA

The morphology of this pleural tumor resembles synovial sarcoma. However, it was negative for t(X;18).

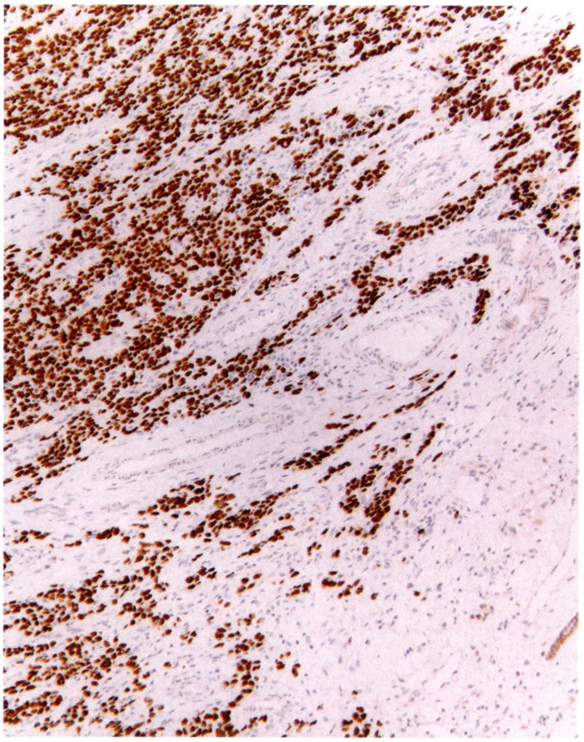

Figure 3-122

SOLITARY FIBROUS TUMOR

STAT6 is strongly positive (nuclear) in this malignant solitary fibrous tumor.

was not significantly improved in a phase 3 trial (283). Radiation therapy may play a role in disease control (284), and multimodality therapy including surgery, chemotherapy, and radiation, have been studied (93).

Peritoneum

Debulking of disease when possible, followed by hyperthermic intraperitoneal chemotherapy (HIPEC) is a standard treatment approach when possible (285). Complete cytoreduction can be achieved in two thirds of patients; the addition of HIPEC with cisplatin improves survival. In a meta-analysis of 1,047 patients, 1-year, 3-year, and 5-year survival rates were 84 percent, 59 percent, and 42 percent, respectively (111). Some patients have prolonged (over 7 years) disease-free survival periods (286). Epithelioid histology, low mitotic rate, negative nodes, and retention of p16 expression by immunohistochemistry have all been reported as potential features associated with a favorable prognosis (227,287). Median survival with successful therapy may be 67 months or longer (227,288,289) in epithelioid subtypes.

Pericardium and Tunica Vaginalis

The treatment approach of diffuse mesothelioma of the pericardium is chemotherapy with cispatin and pemetrexed as the disease is usually not resectable. Disease is usually refractory to therapy overall and portends a fatal outcome (290).

In the tunica vaginalis, surgical removal is recommended (120). The benefits of radiation or chemotherapy are uncertain.

PROGNOSIS

Survival Data

Pleura. Review of large series of patients with mesothelioma since 1980 still shows a poor prognosis, with a survival period of 6 to 12 months (291–294). Survival data from 1,625 patients diagnosed with mesothelioma between

2005 and 2009 in the National Cancer Institute's SEER Medicare database showed a median overall survival period of 8 months (range, 1 to 69 months) (293). A further study showed no difference in survival rates between patients diagnosed between 1973-1999 and 2000-2009 (295). Female gender, younger age, early stage, epithelioid morphology, and treatment with surgery are independent predictors of longer survival (295,296). A variety of other parameters, ranging from blood based (297), imaging (298), histologic patterns and features (99), histologic grading (99,122,171,172), molecular signatures (219,299,300), and tumor microenvironment (301) have all been reported to show prognostic significance, along with various combinations (302,303). However, there is, as yet, no agreed multimodality prognostic algorithm.

Peritoneum. Overall survival is similar to that seen with pleural mesothelioma in some series (5,84,304–306), although better in others (86,285). In one series, patients with peritoneal mesotheliomas had a shorter median survival period than those with pleural tumors, but there was a larger proportion of long-term survivors (306). High blood neutrophil to lymphocyte ratios and hypoalbuminemia are adverse prognostic factors in peritoneal mesothelioma (307).

Pericardium. Data are sparse and mainly limited to case reports and reviews thereof. Occasional localized cases are associated with prolonged survival, if resectable, but survival is otherwise similar to that of pleural tumors (90,116).

Tunica Vaginalis. Most cases report aggressive behavior (91,118–120). In a series of 73 patients, the median survival period was 23 months, which decreased to 14 months in cases of recurrence (120). Recurrence occurred in about 50 percent of patients. In a recent series treated with radical orchiectomy with subsequent surgery, only 5 of 15 patients died at 3.5 years (119).

TNM Staging

Pleura. As for most cancers, TNM staging is a mainstay of prognostication, although pathologic staging is limited to patients who undergo maximal cytoreductive surgery (extended pleurectomy/decortication [EPD] and extrapleural pneumonectomy [EPP]), as only these provide sufficient tissue for a thorough assessment of extent of anatomic involvement by tumor. Smaller samples should be clinically staged (99).

There was no generally accepted staging system for mesothelioma until 1994, when the International Mesothelioma Interest Group (IMIG), in collaboration with the International Association for the Study of Lung Cancer (IASLC), proposed a TNM staging system based on analyses of outcomes in retrospective surgical series and small clinical trials. Subsequently accepted by the Union for International Cancer Control (UICC) and the American Joint Committee on Cancer (AJCC), this became the international staging standard, although it had significant limitations, with some publications showing a lack of prediction of survival (308). Other staging systems have been published showing correlation with survival (309). Therefore, to inform the 8th edition of the UICC/AJCC staging, IMIG in conjunction with the IASLC accrued a database of 3519 patients from which they developed proposals for tumor, node, and metastasis (TNM) categories for mesothelioma (310–313). These proposals were again accepted by both the UICC and the AJCC, although it should be noted that terminology differs between the two groups (314). The 8th edition of staging was published in early 2017, although most institutions only began to change from the 7th edition after 2018.

In the analysis of the data, pathologic staging failed to demonstrate a survival difference between adjacent categories, with the exception of T3 versus T4, and there was better stratification with the merging of T1a and T1b into a single T1 classification (Table 3-7). The clinical and pathologic N1 and N2 classifications of the 7th edition (UICC), comprising ipsilateral, intrathoracic nodal metastases, are also merged into N1 in the 8th edition. Nodes previously classified as N3 in the 7th edition are reclassified as N2 (312). M classification remains unchanged (310). Stage groupings remain unchanged from the 7th edition.

Compared to other thoracic staging systems, namely for lung and thymus cancers, the 8th edition of staging for mesothelioma is relatively unchanged. In fact, it has been simplified in some areas in the hope that this allows better stratification of patients for management decisions.

Peritoneum, Pericardium, and Tunica Vaginalis. There is no accepted staging system for mesothelioma arising at any of these sites,

Table 3-7

TNM STAGING OF PLEURAL MESOTHELIOMA[a]

T Descriptors
- pTX Primary tumor cannot be assessed
- pT0 No evidence of primary tumor
- pT1 Tumor limited to the ipsilateral parietal ± visceral ± mediastinal ± diaphragmatic pleura
- pT2 Tumor involving the ipsilateral pleural surfaces (parietal, mediastinal, diaphragmatic, and visceral pleura) with at least one of the following features:
  - involvement of diaphragmatic muscle
  - extension of tumor from visceral pleura into the underlying pulmonary parenchyma
- pT3 Describes locally advanced but potentially resectable tumor; tumor involving all of the ipsilateral pleural surfaces (parietal, mediastinal, diaphragmatic, and visceral pleura) with at least one of the following features:
  - involvement of the endothoracic fascia
  - extension into the mediastinal fat
  - solitary, completely resectable focus of tumor extending into the soft tissues of the chest wall
  - nontransmural involvement of the pericardium
- pT4 Describes locally advanced technically unresectable tumor

Tumor involving all of the ipsilateral pleural surfaces (parietal, mediastinal, diaphragmatic, and visceral pleura) with at least one of the following features:
- diffuse extension or multifocal masses of tumor in the chest wall, with or without associated rib destruction
- direct transdiaphragmatic extension of tumor to the peritoneum
- direct extension of tumor to the contralateral pleura
- direct extension of tumor to mediastinal organs
- direct extension of tumor into the spine
- tumor extending through to the internal surface of the pericardium with or without a pericardial effusion; or tumor involving the myocardium

N Regional lymph nodes
- pNX Regional lymph nodes cannot be assessed
- pN0 No regional lymph node metastases
- pN1 Metastases in the ipsilateral bronchopulmonary, hilar, or mediastinal (including the internal mammary, peridiaphragmatic, pericardial fat pad, or intercostal lymph nodes) lymph nodes
- pN2 Metastases in the contralateral bronchopulmonary, hilar, or mediastinal lymph nodes or ipsilateral or contralateral supraclavicular lymph nodes

M Distant metastasis
- pM0 No distant metastasis
- pM1 Distant metastasis

Stage Grouping

	N0	N1	N2
T1	IA	II	IIIB
T2	IB	II	IIIB
T3	IB	IIIA	IIIB
T4	IIIB	IIIB	IIIB
M1	IV	IV	IV

[a]Data from UICC TNM classification of malignant tumours, 8th ed. Oxford: Wiley Blackwell; 2017.

although tumors may be subdivided into localized and diffuse, with or without metastatic disease. This subdivision has prognostic value and may be relevant to management (surgery versus other treatments).

Other Prognostic Factors

Pleura. *Clinical and Histologic Features.* Certain clinical features portend a worse prognosis for patients with pleural mesotheliomas, such as older age at diagnosis, poor performance status, anemia, leukocytosis, and the presence of chest pain, dyspnea, and weight loss (293, 295,315,316). Patients with epithelioid mesotheliomas have a better prognosis than those with biphasic mesotheliomas, which in turn confer a better prognosis than sarcomatoid mesotheliomas (296). The extent of the sarcomatoid component in biphasic mesothelioma may also influence prognosis, although the percentage varies between publications (125,165).

Even within subtypes, certain histologic patterns and cytologic and stromal features (Table 3-3) correlate with a worse prognosis, such as solid pattern (over 50 percent of sample), and pleomorphic (156,166), transitional (143,165,167), and rhabdoid (163) cytologic features. Conversely, myxoid stroma, when present in over 50 percent of the sample (155), appears to confer a better prognosis, and lymphohistiocytoid cytology, where cells may be sarcomatoid or epithelioid, confers a prognosis closer to that of epithelioid subtypes (168), possibly reflecting a prominent host response.

Nuclear grading also provides useful prognostic data (see Microscopic Findings) (122,171, 172). Detailed description of subtypes and histologic patterns should therefore be undertaken routinely. Variants such as well-differentiated papillary mesothelial tumor and localized mesothelioma carry a better prognosis and these are described in chapter 4.

Tumor Volume and Maximal Thickness. Tumor volume and maximal thickness of tumor samples may have prognostic significance (317) and, although tumor volume will likely remain the domain of the imaging community, measurement of tumor thickness may become increasingly important. Also, recent data describing differences in immune-genomic phenotyping within resected mesothelioma specimens should encourage pathologists to save and evaluate multiple areas of the resected specimen (125).

Molecular Findings. The last decade has seen a revolution in molecular pathology and immunology in relation to treatment of thoracic cancers, although this has predominantly been in relation to lung cancer. However, as discussed above, mesotheliomas are increasingly becoming a focus for potential new therapies related to molecular and immunologic signatures, with prognostic implications. Trials of new drugs in relation to *BAP1*, *CDKN2A*, and *NF2* are ongoing (318–320). Patients who have a germline *BAP1* mutation have a 7-fold increase in survival, compared to those without (321).

Immunologic Findings. The presence of a dense host lymphocytic reaction to mesothelioma is a favorable prognostic factor, likely due to the host immune system recognizing and killing cells that it views as foreign. With increasing interest in immunomodulatory therapies and several trials ongoing (320), the next decade may well see the introduction of therapies that improve survival. Expression of PDL1 is associated with a poorer prognosis and is more frequently seen in tumors with sarcomatoid morphology (246,322,323). Other immune-related targets by the tumor have not as yet been shown to confer any survival advantage or disadvantage.

Peritoneum. *Clinical and Histologic Features.* Older age at presentation, poor performance status, hypoalbuminemia, high neutrophil to lymphocyte ratios, and abdominal pain are all reported as adverse factors (5,9,110,285,324). Males have a worse prognosis than females although this difference was reported as less pronounced than in the pleura in one series (306). In another series, the survival advantage was limited to pre-menopausal women (325). A solid pattern of growth has been reported as an adverse factor compared to tubulopapillary and micropapillary patterns (287). Grading of peritoneal mesotheliomas is also reported as having prognostic significance (287,326), as has the extent of Ki-67 (327) and MUC1 (328) staining.

Molecular and Immunologic Findings. The presence of *BAP1* mutations and loss of BAP1 staining confer a better prognosis in peritoneal mesotheliomas (87). Mutations in *CDKN2A* and *NF2* are poor prognostic factors (224,237). There is little published data on immunomodulatory therapy, although several trials are ongoing. In one study, PDL1 expression by lymphocytes was associated with better overall and progression-free survival by both univariate and multivariate analyses (329), and this area is likely to burgeon in the next few years.

Pericardium and Tunica Vaginalis. Data on prognostic factors for mesotheliomas arising at these two sites are limited, other than disease being localized or diffuse/metastatic at presentation (90,119,120). For the tunica vaginalis, older age and disseminated disease at presentation are adverse prognostic factors (120).

REFERENCES

1. Marchevsky AM, Khoor A, Walts AE, et al. Localized malignant mesothelioma, an unusual and poorly characterized neoplasm of serosal origin: best current evidence from the literature and the International Mesothelioma Panel. Mod Pathol 2020;33:281-96.
2. Delgermaa V, Takahashi K, Park EK, Le GV, Hara T, Sorahan T. Global mesothelioma deaths reported to the World Health Organization between 1994 and 2008. Bull World Health Organ 2011;89:716-24, 724A-C.
3. Bray F, Ferlay J, Soerjomataram I, Siegel RL, Torre LA, Jemal A. Global cancer statistics 2018: GLOBOCAN estimates of incidence and mortality worldwide for 36 cancers in 185 countries. CA Cancer J Clin 2018;68:394-424.
4. Boffetta P, Malvezzi M, Pira E, Negri E, La Vecchia C. International analysis of age-specific mortality rates from mesothelioma on the basis of the international classification of diseases, 10th revision. J Glob Oncol 2018;4:1-15.
5. Mensi C, De Matteis S, Dallari B, Riboldi L, Bertazzi PA, Consonni D. Incidence of mesothelioma in Lombardy, Italy: exposure to asbestos, time patterns and future projections. Occup Environ Med 2016;73:607-13.
6. Soeberg MJ, Leigh J, Driscoll T, Armstrong B, Young JM, van Zandwijk N. Incidence and survival trends for malignant pleural and peritoneal mesothelioma, Australia, 1982-2009. Occup Environ Med 2016;73:187-94.
7. Metintas S, Batirel HF, Bayram H, et al. Turkey National Mesothelioma Surveillance and Environmental Asbestos Exposure Control Program. Int J Environ Res Public Health 2017;14:1293.
8. Van den Borre L, Deboosere P. Asbestos in Belgium: an underestimated health risk. The evolution of mesothelioma mortality rates (1969-2009). Int J Occup Environ Health 2014;20:134-40.
9. Soeberg MJ, Creighton N, Currow DC, Young JM, van Zandwijk N. Patterns in the incidence, mortality and survival of malignant pleural and peritoneal mesothelioma, New South Wales, 1972-2009. Aust N Z J Public Health 2016;40:255-62.
10. Algranti E, Saito CA, Carneiro AP, Moreira B, Mendonca EM, Bussacos MA. The next mesothelioma wave: mortality trends and forecast to 2030 in Brazil. Cancer Epidemiol 2015;39:687-92.
11. Zhao J, Zuo T, Zheng R, et al. Epidemiology and trend analysis on malignant mesothelioma in China. Chin J Cancer Res 2017;29:361-8.
12. Gilham C, Rake C, Burdett G, et al. Pleural mesothelioma and lung cancer risks in relation to occupational history and asbestos lung burden. Occup Environ Med 2016;73:290-9.
13. Mazurek JM, Syamlal G, Wood JM, Hendricks SA, Weston A. Malignant mesothelioma mortality - United States, 1999-2015. MMWR Morb Mortal Wkly Rep 2017;66:214-8.
14. Reynolds CJ, Minelli C, Darnton A, Cullinan P. Mesothelioma mortality in Great Britain: how much longer will dockyards dominate? Occup Environ Med 2019;76:908-12.
14a. https://gco.iarc.fr/today/home.
14b. Ferlay J, Colombet M, Soerjomataram I, et al. Estimating the global cancer incidence and mortality in 2018: GLOBOCAN sources and methods. Int J Cancer 2019:144:1941-53.
15. Thomas A, Chen Y, Yu T, Gill A, Prasad V. Distinctive clinical characteristics of malignant mesothelioma in young patients. Oncotarget 2015;6:16766-73.
16. Lemen RA. Mesothelioma from asbestos exposures: epidemiologic patterns and impact in the United States. J Toxicol Environ Health B Crit Rev 2016;19:250-65.
17. Marinaccio A, Corfiati M, Binazzi A, et al. The epidemiology of malignant mesothelioma in women: gender differences and modalities of asbestos exposure. Occup Environ Med 2018;75:254-62.
18. Harbour JW, Onken MD, Roberson ED, et al. Frequent mutation of BAP1 in metastasizing uveal melanomas. Science 2010;330:1410-3.
19. Abdel-Rahman MH, Pilarski R, Cebulla CM, et al. Germline BAP1 mutation predisposes to uveal melanoma, lung adenocarcinoma, meningioma, and other cancers. J Med Genet 2011;48:856-9.
20. Testa JR, Cheung M, Pei J, et al. Germline BAP1 mutations predispose to malignant mesothelioma. Nat Genet 2011;43:1022-5.
21. Pastorino S, Yoshikawa Y, Pass HI, et al. A subset of mesotheliomas with improved survival occurring in carriers of BAP1 and other germline mutations. J Clin Oncol 2018;36:JCO2018790352.
22. Walpole S, Pritchard AL, Cebulla CM, et al. Comprehensive study of the clinical phenotype of germline BAP1 variant-carrying families worldwide. J Natl Cancer Inst 2018;110:1328-41.
23. Ohar JA, Cheung M, Talarchek J, et al. Germline BAP1 mutational landscape of asbestos-exposed malignant mesothelioma patients with family history of cancer. Cancer Res 2016;76:206-15.
24. Betti M, Casalone E, Ferrante D, et al. Germline mutations in DNA repair genes predispose asbestos-exposed patients to malignant pleural mesothelioma. Cancer Lett 2017;405:38-45.

25. Ascoli V, Cozzi I, Vatrano S, et al. Mesothelioma families without inheritance of a BAP1 predisposing mutation. Cancer Genet 2016;209:381-7.
26. Betti M, Aspesi A, Biasi A, et al. CDKN2A and BAP1 germline mutations predispose to melanoma and mesothelioma. Cancer Lett 2016;378:120-30.
27. Glass C, Sholl LM, Landgraf JR, Chirieac L, Roggli VL. Molecular analysis of a patient with neurofibromatosis 2 (NF2) and peritoneal malignant mesothelioma. Am J Surg Pathol 2020;44:1290-2.
28. Shih AR, Kradin RL. Malignant mesothelioma in Lynch syndrome: a report of two cases and a review of the literature. Am J Ind Med 2019;62:448-52.
29. Ascoli V, Romeo E, Carnovale Scalzo C, et al. Familial malignant mesothelioma: a population-based study in central Italy (1980-2012). Cancer Epidemiol 2014;38:273-8.
30. Dogan AU, Baris YI, Dogan M, et al. Genetic predisposition to fiber carcinogenesis causes a mesothelioma epidemic in Turkey. Cancer Res 2006;66:5063-8.
31. Tunesi S, Ferrante D, Mirabelli D, et al. Gene-asbestos interaction in malignant pleural mesothelioma susceptibility. Carcinogenesis 2015;36:1129-35.
32. Spirtas R, Heineman EF, Bernstein L, et al. Malignant mesothelioma: attributable risk of asbestos exposure. Occup Environ Med 1994;51:804-11.
32a. Baumann F, Maurizot P, Mangeas M, Ambrosi JP, Douwes J, Robineau B. Pleural mesothelioma in New Caledonia: associations with environmental risk factors. Environ Health Perspect 2011:119:695-700.
32b. Bruno C, Comba P, Zona A. Adverse health effects of fluoro-edenitic fibers: epidemiological evidence and public health priorities. Ann N Y Acad Sci 2006;1076:778-83.
33. Wagner JC, Sleggs CA, Marchand P. Diffuse pleural mesothelioma and asbestos exposure in the North Western Cape Province. Br J Ind Med 1960;17:260-71.
34. McCaughey WT, Wade OL, Elmes PC. Exposure to asbestos dust and diffuse pleural mesotheliomas. Br Med J 1963;1:123.
35. Hourihane DO. The pathology of mesotheliomata and an analysis of their association with asbestos exposure. Thorax 1964;19:268-78.
36. Selikoff IJ, Churg J, Hammond EC. Asbestos exposure and neoplasia. JAMA 1964;188:22-6.
37. McDonald AD, Harper A, McDonald JC, el-Attar OA. Epidemiology of primary malignant mesothelial tumors in Canada. Cancer 1970;26:914-9.
38. Selikoff IJ, Hammond EC, Churg J. Carcinogenicity of amosite asbestos. Arch Environ Health 1972;25:183-6.
39. Marsh GM, Riordan AS, Keeton KA, Benson SM. Non-occupational exposure to asbestos and risk of pleural mesothelioma: review and meta-analysis. Occup Environ Med 2017;74:838-46.
40. Berry G, Pooley F, Gibbs A, Harris JM, McDonald JC. Lung fiber burden in the Nottingham gas mask cohort. Inhal Toxicol 2009;21:168-72.
41. Churg A. Deposition and clearance of chrysotile asbestos. Ann Occup Hyg 1994;38:625-33, 424-5.
42. de Klerk NH, Musk AW, Williams V, Filion PR, Whitaker D, Shilkin KB. Comparison of measures of exposure to asbestos in former crocidolite workers from Wittenoom Gorge, W. Australia. Am J Ind Med 1996;30:579-87.
43. Roggli VL, Sharma A, Butnor KJ, Sporn T, Vollmer RT. Malignant mesothelioma and occupational exposure to asbestos: a clinicopathological correlation of 1445 cases. Ultrastruct Pathol 2002;26:55-65.
44. Goodman M, Teta MJ, Hessel PA, et al. Mesothelioma and lung cancer among motor vehicle mechanics: a meta-analysis. Ann Occup Hyg 2004;48:309-26.
45. Garabrant DH, Alexander DD, Miller PE, et al. Mesothelioma among motor vehicle mechanics: an updated review and meta-analysis. Ann Occup Hyg 2016;60:8-26.
46. Yarborough CM. The risk of mesothelioma from exposure to chrysotile asbestos. Curr Opin Pulm Med 2007;13:334-8.
47. Sichletidis L, Chloros D, Spyratos D, et al. Mortality from occupational exposure to relatively pure chrysotile: a 39-year study. Respiration 2009;78:63-8.
48. Yano E, Wang ZM, Wang XR, Wang MZ, Lan YJ. Cancer mortality among workers exposed to amphibole-free chrysotile asbestos. Am J Epidemiol 2001;154:538-43.
49. Sullivan PA. Vermiculite, respiratory disease, and asbestos exposure in Libby, Montana: update of a cohort mortality study. Environ Health Perspect 2007;115:579-85.
50. McDonald JC, McDonald AD. Chrysotile, tremolite and carcinogenicity. Ann Occup Hyg 1997;41:699-705.
51. Liddell FD, McDonald AD, McDonald JC. The 1891-1920 birth cohort of Quebec chrysotile miners and millers: development from 1904 and mortality to 1992. Ann Occup Hyg 1997;41:13-36.
52. McDonald AD, Case BW, Churg A, et al. Mesothelioma in Quebec chrysotile miners and millers: epidemiology and aetiology. Ann Occup Hyg 1997;41:707-19.
53. Berry G. Mortality and cancer incidence of workers exposed to chrysotile asbestos in the friction-products industry. Ann Occup Hyg 1994;38:539-46, 413.
54. Camus M, Siemiatycki J, Meek B. Nonoccupational exposure to chrysotile asbestos and the risk of lung cancer. N Engl J Med 1998;338:1565-71.

55. Churg A. Chrysotile, tremolite, and malignant mesothelioma in man. Chest 1988;93:621-8.
56. Lanphear BP, Buncher CR. Latent period for malignant mesothelioma of occupational origin. J Occup Med 1992;34:718-21.
57. Nurminen M, Karjalainen A, Takahashi K. Estimating the induction period of pleural mesothelioma from aggregate data on asbestos consumption. J Occup Environ Med 2003;45:1107-15.
58. Frost G. The latency period of mesothelioma among a cohort of British asbestos workers (1978-2005). Br J Cancer 2013;109:1965-73.
59. Ferrante D, Chellini E, Merler E, et al. Italian pool of asbestos workers cohorts: mortality trends of asbestos-related neoplasms after long time since first exposure. Occup Environ Med 2017;74:887-98.
60. Reid A, Franklin P, Olsen N, et al. All-cause mortality and cancer incidence among adults exposed to blue asbestos during childhood. Am J Ind Med 2013;56:133-45.
61. Dalsgaard SB, Wurtz ET, Hansen J, Roe OD, Omland O. Environmental asbestos exposure in childhood and risk of mesothelioma later in life: a long-term follow-up register-based cohort study. Occup Environ Med 2019;76:407-13.
62. Binazzi A, Marinaccio A, Corfiati M, et al. Mesothelioma incidence and asbestos exposure in Italian national priority contaminated sites. Scand J Work Environ Health 2017;43:550-9.
63. Agostin FD, de Michieli P, Negro C. Pleural mesothelioma in household members of asbestos-exposed workers in Friuli Venezia Giulia, Italy. Int J Occup Med Environ Health 2017;30:419-31.
64. Baumann F, Ambrosi JP, Carbone M. Asbestos is not just asbestos: an unrecognised health hazard. Lancet Oncol 2013;14:576-8.
65. Baumann F, Buck BJ, Metcalf RV, McLaurin BT, Merkler DJ, Carbone M. The presence of asbestos in the natural environment is likely related to mesothelioma in young individuals and women from southern Nevada. J Thorac Oncol 2015;10:731-7.
66. Lacourt A, Gramond C, Rolland P, et al. Occupational and non-occupational attributable risk of asbestos exposure for malignant pleural mesothelioma. Thorax 2014;69:532-9.
67. Liu B, van Gerwen M, Bonassi S, Taioli E, International Association for the Study of Lung Cancer Mesothelioma Task F. Epidemiology of environmental exposure and malignant mesothelioma. J Thorac Oncol 2017;12:1031-45.
68. Carbone M, Baris YI, Bertino P, et al. Erionite exposure in North Dakota and Turkish villages with mesothelioma. Proc Natl Acad Sci U S A 2011;108:13618-23.
69. Ortega-Guerrero MA, Carrasco-Nunez G, Barragan-Campos H, Ortega MR. High incidence of lung cancer and malignant mesothelioma linked to erionite fibre exposure in a rural community in Central Mexico. Occup Environ Med 2015;72:216-8.
70. Berry G, Newhouse ML, Antonis P. Combined effect of asbestos and smoking on mortality from lung cancer and mesothelioma in factory workers. Br J Ind Med 1985;42:12-8.
71. Finley BL, Benson SM, Marsh GM. Cosmetic talc as a risk factor for pleural mesothelioma: a weight of evidence evaluation of the epidemiology. Inhal Toxicol 2017;29:179-85.
72. Li X, Brownlee NA, Sporn TA, Mahar A, Roggli VL. Malignant (diffuse) mesothelioma in patients with hematologic malignancies: a clinicopathologic study of 45 cases. Arch Pathol Lab Med 2015;139:1129-36.
73. Cavazza A, Travis LB, Travis WD, et al. Post-irradiation malignant mesothelioma. Cancer 1996; 77:1379-85.
74. Farioli A, Ottone M, Morganti AG, et al. Radiation-induced mesothelioma among long-term solid cancer survivors: a longitudinal analysis of SEER database. Cancer Med 2016;5:950-9.
75. Attanoos RL, Churg A, Galateau-Salle F, Gibbs AR, Roggli VL. Malignant mesothelioma and its non-asbestos causes. Arch Pathol Lab Med 2018; 142:753-60.
75a. Gibb H, Fulcher K, Nagarajan S, et al. Analyses of radiation and mesothelioma in the US Transuranium and Uranium Registries. Am J Public Health. 2013;103:710-6.
76. Brown LM, Chen BE, Pfeiffer RM, et al. Risk of second non-hematological malignancies among 376,825 breast cancer survivors. Breast Cancer Res Treat 2007;106:439-51.
77. Deutsch M, Land SR, Begovic M, Cecchini R, Wolmark N. An association between postoperative radiotherapy for primary breast cancer in 11 National Surgical Adjuvant Breast and Bowel Project (NSABP) studies and the subsequent appearance of pleural mesothelioma. Am J Clin Oncol 2007;30:294-6.
78. Travis LB, Fossa SD, Schonfeld SJ, et al. Second cancers among 40,576 testicular cancer patients: focus on long-term survivors. J Natl Cancer Inst 2005;97:1354-65.
79. Tward JD, Wendland MM, Shrieve DC, Szabo A, Gaffney DK. The risk of secondary malignancies over 30 years after the treatment of non-Hodgkin lymphoma. Cancer 2006;107:108-15.
80. De Bruin ML, Burgers JA, Baas P, et al. Malignant mesothelioma after radiation treatment for Hodgkin lymphoma. Blood 2009;113:3679-81.
81. Hodgson DC, Gilbert ES, Dores GM, et al. Long-term solid cancer risk among 5-year survivors of Hodgkin's lymphoma. J Clin Oncol 2007; 25:1489-97.
82. Boffetta P. Epidemiology of peritoneal mesothelioma: a review. Ann Oncol 2007;18:985-90.

83. Salo SA, Ilonen I, Laaksonen S, Myllarniemi M, Salo JA, Rantanen T. Epidemiology of malignant peritoneal mesothelioma: a population-based study. Cancer Epidemiol 2017;51:81-6.
84. Conti S, Minelli G, Ascoli V, et al. Peritoneal mesothelioma in Italy: trends and geography of mortality and incidence. Am J Ind Med 2015;58:1050-8.
85. Berry G, Reid A, Aboagye-Sarfo P, et al. Malignant mesotheliomas in former miners and millers of crocidolite at Wittenoom (Western Australia) after more than 50 years follow-up. Br J Cancer 2012;106:1016-20.
86. Malpica A, Euscher ED, Marques-Piubelli ML, et al. Malignant mesothelioma of the peritoneum in women: a clinicopathologic study of 164 cases. Am J Surg Pathol 2021;45:45-58.
87. Leblay N, Lepretre F, Le Stang N, et al. BAP1 is altered by copy number loss, mutation, and/or loss of protein expression in more than 70% of malignant peritoneal mesotheliomas. J Thorac Oncol 2017;12:724-33.
88. Lowry SJ, Weiss NS. Geographic distribution of incidence of pericardial and paratesticular mesotheliomas in the USA. Cancer Causes Control 2016;27:1487-9.
89. Mezei G, Chang ET, Mowat FS, Moolgavkar SH. Epidemiology of mesothelioma of the pericardium and tunica vaginalis testis. Ann Epidemiol 2017;27:348-59.e311.
90. Cao S, Jin S, Cao J, et al. Malignant pericardial mesothelioma : a systematic review of current practice. Herz 2018;43:61-8.
91. Butnor KJ, Pavlisko EN, Sporn TA, Roggli VL. Mesothelioma of the tunica vaginalis testis. Hum Pathol 2019;92:48-58.
92. Baqui AA, Boire NA, Baqui TT, Etwaru DJ. Mlignant mesothelioma of the tunica vaginalis testis-a malignancy associated with asbestos exposure and trauma: a case report and literature review. J Investig Med High Impact Case Rep 2019;7:2324709619827335.
93. Zellos LS, Sugarbaker DJ. Diffuse malignant mesothelioma of the pleural space and its management. Oncology (Williston Park) 2002;16:907-13.
94. Sheard JD, Taylor W, Soorae A, Pearson MG. Pneumothorax and malignant mesothelioma in patients over the age of 40. Thorax 1991;46:584-5.
95. Situnayake RD, Middleton WG. Recurrent pneumothorax and malignant pleural mesothelioma. Respir Med 1991;85:255-6.
96. Sussman J, Rosai J. Lymph node metastasis as the initial manifestation of malignant mesothelioma. Report of six cases. Am J Surg Pathol 1990;14:819-28.
97. Larsen BT, Klein JR, Hornychova H, et al. Diffuse intrapulmonary malignant mesothelioma masquerading as interstitial lung disease: a distinctive variant of mesothelioma. Am J Surg Pathol 2013;37:1555-64.
98. Marom EM, Erasmus JJ, Pass HI, Patz EF Jr. The role of imaging in malignant pleural mesothelioma. Semin Oncol 2002;29:26-35.
99. Nicholson AG, Sauter JL, Nowak AK, et al. EURACAN/IASLC proposals for updating the histologic classification of pleural mesothelioma: towards a more multidisciplinary approach. J Thorac Oncol 2020;15:29-49.
100. Gill RR, Gerbaudo VH, Sugarbaker DJ, Hatabu H. Current trends in radiologic management of malignant pleural mesothelioma. Semin Thorac Cardiovasc Surg 2009;21:111-20.
101. Gerbaudo VH, Katz SI, Nowak AK, Francis RJ. Multimodality imaging review of malignant pleural mesothelioma diagnosis and staging. PET Clin 2011;6:275-97.
102. Gill RR, Tsao AS, Kindler HL, et al. Radiologic considerations and standardization of malignant pleural mesothelioma imaging within clinical trials: consensus statement from the NCI Thoracic Malignancy Steering Committee - International Association for the Study of Lung Cancer - Mesothelioma Applied Research Foundation Clinical Trials Planning Meeting. J Thorac Oncol 2019;14:1718-31.
103. Armato SG 3rd, Blyth KG, Keating JJ, et al. Imaging in pleural mesothelioma: a review of the 13th International Conference of the International Mesothelioma Interest Group. Lung Cancer 2016;101:48-58.
104. Liu F, Zhao B, Krug LM, et al. Assessment of therapy responses and prediction of survival in malignant pleural mesothelioma through computer-aided volumetric measurement on computed tomography scans. J Thorac Oncol 2010;5:879-84.
105. Archer HA, Panopoulou A, Bhatt N, Edey AJ, Giffin NJ. Mesothelioma and anti-Ma paraneoplastic syndrome; heterogeneity in immunogenic tumours increases. Pract Neurol 2014;14:33-5.
106. Tanriverdi O, Meydan N, Barutca S, Ozsan N, Gurel D, Veral A. Anti-Yo antibody-mediated paraneoplastic cerebellar degeneration in a female patient with pleural malignant mesothelioma. Jpn J Clin Oncol 2013;43:563-8.
107. Venzano C, Di Marco E, Garbero M, Forno P, Marchese A, Borghesi R. Nephrotic syndrome associated with pleural mesothelioma. An unusual paraneoplastic event. Recenti Prog Med 1990;81:325-6.
108. Perks WH, Crow JC, Green M. Mesothelioma associated with the syndrome of inappropriate secretion of antidiuretic hormone. Am Rev Respir Dis 1978;117:789-94.
109. De Pangher Manzini V, Brollo A, Bianchi C. Thrombocytosis in malignant pleural mesothelioma. Tumori 1990;76:576-8.

110. Garcia-Fadrique A, Mehta A, Mohamed F, Dayal S, Cecil T, Moran BJ. Clinical presentation, diagnosis, classification and management of peritoneal mesothelioma: a review. J Gastrointest Oncol 2017;8:915-24.
111. Helm JH, Miura JT, Glenn JA, et al. Cytoreductive surgery and hyperthermic intraperitoneal chemotherapy for malignant peritoneal mesothelioma: a systematic review and meta-analysis. Ann Surg Oncol 2015;22:1686-93.
112. Nicholson CP, Donohue JH, Thompson GB, Lewis JE. A study of metastatic cancer found during inguinal hernia repair. Cancer 1992;69:3008-11.
113. Istomin V, Blondheim DS, Meisel SR, Frimerman A, Lapidot M, Rachmilevitch R. Pericardial effusion due to primary malignant pericardial mesothelioma: a common finding but an uncommon cause. Case Rep Med 2016;2016:4810901.
114. Gong W, Ye X, Shi K, Zhao Q. Primary malignant pericardial mesothelioma-a rare cause of superior vena cava thrombosis and constrictive pericarditis. J Thorac Dis 2014;6:E272-5.
115. Barroso AS, Leite S, Friões F, et al. Pericardial mesothelioma presenting as a suspected ST-elevation myocardial infarction. Rev Port Cardiol 2017;36:307.e1-307.e5.
116. Sardar MR, Kuntz C, Patel T, et al. Primary pericardial mesothelioma unique case and literature review. Tex Heart Inst J 2012;39:261-4.
117. Miyamoto Y, Nakano S, Shimazaki Y, Matsuda H, Fukuda H. Pericardial mesothelioma presenting as left atrial thrombus in a patient with mitral stenosis. Cardiovasc Surg 1996;4:51-2.
118. An JY, Kim D, Tanakchi S, et al. Clinical features and outcomes of tunica vaginalis mesothelioma: a case series from the National Institutes of Health. Clin Genitourin Cancer 2017;15:e871-5.
119. Recabal P, Rosenzweig B, Bazzi WM, Carver BS, Sheinfeld J. Malignant mesothelioma of the tunica vaginalis testis: outcomes following surgical management beyond radical orchiectomy. Urology 2017;107:166-70.
120. Plas E, Riedl CR, Pfluger H. Malignant mesothelioma of the tunica vaginalis testis: review of the literature and assessment of prognostic parameters. Cancer 1998;83:2437-46.
121. Rice D, Rusch V, Pass H, et al. Recommendations for uniform definitions of surgical techniques for malignant pleural mesothelioma: a consensus report of the international association for the study of lung cancer international staging committee and the international mesothelioma interest group. J Thorac Oncol 2011;6:1304-12.
122. Zhang YZ, Brambilla C, Molyneaux PL, et al. Utility of nuclear grading system in epithelioid malignant pleural mesothelioma in biopsy-heavy setting: an external validation study of 563 cases. Am J Surg Pathol 2020;44:347-56.
123. Tsao MS, Carbone M, Galateau-Salle F, et al. Pathologic considerations and standardization in mesothelioma clinical trials. J Thorac Oncol 2019;14:1704-17.
124. Chirieac LR, Hung YP, Foo WC, et al. Diagnostic value of biopsy sampling in predicting histology in patients with diffuse malignant pleural mesothelioma. Cancer 2019;125:4164-71.
125. Vigneswaran WT, Kircheva DY, Ananthanarayanan V, et al. Amount of epithelioid differentiation is a predictor of survival in malignant pleural mesothelioma. Ann Thorac Surg 2017;103:962-6.
126. Kao SC, Yan TD, Lee K, et al. Accuracy of diagnostic biopsy for the histological subtype of malignant pleural mesothelioma. J Thorac Oncol 2011;6:602-5.
127. Greillier L, Cavailles A, Fraticelli A, et al. Accuracy of pleural biopsy using thoracoscopy for the diagnosis of histologic subtype in patients with malignant pleural mesothelioma. Cancer 2007;110:2248-52.
128. Brims FJ, Arif M, Chauhan AJ. Outcomes and complications following medical thoracoscopy. Clin Respir J 2012;6:144-9.
129. Walters J, Maskell NA. Biopsy techniques for the diagnosis of mesothelioma. Recent Results Cancer Res 2011;189:45-55.
130. Maskell NA, Gleeson FV, Davies RJ. Standard pleural biopsy versus CT-guided cutting-needle biopsy for diagnosis of malignant disease in pleural effusions: a randomised controlled trial. Lancet 2003;361:1326-30.
131. Boutin C, Rey F, Gouvernet J, Viallat JR, Astoul P, Ledoray V. Thoracoscopy in pleural malignant mesothelioma: a prospective study of 188 consecutive patients. Part 2: prognosis and staging. Cancer 1993;72:394-404.
132. Metintas M, Ak G, Dundar E, et al. Medical thoracoscopy vs CT scan-guided Abrams pleural needle biopsy for diagnosis of patients with pleural effusions: a randomized, controlled trial. Chest 2010;137:1362-8.
133. Agarwal PP, Seely JM, Matzinger FR, et al. Pleural mesothelioma: sensitivity and incidence of needle track seeding after image-guided biopsy versus surgical biopsy. Radiology 2006;241:589-94.
134. Koegelenberg CF, Diacon AH. Pleural controversy: close needle pleural biopsy or thoracoscopy-which first? Respirology 2011;16:738-46.
135. Churg A, Hwang H, Tan L, et al. Malignant mesothelioma in situ. Histopathology 2018;72:1033-8.

136. Churg A, Galateau-Salle F, Roden AC, et al. Malignant mesothelioma in situ: morphologic features and clinical outcome. Mod Pathol 2020;33:297-30.
137. Minami K, Jimbo N, Tanaka Y, et al. Malignant mesothelioma in situ diagnosed by methylthioadenosine phosphorylase loss and homozygous deletion of CDKN2A: a case report. Virchows Arch 2020;476:469-73.
138. Allen TC, Cagle PT, Churg AM, et al. Localized malignant mesothelioma. Am J Surg Pathol 2005;29:866-73.
139. Finn RS, Brims FJH, Gandhi A, et al. Postmortem findings of malignant pleural mesothelioma: a two-center study of 318 patients. Chest 2012; 142:1267-73.
140. Travis WD, Brambilla E, Burke AP, Marx A, Nicholson AG, eds. WHO classification of tumours of the lung, pleura, thymus and heart, 4th ed. Lyons: IARC; 2015.
140a. WHO Classification of Tumours Editorial Board. Thoracic tumours, 5th ed. WHO classification of tumour series. Lyon: IARC; 2021.
141. Brcic L, Vlacic G, Quehenberger F, Kern I. Reproducibility of malignant pleural mesothelioma histopathologic subtyping. Arch Pathol Lab Med 2018;142:747-52.
142. Brcic L, Jakopovic M, Brcic I, et al. Reproducibility of histological subtyping of malignant pleural mesothelioma. Virchows Arch 2014;465:679-85.
143. Dacic S, Le Stang N, Husain A, et al. Interobserver variation in the assessment of the sarcomatoid and transitional components in biphasic mesotheliomas. Mod Pathol 2020;33:255-62.
144. Whitaker D, Henderson DW, Shilkin KB. The concept of mesothelioma in situ: implications for diagnosis and histogenesis. Semin Diagn Pathol 1992;9:151-61.
145. Churg A, Sheffield BS, Galateau-Salle F. New markers for separating benign from malignant mesothelial proliferations: are we there yet? Arch Pathol Lab Med 2016;140:318-21.
146. Sheffield BS, Hwang HC, Lee AF, et al. BAP1 immunohistochemistry and p16 FISH to separate benign from malignant mesothelial proliferations. Am J Surg Pathol 2015;39:977-82.
147. McGregor SM, McElherne J, Minor A, et al. BAP1 immunohistochemistry has limited prognostic utility as a complement of CDKN2A (p16) fluorescence in situ hybridization in malignant pleural mesothelioma. Hum Pathol 2017;60:86-94.
148. Hida T, Hamasaki M, Matsumoto S, et al. BAP1 immunohistochemistry and p16 FISH results in combination provide higher confidence in malignant pleural mesothelioma diagnosis: ROC analysis of the two tests. Pathol Int 2016;66:563-70.
149. Walts AE, Hiroshima K, McGregor SM, Wu D, Husain AN, Marchevsky AM. BAP1 immunostain and CDKN2A (p16) FISH analysis: clinical applicability for the diagnosis of malignant mesothelioma in effusions. Diagn Cytopathol 2016;44:599-606.
150. Bruno R, Ali G, Fontanini G. Molecular markers and new diagnostic methods to differentiate malignant from benign mesothelial pleural proliferations: a literature review. J Thorac Dis 2018;10(Suppl 2):S342-52.
151. Chapel DB, Schulte JJ, Berg K, et al. MTAP immunohistochemistry is an accurate and reproducible surrogate for CDKN2A fluorescence in situ hybridization in diagnosis of malignant pleural mesothelioma. Mod Pathol 2020;33:245-54.
152. Hamasaki M, Kinoshita Y, Yoshimura M, et al. Cytoplasmic MTAP expression loss detected by immunohistochemistry correlates with 9p21 homozygous deletion detected by FISH in pleural effusion cytology of mesothelioma. Histopathology 2019;75:153-5.
153. Berg KB, Dacic S, Miller C, Cheung S, Churg A. Utility of methylthioadenosine phosphorylase compared with BAP1 immunohistochemistry, and CDKN2A and NF2 fluorescence in situ hybridization in separating reactive mesothelial proliferations from epithelioid malignant mesotheliomas. Arch Pathol Lab Med 2018; 142:1549-53.
154. Galateau-Sallé F, Vignaud JM, Burke L, et al. Well-differentiated papillary mesothelioma of the pleura: a series of 24 cases. Am J Surg Pathol 2004;28:534-40.
155. Krasinskas AM, Borczuk AC, Hartman DJ, et al. Prognostic significance of morphological growth patterns and mitotic index of epithelioid malignant peritoneal mesothelioma. Histopathology 2016;68:729-37.
156. Kadota K, Suzuki K, Sima CS, Rusch VW, Adusumilli PS, Travis WD. Pleomorphic epithelioid diffuse malignant pleural mesothelioma: a clinicopathological review and conceptual proposal to reclassify as biphasic or sarcomatoid mesothelioma. J Thorac Oncol 2011;6:896-904.
157. Alchami FS, Attanoos RL, Bamber AR. Myxoid variant epithelioid pleural mesothelioma defines a favourable prognosis group: an analysis of 191 patients with pleural malignant mesothelioma. J Clin Pathol 2017;70:179-82.
158. Ordonez NG. Mesothelioma with signet-ring cell features: report of 23 cases. Mod Pathol 2013;26:370-84.
159. Ordonez NG. Deciduoid mesothelioma: report of 21 cases with review of the literature. Mod Pathol 2012;25:1481-95.

160. Shanks JH, Harris M, Banerjee SS, et al. Mesotheliomas with deciduoid morphology: a morphologic spectrum and a variant not confined to young females. Am J Surg Pathol 2000;24:285-94.
161. Nascimento AG, Keeney GL, Fletcher CD. Deciduoid peritoneal mesothelioma. An unusual phenotype affecting young females. Am J Surg Pathol 1994;18:439-45.
162. Ordonez NG. Mesotheliomas with small cell features: report of eight cases. Mod Pathol 2012;25:689-98.
163. Ordonez NG. Mesothelioma with rhabdoid features: an ultrastructural and immunohistochemical study of 10 cases. Mod Pathol 2006; 19:373-83.
164. Ahadi MS, Gill AJ. SMARCA4 loss is very rare in thoracic mesothelioma. Am J Surg Pathol 2019;43:1154-5.
165. Galateau Salle F, Le Stang N, Nicholson AG, et al. New insights on diagnostic reproducibility of biphasic mesotheliomas: a multi-institutional evaluation by the international mesothelioma panel from the MESOPATH reference center. J Thorac Oncol 2018;13:1189-203.
166. Ordonez NG. Pleomorphic mesothelioma: report of 10 cases. Mod Pathol 2012;25:1011-22.
167. Zhang YZ, Brambilla C, Molyneaux PL, et al. Presence of pleomorphic features but not growth patterns improves prognostic stratification of epithelioid malignant pleural mesothelioma by 2-tier nuclear grade. Histopathology 2020;77:423-36.
168. Galateau-Salle F, Le Stang N, Tirode F, et al. Comprehensive molecular and pathological evaluation of transitional mesothelioma assisted by deep learning approach: a multi institutional study of the International Mesothelioma Panel from MESOPATH Reference Center. J Thorac Oncol 2020;15:1037-53.
169. Galateau-Salle F, Attanoos R, Gibbs AR, et al. Lymphohistiocytoid variant of malignant mesothelioma of the pleura: a series of 22 cases. Am J Surg Pathol 2007;31:711-6.
170. Nind NR, Attanoos RL, Gibbs AR. Unusual intraparenchymal growth patterns of malignant pleural mesothelioma. Histopathology 2003;42:150-5.
171. Kadota K, Suzuki K, Colovos C, et al. A nuclear grading system is a strong predictor of survival in epitheloid diffuse malignant pleural mesothelioma. Mod Pathol 2012;25:260-71.
172. Rosen LE, Karrison T, Ananthanarayanan V, et al. Nuclear grade and necrosis predict prognosis in malignant epithelioid pleural mesothelioma: a multi-institutional study. Mod Pathol 2018;31:598-606.
173. Pelosi G, Papotti M, Righi L, et al. Pathologic grading of malignant pleural mesothelioma: an evidence-based proposal. J Thorac Oncol 2018; 13:1750-61.
174. Klebe S, Brownlee NA, Mahar A, et al. Sarcomatoid mesothelioma: a clinical-pathologic correlation of 326 cases. Mod Pathol 2010;23:470-9.
175. Klebe S, Mahar A, Henderson DW, Roggli VL. Malignant mesothelioma with heterologous elements: clinicopathological correlation of 27 cases and literature review. Mod Pathol 2008;21:1084-94.
176. Klabatsa A, Nicholson AG, Dulay K, Rudd RM, Sheaff MT. Diffuse pleural mesothelioma with epithelioid and angiosarcomatous components—a hitherto undescribed pattern of differentiation. Histopathology 2012;60:1164-6.
177. McGregor SM, Dunning R, Hyjek E, Vigneswaran W, Husain AN, Krausz T. BAP1 facilitates diagnostic objectivity, classification, and prognostication in malignant pleural mesothelioma. Hum Pathol 2015;46:1670-8.
178. Kinoshita Y, Hamasaki M, Yoshimura M, et al. A combination of MTAP and BAP1 immunohistochemistry is effective for distinguishing sarcomatoid mesothelioma from fibrous pleuritis. Lung Cancer 2018;125:198-204.
179. Galateau-Salle F, Churg A, Roggli V, Travis WD, WHO Committee for Tumors of the Pleura. The 2015 World Health Organization classification of tumors of the pleura: advances since the 2004 classification. J Thorac Oncol 2016; 11:142-54.
180. Husain AN, Colby TV, Ordonez NG, et al. Guidelines for pathologic diagnosis of malignant mesothelioma 2017 update of the consensus statement from the International Mesothelioma Interest Group. Arch Pathol Lab Med 2018; 142:89-108.
181. Berg KB, Churg A. GATA3 Immunohistochemistry for distinguishing sarcomatoid and desmoplastic mesothelioma from sarcomatoid carcinoma of the lung. Am J Surg Pathol 2017;41:1221-5.
182. Light RW, Macgregor MI, Luchsinger PC, Ball WC Jr. Pleural effusions: the diagnostic separation of transudates and exudates. Ann Intern Med 1972;77:507-13.
183. Asciak R, Rahman NM. Malignant pleural effusion: from diagnostics to therapeutics. Clin Chest Med 2018;39:181-93.
184. Garcia LW, Ducatman BS, Wang HH. The value of multiple fluid specimens in the cytological diagnosis of malignancy. Mod Pathol 1994;7:665-8.
185. Sears D, Hajdu SI. The cytologic diagnosis of malignant neoplasms in pleural and peritoneal effusions. Acta Cytol 1987;31:85-97.

186. Tabatabai ZL, Nayar R, Souers RJ, Crothers BA, Davey DD. Performance characteristics of body fluid cytology analysis of 344 380 responses from the College of American Pathologists Interlaboratory Comparison Program in Nongynecologic Cytopathology. Arch Pathol Lab Med 2018;142:53-8.
187. Hyun TS, Barnes M, Tabatabai ZL. The diagnostic utility of D2-40, calretinin, CK5/6, desmin and MOC-31 in the differentiation of mesothelioma from adenocarcinoma in pleural effusion cytology. Acta Cytol 2012;56:527-32.
188. Husain AN, Colby T, Ordonez N, et al. Guidelines for pathologic diagnosis of malignant mesothelioma: 2012 update of the consensus statement from the International Mesothelioma Interest Group. Arch Pathol Lab Med 2013;137:647-67.
189. Ehya H. The cytologic diagnosis of mesothelioma. Semin Diagn Pathol 1986;3:196-203.
190. Tao LC. The cytopathology of mesothelioma. Acta Cytol 1979;23:209-13.
191. Triol JH, Conston AS, Chandler SV. Malignant mesothelioma. Cytopathology of 75 cases seen in a New Jersey community hospital. Acta Cytol 1984;28:37-45.
192. Henderson DW, Reid G, Kao SC, van Zandwijk N, Klebe S. Challenges and controversies in the diagnosis of mesothelioma: part 1. Cytology-only diagnosis, biopsies, immunohistochemistry, discrimination between mesothelioma and reactive mesothelial hyperplasia, and biomarkers. J Clin Pathol 2013;66:847-53.
193. Paintal A, Raparia K, Zakowski MF, Nayar R. The diagnosis of malignant mesothelioma in effusion cytology: a reappraisal and results of a multi-institution survey. Cancer Cytopathol 2013;121:703-7.
194. Hwang HC, Sheffield BS, Rodriguez S, et al. Utility of BAP1 immunohistochemistry and p16 (CDKN2A) FISH in the diagnosis of malignant mesothelioma in effusion cytology specimens. Am J Surg Pathol 2016;40:120-6.
195. Monaco S, Mehrad M, Dacic S. Recent advances in the diagnosis of malignant mesothelioma: focus on approach in challenging cases and in limited tissue and cytologic samples. Adv Anat Pathol 2018;25:24-30.
196. Hjerpe A, Ascoli V, Bedrossian CW, et al. Guidelines for the cytopathologic diagnosis of epithelioid and mixed-type malignant mesothelioma: complementary statement from the International Mesothelioma Interest Group, also endorsed by the International Academy of Cytology and the Papanicolaou Society of Cytopathology. Diagn Cytopathol 2015;43:563-76.
197. Hiroshima K, Wu D, Hamakawa S, et al. HEG1, BAP1, and MTAP are useful in cytologic diagnosis of malignant mesothelioma with effusion. Diagn Cytopathol 2021;49:622-32.
198. Chevrier M, Monaco SE, Jerome JA, Galateau-Salle F, Churg A, Dacic S. Testing for BAP1 loss and CDKN2A/p16 homozygous deletion improves the accurate diagnosis of mesothelial proliferations in effusion cytology. Cancer Cytopathol 2020;128:939-47.
199. Sharyn I, Katz SI, Roshkovan L, Berger I, et al. Serum soluble mesothelin-related protein (SMRP) and fibulin-3 levels correlate with baseline malignant pleural mesothelioma (MPM) tumor volumes but are not useful as biomarkers of response in an immunotherapy trial. Lung Cancer 2021;154:5-12.
200. Pantazopoulos I, Boura P, Xanthos T, Syrigos K. Effectiveness of mesothelin family proteins and osteopontin for malignant mesothelioma. Eur Respir J 2013;41:706-15.
201. Creaney J, Segal A, Olsen N, et al. Pleural fluid mesothelin as an adjunct to the diagnosis of pleural malignant mesothelioma. Dis Markers 2014;2014:413946.
202. Hollevoet K, Reitsma JB, Creaney J, et al. Serum mesothelin for diagnosing malignant pleural mesothelioma: an individual patient data meta-analysis. J Clin Oncol 2012;30:1541-9.
203. Hassan R, Kindler HL, Jahan T, et al. Phase II clinical trial of amatuximab, a chimeric antimesothelin antibody with pemetrexed and cisplatin in advanced unresectable pleural mesothelioma. Clin Cancer Res 2014;20:5927-36.
204. Cherkassky L, Morello A, Villena-Vargas J, et al. Human CAR T cells with cell-intrinsic PD-1 checkpoint blockade resist tumor-mediated inhibition. J Clin Invest 2016;126:3130-44.
205. Pass HI, Wali A, Tang N, et al. Soluble mesothelin-related peptide level elevation in mesothelioma serum and pleural effusions. Ann Thorac Surg 2008;85:265-72.
206. Demir M, Kaya H, Taylan M, et al. Evaluation of new biomarkers in the prediction of malignant mesothelioma in subjects with environmental asbestos exposure. Lung 2016;194:409-17.
207. Wheatley-Price P, Yang B, Patsios D, et al. Soluble mesothelin-related peptide and osteopontin as markers of response in malignant mesothelioma. J Clin Oncol 2010;28:3316-22.
208. Creaney J, Yeoman D, Demelker Y, et al. Comparison of osteopontin, megakaryocyte potentiating factor, and mesothelin proteins as markers in the serum of patients with malignant mesothelioma. J Thorac Oncol 2008;3:851-7.

209. Pass HI, Lott D, Lonardo F, et al. Asbestos exposure, pleural mesothelioma, and serum osteopontin levels. N Engl J Med 2005;353:1564-73.
210. Pass HI, Goparaju C, Espin-Garcia O, et al. Plasma biomarker enrichment of clinical prognostic indices in malignant pleural mesothelioma. J Thorac Oncol 2016;11:900-9.
211. Napolitano A, Antoine DJ, Pellegrini L, et al. HMGB1 and its hyperacetylated isoform are sensitive and specific serum biomarkers to detect asbestos exposure and to identify mesothelioma patients. Clin Cancer Res 2016;22:3087-96.
212. Kirschner MB, Pulford E, Hoda MA, et al. Fibulin-3 levels in malignant pleural mesothelioma are associated with prognosis but not diagnosis. Br J Cancer 2015;113:963-9.
213. Kovac V, Dodic-Fikfak M, Arneric N, Dolzan V, Franko A. Fibulin-3 as a biomarker of response to treatment in malignant mesothelioma. Radiol Oncol 2015;49:279-85.
214. Pass HI, Levin SM, Harbut MR, et al. Fibulin-3 as a blood and effusion biomarker for pleural mesothelioma. N Engl J Med 2012;367:1417-27.
215. Tabata C, Terada T, Tabata R, et al. Serum thioredoxin-1 as a diagnostic marker for malignant peritoneal mesothelioma. J Clin Gastroenterol 2013;47:e7-11.
216. Johnen G, Gawrych K, Raiko I, et al. Calretinin as a blood-based biomarker for mesothelioma. BMC Cancer 2017;17:386.
217. Jimenez-Ramirez C, Casjens S, Juarez-Perez CA, et al. Mesothelin, calretinin, and megakaryocyte potentiating factor as biomarkers of malignant pleural mesothelioma. Lung 2019;197:641-9.
218. Guo G, Chmielecki J, Goparaju C, et al. Whole-exome sequencing reveals frequent genetic alterations in BAP1, NF2, CDKN2A, and CUL1 in malignant pleural mesothelioma. Cancer Res 2015;75:264-9.
219. Hmeljak J, Sanchez-Vega F, Hoadley KA, et al. Integrative molecular characterization of malignant pleural mesothelioma. Cancer Discov 2018;8:1548-65.
220. Muller S, Lai WV, Adusumilli PS, et al. V-domain Ig-containing suppressor of T-cell activation (VISTA), a potentially targetable immune checkpoint molecule, is highly expressed in epithelioid malignant pleural mesothelioma. Mod Pathol 2020;33:303-11.
221. Illei PB, Ladanyi M, Rusch VW, Zakowski MF. The use of CDKN2A deletion as a diagnostic marker for malignant mesothelioma in body cavity effusions. Cancer 2003;99:51-6.
222. Monaco SE, Shuai Y, Bansal M, Krasinskas AM, Dacic S. The diagnostic utility of p16 FISH and GLUT-1 immunohistochemical analysis in mesothelial proliferations. Am J Clin Pathol 2011;135:619-27.
223. Chiosea S, Krasinskas A, Cagle PT, Mitchell KA, Zander DS, Dacic S. Diagnostic importance of 9p21 homozygous deletion in malignant mesotheliomas. Mod Pathol 2008;21:742-7.
224. Krasinskas AM, Bartlett DL, Cieply K, Dacic S. CDKN2A and MTAP deletions in peritoneal mesotheliomas are correlated with loss of p16 protein expression and poor survival. Mod Pathol 2010;23:531-8.
225. Guled M, Lahti L, Lindholm PM, et al. CDKN2A, NF2, and JUN are dysregulated among other genes by miRNAs in malignant mesothelioma -a miRNA microarray analysis. Genes Chromosomes Cancer 2009;48:615-23.
226. Lopez-Rios F, Chuai S, Flores R, et al. Global gene expression profiling of pleural mesotheliomas: overexpression of aurora kinases and P16/CDKN2A deletion as prognostic factors and critical evaluation of microarray-based prognostic prediction. Cancer Res 2006;66:2970-9.
227. Borczuk AC, Taub RN, Hesdorffer M, et al. P16 loss and mitotic activity predict poor survival in patients with peritoneal malignant mesothelioma. Clin Cancer Res 2005;11:3303-8.
228. Shahi M, Eguchi T, Zauderer MG, et al. Comparison of MTAP immunohistochemistry (IHC) with P16/CDKN2A deletion/loss and utility of MTAP and BAP1 IHC in the diagnosis of malignant pleural mesothelioma (MPM). Laboratory Investigation 2018;98(Suppl 1):750-1.
229. Hida T, Hamasaki M, Matsumoto S, et al. Immunohistochemical detection of MTAP and BAP1 protein loss for mesothelioma diagnosis: comparison with 9p21 FISH and BAP1 immunohistochemistry. Lung Cancer 2017;104:98-105.
230. Ladanyi M, Zauderer MG, Krug LM, et al. New strategies in pleural mesothelioma: BAP1 and NF2 as novel targets for therapeutic development and risk assessment. Clin Cancer Res 2012;18:4485-90.
231. Bott M, Brevet M, Taylor BS, et al. The nuclear deubiquitinase BAP1 is commonly inactivated by somatic mutations and 3p21.1 losses in malignant pleural mesothelioma. Nat Genet 2011;43:668-72.
232. Nasu M, Emi M, Pastorino S, et al. High incidence of somatic BAP1 alterations in sporadic malignant mesothelioma. J Thorac Oncol 2015;10:565-76.
233. Zauderer MG, Bott M, McMillan R, et al. Clinical characteristics of patients with malignant pleural mesothelioma harboring somatic BAP1 mutations. J Thorac Oncol 2013;8:1430-3.

234. Borczuk AC, Pei J, Taub RN, et al. Genome-wide analysis of abdominal and pleural malignant mesothelioma with DNA arrays reveals both common and distinct regions of copy number alteration. Cancer Biol Ther 2016;17:328-35.
235. Hung YP, Dong F, Torre M, Crum CP, Bueno R, Chirieac LR. Molecular characterization of diffuse malignant peritoneal mesothelioma. Mod Pathol 2020;33:2269-79.
236. Cigognetti M, Lonardi S, Fisogni S, et al. BAP1 (BRCA1-associated protein 1) is a highly specific marker for differentiating mesothelioma from reactive mesothelial proliferations. Mod Pathol 2015;28:1043-57.
237. Singhi AD, Krasinskas AM, Choudry HA, et al. The prognostic significance of BAP1, NF2, and CDKN2A in malignant peritoneal mesothelioma. Mod Pathol 2016;29:14-24.
238. Farzin M, Toon CW, Clarkson A, et al. Loss of expression of BAP1 predicts longer survival in mesothelioma. Pathology 2015;47:302-7.
239. Tandon RT, Jimenez-Cortez Y, Taub R, Borczuk AC. Immunohistochemistry in peritoneal mesothelioma: a single-center experience of 244 cases. Arch Pathol Lab Med 2018;142:236-42.
240. Sheffield BS, Lorette J, Shen Y, Marra MA, Churg A. Immunohistochemistry for NF2, LATS1/2, and YAP/TAZ fails to separate benign from malignant mesothelial proliferations. Arch Pathol Lab Med 2016;140:391.
241. Tallet A, Nault JC, Renier A, et al. Overexpression and promoter mutation of the TERT gene in malignant pleural mesothelioma. Oncogene 2014;33:3748-52.
242. Sekido Y. Molecular biology of malignant mesothelioma. Environ Health Prev Med 2008;13:65-70.
243. Bueno R, Stawiski EW, Goldstein LD, et al. Comprehensive genomic analysis of malignant pleural mesothelioma identifies recurrent mutations, gene fusions and splicing alterations. Nat Genet 2016;48:407-16.
244. Murakami H, Mizuno T, Taniguchi T, et al. LATS2 is a tumor suppressor gene of malignant mesothelioma. Cancer Res 2011;71:873-83.
245. Ou WB, Hubert C, Corson JM, et al. Targeted inhibition of multiple receptor tyrosine kinases in mesothelioma. Neoplasia 2011;13:12-22.
246. Cedres S, Ponce-Aix S, Zugazagoitia J, et al. Analysis of expression of programmed cell death 1 ligand 1 (PD-L1) in malignant pleural mesothelioma (MPM). PLoS One 2015;10:e0121071.
247. Combaz-Lair C, Galateau-Salle F, McLeer-Florin A, et al. Immune biomarkers PD-1/PD-L1 and TLR3 in malignant pleural mesotheliomas. Hum Pathol 2016;52:9-18.
248. Alley EW, Lopez J, Santoro A, et al. Clinical safety and activity of pembrolizumab in patients with malignant pleural mesothelioma (KEYNOTE-028): preliminary results from a non-randomised, open-label, phase 1b trial. Lancet Oncol 2017;18:623-30.
249. Calabro L, Morra A, Fonsatti E, et al. Efficacy and safety of an intensified schedule of tremelimumab for chemotherapy-resistant malignant mesothelioma: an open-label, single-arm, phase 2 study. Lancet Respir Med 2015;3:301-9.
250. Maio M, Scherpereel A, Calabro L, et al. Tremelimumab as second-line or third-line treatment in relapsed malignant mesothelioma (DETERMINE): a multicentre, international, randomised, double-blind, placebo-controlled phase 2b trial. Lancet Oncol 2017;18:1261-73.
251. Loharamtaweethong K, Puripat N, Aoonjai N, Sutepvarnon A, Bandidwattanawong C. Anaplastic lymphoma kinase (ALK) translocation in paediatric malignant peritoneal mesothelioma: a case report of novel ALK-related tumour spectrum. Histopathology 2016;68:603-7.
252. Desmeules P, Joubert P, Zhang L, et al. A subset of malignant mesotheliomas in young adults are associated with recurrent ewsr1/fus-atf1 fusions. Am J Surg Pathol 2017;41:980-8.
253. Burns TR, Greenberg SD, Mace ML, Johnson EH. Ultrastructural diagnosis of epithelial malignant mesothelioma. Cancer 1985;56:2036-40.
254. Dardick I, Al-Jabi M, McCaughey WT, Srigley JR, van Nostrand AW, Ritchie AC. Ultrastructure of poorly differentiated diffuse epithelial mesotheliomas. Ultrastruct Pathol 1984;7:151-60.
255. Dardick I, Jabi M, McCaughey WT, Deodhare S, van Nostrand AW, Srigley JR. Diffuse epithelial mesothelioma: a review of the ultrastructural spectrum. Ultrastruct Pathol 1987;11:503-3.
256. Oury TD, Hammar SP, Roggli VL. Ultrastructural features of diffuse malignant mesotheliomas. Hum Pathol 1998;29:1382-92.
257. Wang NS. The regional difference of pleural mesothelial cells in rabbits. Am Rev Respir Dis 1974;110:623-33.
258. Warhol MJ, Hickey WF, Corson JM. Malignant mesothelioma: ultrastructural distinction from adenocarcinoma. Am J Surg Pathol 1982;6:307-14.
259. Warhol MJ, Corson JM. An ultrastructural comparison of mesotheliomas with adenocarcinomas of the lung and breast. Hum Pathol 1985;16:50-5.
260. Jandik WR, Landas SK, Bray CK, Lager DJ. Scanning electron microscopic distinction of pleural mesotheliomas from adenocarcinomas. Mod Pathol 1993;6:761-4.

261. Churg A, Cagle PT, Roggli VL. Tumors of the serosal membranes. AFIP Atlas of Tumor Pathology, 4th Series, Fascicle 3. Washington, DC: American Registry of Pathology; 2006.
262. Yoshimura M, Kinoshita Y, Hamasaki M, et al. Highly expressed EZH2 in combination with BAP1 and MTAP loss, as detected by immunohistochemistry, is useful for differentiating malignant pleural mesothelioma from reactive mesothelial hyperplasia. Lung Cancer 2019;130:187-93.
263. Shinozaki-Ushiku A, Ushiku T, Morita S, Anraku M, Nakajima J, Fukayama M. Diagnostic utility of BAP1 and EZH2 expression in malignant mesothelioma. Histopathology 2017;70:722-33.
264. Pors J, Naso J, Berg K, Churg A. Cyclin D1 immunohistochemical staining to separate benign from malignant mesothelial proliferations. Mod Pathol 2020;33:312-8.
265. Churg A, Cagle P, Colby TV, et al. The fake fat phenomenon in organizing pleuritis: a source of confusion with desmoplastic malignant mesotheliomas. Am J Surg Pathol 2011;35:1823-9.
266. Chirac P, Maillet D, Lepretre F, et al. Genomic copy number alterations in 33 malignant peritoneal mesothelioma analyzed by comparative genomic hybridization array. Hum Pathol 2016;55:72-82.
267. Yoshikawa Y, Emi M, Hashimoto-Tamaoki T, et al. High-density array-CGH with targeted NGS unmask multiple noncontiguous minute deletions on chromosome 3p21 in mesothelioma. Proc Natl Acad Sci U S A 2016;113:13432-7.
268. Derakhshan F, Ionescu D, Cheung S, Churg A. Use of programmed death ligand-1 (pd-l1) staining to separate sarcomatoid malignant mesotheliomas from benign mesothelial reactions. Arch Pathol Lab Med 2020;144:185-8.
269. Papotti M, Nicholson AG, Dacic S. Mesothelioma and immunohistochemsitry. In: Yatabe Y, Borczuk AC, Cooper WA, et al, eds. IASLC atlas of diagnostic immunohistochemistry. Denver: IASLC; 2020:157-66.
270. Ordonez NG, Sahin AA. Diagnostic utility of immunohistochemistry in distinguishing between epithelioid pleural mesotheliomas and breast carcinomas: a comparative study. Hum Pathol 2014;45:1529-40.
271. Miettinen M, McCue PA, Sarlomo-Rikala M, et al. GATA3: a multispecific but potentially useful marker in surgical pathology: a systematic analysis of 2500 epithelial and nonepithelial tumors. Am J Surg Pathol 2014;38:13-22.
272. Osborn M, Pelling N, Walker MM, Fisher C, Nicholson AG. The value of 'mesothelium-associated' antibodies in distinguishing between metastatic renal cell carcinomas and mesotheliomas. Histopathology 2002;41:301-7.
273. Baranov E, McBride MJ, Bellizzi AM, et al. A novel SS18-SSX fusion-specific antibody for the diagnosis of synovial sarcoma. Am J Surg Pathol 2020;44:922-33.
274. Burt BM, Cameron RB, Mollberg NM, et al. Malignant pleural mesothelioma and the Society of Thoracic Surgeons Database: an analysis of surgical morbidity and mortality. J Thorac Cardiovasc Surg 2014;148:30-5.
275. Miyazaki T, Yamasaki N, Tsuchiya T, et al. Is pleurectomy/decortication superior to extrapleural pneumonectomy for patients with malignant pleural mesothelioma? A single-institutional experience. Ann Thorac Cardiovasc Surg 2018;24:81-8.
276. Treasure T, Lang-Lazdunski L, Waller D, et al. Extra-pleural pneumonectomy versus no extra-pleural pneumonectomy for patients with malignant pleural mesothelioma: clinical outcomes of the mesothelioma and radical surgery (MARS) randomised feasibility study. Lancet Oncol 2011;12:763-72.
277. Friedberg JS, Culligan MJ, Tsao AS, et al. A proposed system toward standardizing surgical-based treatments for malignant pleural mesothelioma, from the joint national cancer institute-international association for the study of lung cancer-mesothelioma applied research foundation taskforce. J Thorac Oncol 2019;14:1343-53.
278. Vogelzang NJ, Rusthoven JJ, Symanowski J, et al. Phase III study of pemetrexed in combination with cisplatin versus cisplatin alone in patients with malignant pleural mesothelioma. J Clin Oncol 2003;21:2636-44.
279. Zalcman G, Mazieres J, Margery J, et al. Bevacizumab for newly diagnosed pleural mesothelioma in the Mesothelioma Avastin Cisplatin Pemetrexed Study (MAPS): a randomised, controlled, open-label, phase 3 trial. Lancet 2016; 387:1405-14.
280. Scherpereel A, Mazieres J, Greillier L, et al. Nivolumab or nivolumab plus ipilimumab in patients with relapsed malignant pleural mesothelioma (IFCT-1501 MAPS2): a multicentre, open-label, randomised, non-comparative, phase 2 trial. Lancet Oncol 2019;20:239-53.
281. Disselhorst MJ, Quispel-Janssen J, Lalezari F, et al. Ipilimumab and nivolumab in the treatment of recurrent malignant pleural mesothelioma (INITIATE): results of a prospective, single-arm, phase 2 trial. Lancet Respir Med 2019;7:260-70.
282. Grosso F, Steele N, Novello S, et al. Nintedanib plus pemetrexed/cisplatin in patients with malignant pleural mesothelioma: phase ii results from the randomized, placebo-controlled LUME-Meso Trial. J Clin Oncol 2017;35:3591-600.

283. Scagliotti GV, Gaafar R, Nowak AK, et al. Nintedanib in combination with pemetrexed and cisplatin for chemotherapy-naive patients with advanced malignant pleural mesothelioma (LUME-Meso): a double-blind, randomised, placebo-controlled phase 3 trial. Lancet Respir Med 2019;7:569-80.
284. Gomez DR, Rimner A, Simone CB 2nd, et al. The use of radiation therapy for the treatment of malignant pleural mesothelioma: expert opinion from the National Cancer Institute Thoracic Malignancy Steering Committee, International Association for the Study of Lung Cancer, and Mesothelioma Applied Research Foundation. J Thorac Oncol 2019;14:1172-83.
285. Magge D, Zenati MS, Austin F, et al. Malignant peritoneal mesothelioma: prognostic factors and oncologic outcome analysis. Ann Surg Oncol 2014;21:1159-65.
286. Baratti D, Kusamura S, Cabras AD, Bertulli R, Hutanu I, Deraco M. Diffuse malignant peritoneal mesothelioma: long-term survival with complete cytoreductive surgery followed by hyperthermic intraperitoneal chemotherapy (HIPEC). Eur J Cancer 2013;49:3140-8.
287. Krasinskas AM, Borczuk AC, Hartman DJ, et al. Prognostic significance of morphological growth patterns and mitotic index of epithelioid malignant peritoneal mesothelioma. Histopathology 2016;68:729-37.
288. Kluger MD, Taub RN, Hesdorffer M, Jin Z, Chabot JA. Two-stage operative cytoreduction and intraperitoneal chemotherapy for diffuse malignant peritoneal mesothelioma: operative morbidity and mortality in phase I and II trials. Eur J Surg Oncol 2010;36:997-1003.
289. Sugarbaker PH, Welch LS, Mohamed F, Glehen O. A review of peritoneal mesothelioma at the Washington Cancer Institute. Surg Oncol Clin N Am 2003;12:605-21.
290. Kim JS, Lim SY, Hwang J, Kang EJ, Choi YJ. A case report of primary pericardial malignant mesothelioma treated with pemetrexed and cisplatin. J Korean Med Sci 2017;32:1879-84.
291. Alberts AS, Falkson G, Goedhals L, Vorobiof DA, Van der Merwe CA. Malignant pleural mesothelioma: a disease unaffected by current therapeutic maneuvers. J Clin Oncol 1988;6:527-35.
292. Rusch VW. A proposed new international TNM staging system for malignant pleural mesothelioma. From the International Mesothelioma Interest Group. Chest 1995;108:1122-8.
293. Spirtas R, Connelly RR, Tucker MA. Survival patterns for malignant mesothelioma: the SEER experience. Int J Cancer 1988;41:525-30.
294. Milano MT, Zhang H. Malignant pleural mesothelioma: a population-based study of survival. J Thorac Oncol 2010;5:1841-8.
295. Taioli E, Wolf AS, Camacho-Rivera M, et al. Determinants of survival in malignant pleural mesothelioma: a surveillance, epidemiology, and end results (SEER) study of 14,228 patients. PLoS One 2015;10:e0145039.
296. Meyerhoff RR, Yang CF, Speicher PJ, et al. Impact of mesothelioma histologic subtype on outcomes in the Surveillance, Epidemiology, and End Results database. J Surg Res 2015;196:23-32.
297. Brims FJ, Meniawy TM, Duffus I, et al. A novel clinical prediction model for prognosis in malignant pleural mesothelioma using decision tree analysis. J Thorac Oncol 2016;11:573-82.
298. Gill RR, Yeap BY, Bueno R, Richards WG. Quantitative clinical staging for patients with malignant pleural mesothelioma. J Natl Cancer Inst 2018;110:258-64.
299. Gordon GJ, Dong L, Yeap BY, et al. Four-gene expression ratio test for survival in patients undergoing surgery for mesothelioma. J Natl Cancer Inst 2009;101:678-86.
300. Alcala N, Mangiante L, Le-Stang N, et al. Redefining malignant pleural mesothelioma types as a continuum uncovers immune-vascular interactions. EBioMedicine 2019;48:191-202.
301. Suzuki K, Kadota K, Sima CS, et al. Chronic inflammation in tumor stroma is an independent predictor of prolonged survival in epithelioid malignant pleural mesothelioma patients. Cancer Immunol Immunother 2011;60:1721-8.
302. Doi H, Kuribayashi K, Kitajima K, Yamakado K, Kijima T. Development of a novel prognostic risk classification system for malignant pleural mesothelioma. Clin Lung Cancer 2020;21:66-74.e2.
303. Forest F, Patoir A, Dal Col P, et al. Nuclear grading, BAP1, mesothelin and PD-L1 expression in malignant pleural mesothelioma: prognostic implications. Pathology 2018;50:635-41.
304. Shavelle R, Vavra-Musser K, Lee J, Brooks J. Life expectancy in pleural and peritoneal mesothelioma. Lung Cancer Int 2017;2017:2782590.
305. Beebe-Dimmer JL, Fryzek JP, Yee CL, et al. Mesothelioma in the United States: a surveillance, epidemiology, and end results (SEER)-Medicare investigation of treatment patterns and overall survival. Clin Epidemiol 2016;8:743-50.
306. Mirabelli D, Roberti S, Gangemi M, et al. Survival of peritoneal malignant mesothelioma in Italy: a population-based study. Int J Cancer 2009;124:194-200.
307. Yin W, Zheng G, Yang K, Song H, Liang Y. Analysis of prognostic factors of patients with malignant peritoneal mesothelioma. World J Surg Oncol 2018;16:44.

308. Zellos LS, Sugarbaker DJ. Multimodality treatment of diffuse malignant pleural mesothelioma. Semin Oncol 2002;29:41-50.
309. Sugarbaker DJ, Flores RM, Jaklitsch MT, et al. Resection margins, extrapleural nodal status, and cell type determine postoperative long-term survival in trimodality therapy of malignant pleural mesothelioma: results in 183 patients. J Thorac Cardiovasc Surg 1999;117:54-63.
309a. UICC TNM classification of malignant tumours, 8th ed. Oxford: Wiley Blackwell; 2017.
310. Rusch VW, Chansky K, Kindler HL, et al. The IASLC Mesothelioma Staging Project: proposals for the M descriptors and for revision of the TNM stage groupings in the forthcoming (eighth) edition of the TNM Classification for Mesothelioma. J Thorac Oncol 2016;11:2112-9.
311. Nowak AK, Chansky K, Rice DC, et al. The IASLC Mesothelioma Staging Project: proposals for revisions of the T descriptors in the forthcoming eighth edition of the TNM Classification for Pleural Mesothelioma. J Thorac Oncol 2016;11:2089-99.
312. Rice D, Chansky K, Nowak A, et al. The IASLC Mesothelioma Staging Project: proposals for revisions of the n descriptors in the forthcoming eighth edition of the TNM Classification for Pleural Mesothelioma. J Thorac Oncol 2016; 11:2100-11.
313. Pass H, Giroux D, Kennedy C, et al. The IASLC Mesothelioma Staging Project: improving staging of a rare disease through international participation. J Thorac Oncol 2016;11:2082-8.
314. Nicholson AG, Tsao MS, Travis WD, et al. Eighth edition staging of thoracic malignancies: implications for the reporting pathologist. Arch Pathol Lab Med 2018;142:645-61.
315. Herndon JE, Green MR, Chahinian AP, Corson JM, Suzuki Y, Vogelzang NJ. Factors predictive of survival among 337 patients with mesothelioma treated between 1984 and 1994 by the Cancer and Leukemia Group B. Chest 1998; 113:723-31.
316. Edwards JG, Abrams KR, Leverment JN, Spyt TJ, Waller DA, O'Byrne KJ. Prognostic factors for malignant mesothelioma in 142 patients: validation of CALGB and EORTC prognostic scoring systems. Thorax 2000;55:731-5.
317. Rusch VW, Gill R, Mitchell A, et al. A multicenter study of volumetric computed tomography for staging malignant pleural mesothelioma. Ann Thorac Surg 2016;102:1059-66.
318. Carbone M, Gaudino G, Yang H. Recent insights emerging from malignant mesothelioma genome sequencing. J Thorac Oncol 2015;10:409-11.
319. Panou V, Vyberg M, Weinreich UM, Meristoudis C, Falkmer UG, Roe OD. The established and future biomarkers of malignant pleural mesothelioma. Cancer Treat Rev 2015;41:486-95.
320. McCambridge AJ, Napolitano A, Mansfield AS, et al. State of the art: advances in malignant pleural mesothelioma in 2017. J Thorac Oncol 2018.
321. Baumann F, Flores E, Napolitano A, et al. Mesothelioma patients with germline BAP1 mutations have 7-fold improved long-term survival. Carcinogenesis 2015;36:76-81.
322. Mansfield AS, Roden AC, Peikert T, et al. B7-H1 expression in malignant pleural mesothelioma is associated with sarcomatoid histology and poor prognosis. J Thorac Oncol 2014;9:1036-40.
323. Nguyen BH, Montgomery R, Fadia M, Wang J, Ali S. PD-L1 expression associated with worse survival outcome in malignant pleural mesothelioma. Asia Pac J Clin Oncol 2018;14:69-73.
324. Kaya H, Sezgi C, Tanrikulu AC, et al. Prognostic factors influencing survival in 35 patients with malignant peritoneal mesothelioma. Neoplasma 2014;61:433-8.
325. Huang Y, Alzahrani NA, Liauw W, Morris DL. Effects of sex hormones on survival of peritoneal mesothelioma. World J Surg Oncol 2015;13:210.
326. Valente K, Blackham AU, Levine E, et al. A histomorphologic grading system that predicts overall survival in diffuse malignant peritoneal mesothelioma with epithelioid subtype. Am J Surg Pathol 2016;40:1243-8.
327. Pillai K, Pourgholami MH, Chua TC, Morris DL. Prognostic significance of Ki67 expression in malignant peritoneal mesothelioma. Am J Clin Oncol 2015;38:388-94.
328. Pillai K, Pourgholami MH, Chua TC, Morris DL. MUC1 has prognostic significance in malignant peritoneal mesothelioma. Int J Biol Markers 2013;28:303-12.
329. Valmary-Degano S, Colpart P, Villeneuve L, et al. Immunohistochemical evaluation of two antibodies against PD-L1 and prognostic significance of PD-L1 expression in epithelioid peritoneal malignant mesothelioma: a RENAPE study. Eur J Surg Oncol 2017;43:1915-23.

4 OTHER MESOTHELIAL TUMORS

Tumors of mesothelial origin that are distinguished from diffuse mesothelioma either by morphologic differences, anatomic localization, or both, lead to differences in management and prognosis, thus warranting separate classification. The entity "mesothelioma of the atrioventricular node" is now viewed as an epithelial endodermal embryologic rest rather than a neoplastic mesothelial proliferation and is discussed in chapter 8.

WELL-DIFFERENTIATED PAPILLARY MESOTHELIAL TUMOR

Definition. *Well-differentiated papillary mesothelial tumor* (WDPMT) is a neoplastic mesothelial tumor characterized by bland cells lining distinct fibrovascular cores with a papillary architecture. The term *well-differentiated papillary mesothelioma* has been commonly used, but the above term is preferred to highlight its distinction from diffuse and localized mesotheliomas (1).

Clinical Features. WDPMT was first described in a series of eight cases as an unusually well-differentiated mesothelioma (2). This was followed by a larger series of 22 cases in the peritoneum in which most were in women, and two were siblings (3). Some patients had a history of exposure to asbestos. Importantly, these were mesothelial tumors with a good prognosis. WDPMT occurs most commonly in the peritoneum (3), although cases have been described in the tunica vaginalis (4–6), pleura (7), and pericardium (8). While peritoneal cases occur more commonly in women (9), other sites (except tunica vaginalis) have a closer male to female ratio (7,10).

The patient's age at presentation ranges from the 20s to 70s in most series. The percentage of cases presenting incidentally varies between series, some identified on imaging and others at surgery, with pain as a presenting symptom in some cases (7,9). Ascites is also reported at presentation (11). In a series that included cases from sites other than the peritoneum, asbestos exposure has been reported (7,10).

Gross Findings. WDPMT can be a single area of nodularity but often, multiple nodules of a few millimeters to several centimeters are seen (fig. 4-1) (9). Size criteria have been proposed as upper limits, but these are a guideline rather than an absolute cutoff value. Large examples require extensive sampling to rule out areas of diffuse or localized mesothelioma, especially when there is prominent papillary architecture. WDPMT combined with multicystic mesothelioma and adenomatoid tumor has been reported (9,12,13).

Microscopic Findings. The key histologic features of these tumors are papillary architecture (fig. 4-2), lack of stratification, and most importantly, low nuclear grade. Nuclei should be bland, with smooth nuclear contours and absent to small nucleoli (fig. 4-2). Mitoses are rare.

Figure 4-1

WELL-DIFFERENTIATED PAPILLARY MESOTHELIAL TUMOR (WDPMT)

Diffuse excrescences are seen in the omentum.

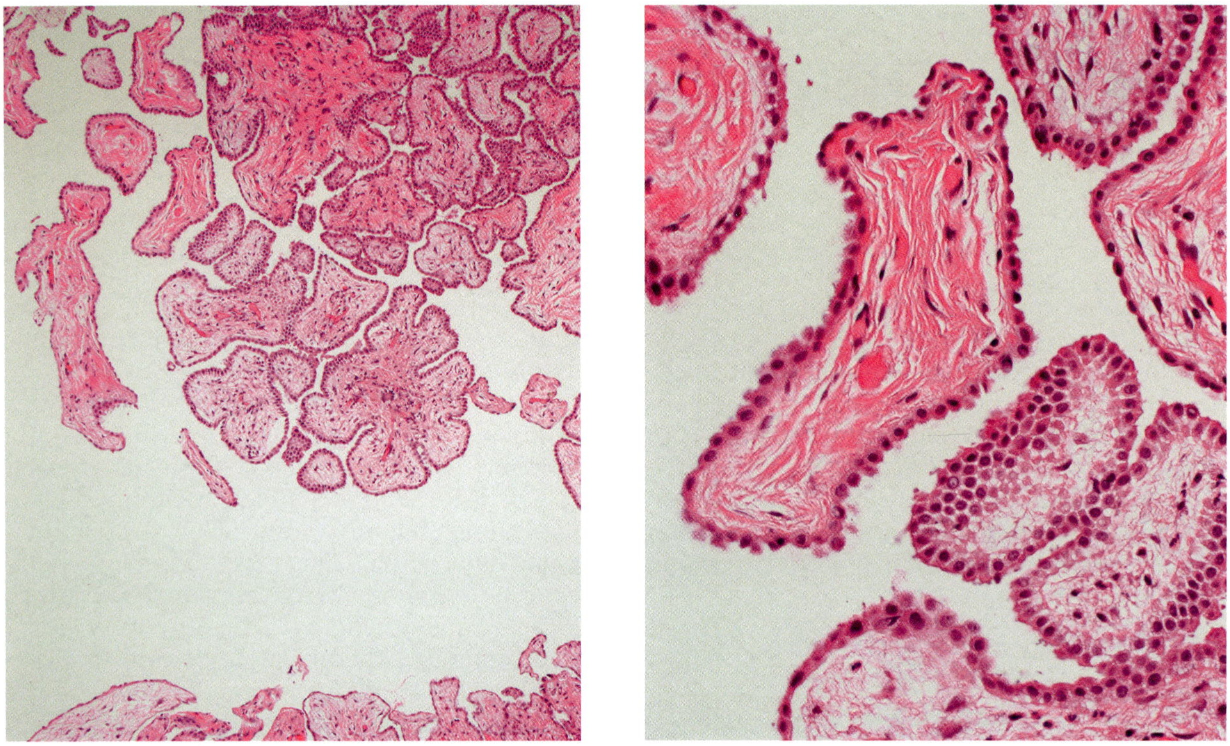

Figure 4-2

WELL-DIFFERENTIATED PAPILLARY MESOTHELIAL TUMOR

Left: Arborizing papillary structures are lined by a single layer of bland mesothelial cells.
Right: The nuclear grade is low, with uniform nuclei without nucleoli.

The cells line fibrovascular cores without stratification. Psammoma bodies may be present, and in some cases, papillary cores become hyalinized. Areas of stromal sclerosis can mimic invasion, and invasion of underlying stroma has been reported.

In a series classified as WDPMTs with invasion, 13 were peritoneal, 3 pleural, 3 hydroceles, and 1 in hernia sac. Invasive foci did not seem to predict negative outcomes as a sole criterion (14). It remains uncertain whether these represent true WDPMTs or diffuse/localized epithelioid mesothelioma with prominent papillary architecture and early invasion. Recent molecular and immunohistochemical advances may aid in making this distinction (9,15).

In the tunica vaginalis, a category of uncertain malignant potential has been proposed for cases with a papillary pattern but in which there is localized disease and some nuclear atypia (16,17). Whether these correspond to WDPMT, localized mesothelioma or an intermediate group requires further study.

Immunohistochemical Findings. WDPMT has an immunophenotype similar to that of classic epithelioid mesothelioma, although proliferation rates are lower (18) and loss of BAP1 and/or MTAP staining is not observed (19,20). Sixty percent of WDPMTs stain for PAX8 while twelve percent of peritoneal mesotheliomas stain for PAX8, which may help distinguish WDPMTs from early diffuse or localized mesothelioma (9,21–23).

Cytologic Findings. Tubulopapillary architecture, monolayered pavement-like sheets, and bland ovoid to round cells are seen (24). Spherical clusters with smooth contour and collagenous balls have also been described (25).

Molecular Genetic Findings. Abnormal karyotypes support a neoplastic proliferation (14). One reported case had a *NF2* heterozygous deletion (26), and one case an *E2F1* mutation (27). In another series, WDPMTs harbored somatic missense mutations in either the *TRAF7* or *CDC42* genes, and lacked alterations involving *BAP1, NF2, CDKN2A, DDX3X, SETD2, LASTS1/2, PBRM1, SMARCC1,* and *ALK* that are found in diffuse mesothelioma, a finding that may aid in diagnosis (15,28). A whole exome sequencing analysis of five cases also showed distinct mutations in *EHD1, ATM, FBXO10, SH2D2A, CDH5, MAGED1,* and *TP73* and that WDPMTs were strongly enriched with C > A transversion substitution mutations, features not found in diffuse mesotheliomas (28). Two siblings with a germline *BAP1* mutation were reported to have WDPMT, although it is difficult to be certain that these are not indolent epithelioid mesotheliomas with prominent papillary architecture (29).

Differential Diagnosis. Distinction from papillary mesothelial hyperplasia is based on presentation as nodules, presence of distinct papillary architecture, lack of cause for mesothelial proliferation, and lack of inflammation. During abdominal surgery, in particular, nodules of WDMPT are encountered and often sent for frozen section; therefore, the differential diagnosis includes primary tumors that may be the target of such surgery. The architecture, size of the lesion, and presence of bland nuclei should help when the differential diagnosis is adenocarcinoma (7,10,30). However, larger examples or areas of sclerosis can be challenging, so focus on features in the cellular population and the architecture is critical.

When WDPMT is considered at frozen section, the pathologist should ask the surgeon to sample additional foci if present. Additional sampling is intended to address the main differential diagnosis of mesothelioma with a papillary pattern, diffuse or localized. Distinguishing multifocal WDPMT from diffuse mesothelioma may also be problematic on small biopsies, although the latter usually displays greater nuclear atypia, variation in histologic patterns, and a greater extent of stromal invasion. This emphasizes the need for thorough sampling (fig. 4-3).

WDPMTs stain for mesothelial markers, which aid in distinction from metastatic carcinomas. BAP1 staining is maintained in WDPMT, which may help in distinguishing it from diffuse and localized mesotheliomas (19,20). WDPMTs stain for PAX8, so use of this antibody is limited in relation to distinguishing WDPMT from metastatic carcinomas from the gynecologic tract (22), but may be of value in differentiation from diffuse/localized mesothelioma (9,21,23). WDPMTs also harbor mutations in either the *TRAF7* or *CDC42* genes, and lack alterations involving *BAP1, NF2, CDKN2A, DDX3X, SETD2,* and *ALK,* which are frequent in diffuse mesothelioma (15).

Treatment and Prognosis. Removal of the visible lesions when feasible is the first line of therapy, but other therapies are more controversial as complications of therapy may exceed the risks conferred by the tumor. Therefore, conservative approaches to this disease have been favored. However, methods of complete cytoreduction for multifocal disease have been used, as well as hyperthermic intraperitoneal chemotherapy, with possible reduction in recurrence rate. Patient selection may be a critical parameter, with extensive or symptomatic disease requiring more radical cytoreduction or chemotherapy (9,18,31,32).

The prognosis of patients with this tumor is favorable. The majority of cases are indolent, but recurrences and persistent disease occur (3,7,9,10,12,31,32). Examples have been reported of diffuse mesothelioma following a prior diagnosis of WDPMT, although it is uncertain whether this is true progression from WDPMT to mesothelioma, separate tumors, or dedifferentiation within an indolent mesothelioma (33,34). In cases arising in the tunica vaginalis, a borderline uncertain potential category has been proposed (17).

LOCALIZED MESOTHELIOMA

Definition. *Localized mesothelioma* is an unusual mesothelioma that has a distinctive presentation as a single circumscribed mass. The definition needs to be applied strictly, and satellite lesions, even if smaller than the main mass, exclude this diagnosis. Diagnosis, therefore, requires thorough review of clinical and imaging data. Tumors are microscopically, immunohistochemically, and ultrastructurally the same as diffuse mesotheliomas, other than being unifocal. The term solitary mesothelioma should no longer be used.

Clinical Findings. Some cases are incidental findings, while others present with chest pain, dyspnea, and systemic symptoms. Age range is variable: in one series it ranged from 6 to 82

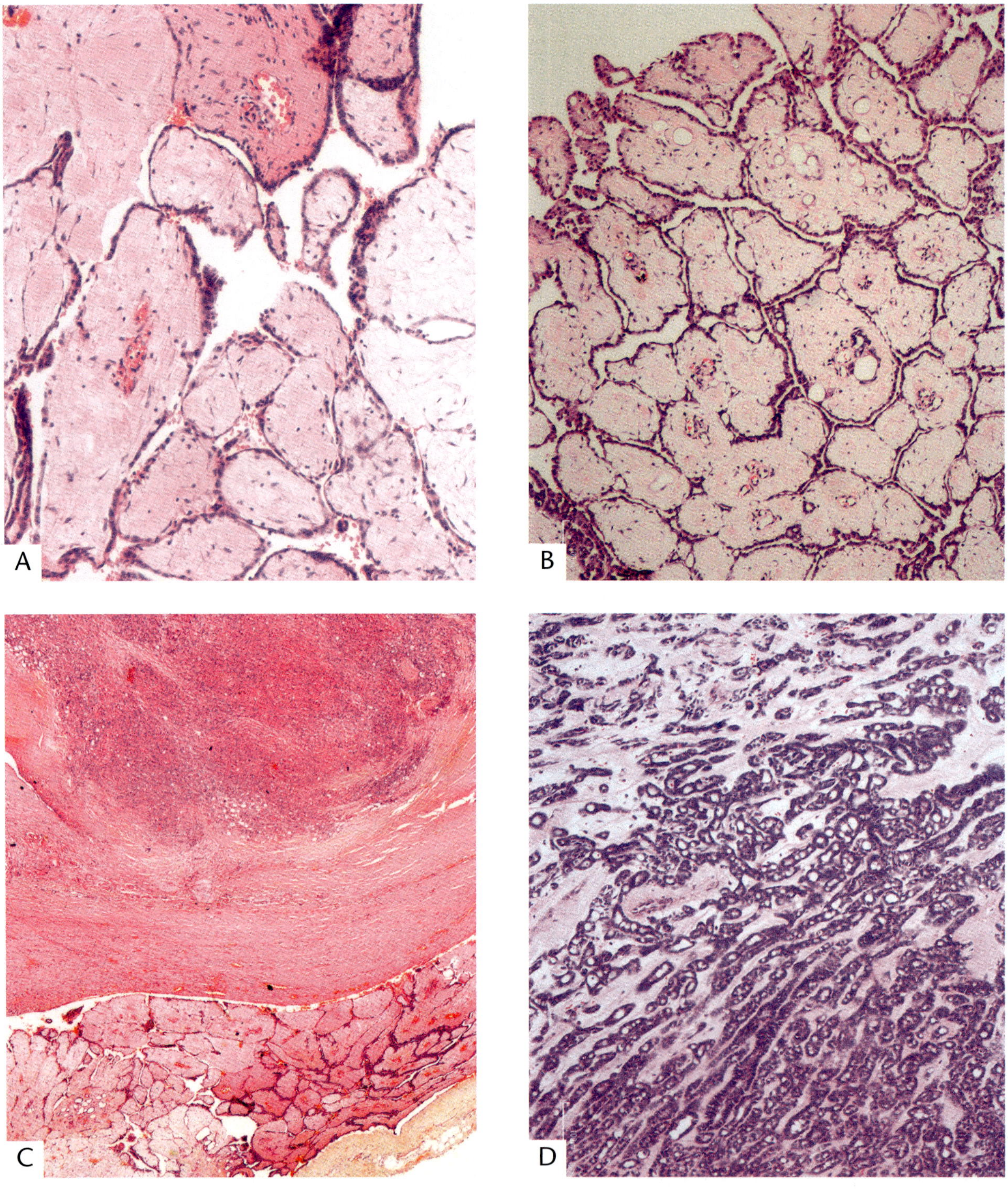

Figure 4-3

DIFFUSE MESOTHELIOMA MIMICKING A WELL-DIFFERENTIATED PAPILLARY MESOTHELIAL TUMOR

A: In this area, the features are virtually indistinguishable from a WDPMT.

B: A few papillae show possible microcystic foci within stromal cores.

C,D: At low power, unequivocal diffuse mesothelioma is seen (C), with abundant tubulopapillary areas (D). (Courtesy of Dr. Y. Z. Zhang, London, United Kingdom)

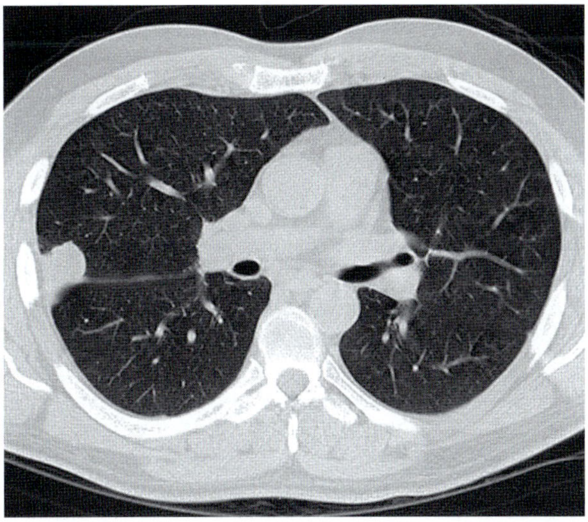

Figure 4-4

LOCALIZED MESOTHELIOMA

Computerized tomography (CT) shows a localized mass involving the pleura and lung, indistinguishable from a peripheral nonsmall cell lung carcinoma.

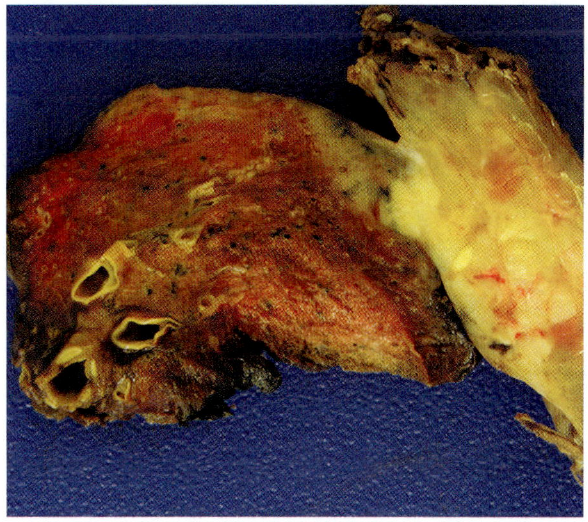

Figure 4-5

LOCALIZED MESOTHELIOMA

A localized mesothelioma involves the chest wall and peripheral lung.

years, with 75 percent in males (35,36). Pleural and peritoneal examples are described, as well as rare examples in the liver (35), pancreas (37), and spleen (38). It may be impossible to distinguish cases from peripheral lung cancers on imaging (fig. 4-4), especially when there is lung involvement. Asbestos exposure is reported in some cases. The incidence of localized mesothelioma is 0.5 to 1.6 percent of diagnosed mesotheliomas.

Gross Findings. Localized mesotheliomas are typically solid masses centered on the pleura that extend into the chest wall, lung, or both (fig. 4-5). Bony destruction may occur. Although there is little evidence as yet, a pleural effusion does not preclude the diagnosis as long as cytology samples are negative (35).

Microscopic Findings. Histologic subtypes, immunohistochemical findings, and differential diagnoses are essentially the same as for diffuse mesotheliomas (see chapter 3). Epithelioid, biphasic, and sarcomatoid examples are all seen, although epithelioid variants are the most common (36).

Molecular Genetic Findings. Localized mesotheliomas have genetic features both similar and different than diffuse pleural mesothelioma, including *BAP1*-mutated, *TRAF7*-mutated, and near-haploid subgroups (39). These, therefore, overlap with the mutation profiles of diffuse mesotheliomas (*BAP1* mutated) and WDPMTs (*TRAF7* mutated). Genomic near-haploidization has also been recently identified in a subset of diffuse pleural mesotheliomas (40,41).

Treatment and Prognosis. The localization of this tumor gives potential for complete excision, and this can be curative. Survival is better compared to diffuse mesotheliomas of the same histology, and surgery is the treatment of choice (42). This is still, however, a malignant tumor associated with recurrence and death from disease. Some patients receive chemotherapy or radiation therapy. In a recent series, the median survival period was 134 months, compared to 29 months in a review of the preceding literature, although this may reflect better patient selection (35).

MULTICYSTIC MESOTHELIOMA

Definition. *Multicystic mesothelioma* is a multicystic mesothelial lesion, usually of the pelvis, that can involve uterus, ovaries, bladder, and rectum, and can extend into the abdomen and retroperitoneum. Terms including *peritoneal inclusion cyst* and *benign cystic mesothelioma* have been used (43). The prevailing theory is that most cases are the result of an excessive inflammatory process, but they are discussed in this chapter because other cases are reported as overlapping with adenomatoid tumors and WDPMTs.

Clinical Findings. Multicystic mesotheliomas occur in women with a mean age of 37 years (44), but also occur in the pediatric population (45,46). Similar lesions have been described in the pleural cavity, pericardium (47,48), and spermatic cord (49). In a series of 17 cases, 13 were in women, and 5 cases were associated with recurrence (50).

Patients with these tumors can be asymptomatic but also have associated pelvic or abdominal pain. The variation in size and distribution can influence clinical symptoms. Lesions may be solitary, but may also be multifocal (44). Computerized tomography (CT) shows thin-walled cysts suggestive of the lesion (51).

Gross Findings. Grossly, lesions are multicystic, thin-walled translucent cysts in clusters (fig. 4-6), with very thin to somewhat thicker fibrous walls. Intracystic fluid can be clear or hemorrhagic, and can be viscous. These lesions range from small (millimeters) to very large (20 cm). They may be seen on the surface of intra-abdominal organs that have a serosal surface (fig. 4-7).

Microscopic Findings. Multicystic mesotheliomas are thin-walled cysts lined by a flat layer of mesothelial cells (fig. 4-8). Mesothelial hyperplasia can be present, as well as squamous metaplasia. Fibrous septal walls may be thin and collagenous, with sparse chronic inflammation. In thicker walled lesions, fibroblastic (fig. 4-9) and smooth muscle cells may be present. Some multicystic mesotheliomas show morphologic overlap with adenomatoid tumors (9) as well as WDPMTs (fig. 4-10) (9), although the significance of this is uncertain (see Differential Diagnosis).

Immunohistochemical Findings. The lining cells stain for calretinin, and a subset of cases are estrogen receptor (ER) and progesterone receptor (PR) positive (50). CA19-9 is also reported as positive in serum and tissue (52).

Cytologic Findings. Rarely, cytology is an adjunct to diagnosis. Numerous mesothelial cells without atypia, in sheets and in two cell-thick strands, are reported as support for multicystic mesothelioma in a clinically cystic lesion (53,54).

Molecular Genetic Findings. Little is known of the molecular characteristics of these tumors, other than a case report of *TNS3-MAP3K3* and *ZFPM2-ELF5* fusion genes identified by RNA

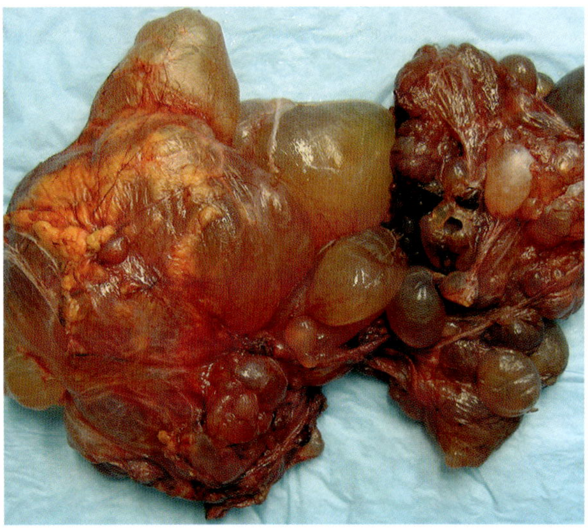

Figure 4-6

MULTICYSTIC MESOTHELIOMA

Multiple thin-walled, variably sized fluid-filled cysts are characteristic of this lesion.

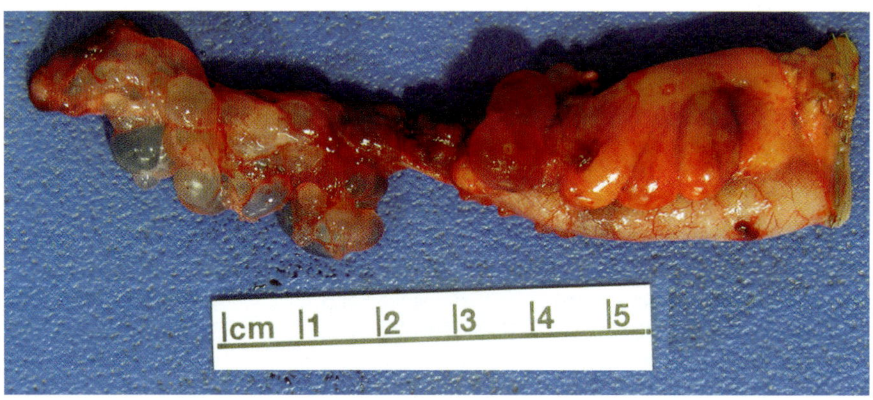

Figure 4-7

MULTICYSTIC MESOTHELIOMA

This example in the appendix shows multiple, smaller, clear fluid-filled cysts involving the appendiceal tip.

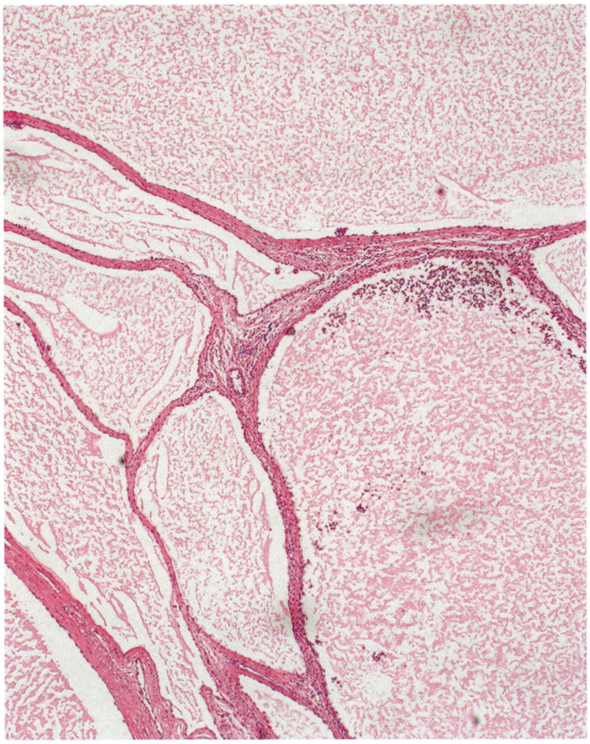

Figure 4-8

MULTICYSTIC MESOTHELIOMA

Multiloculated cyst has a flat nonstratified lining.

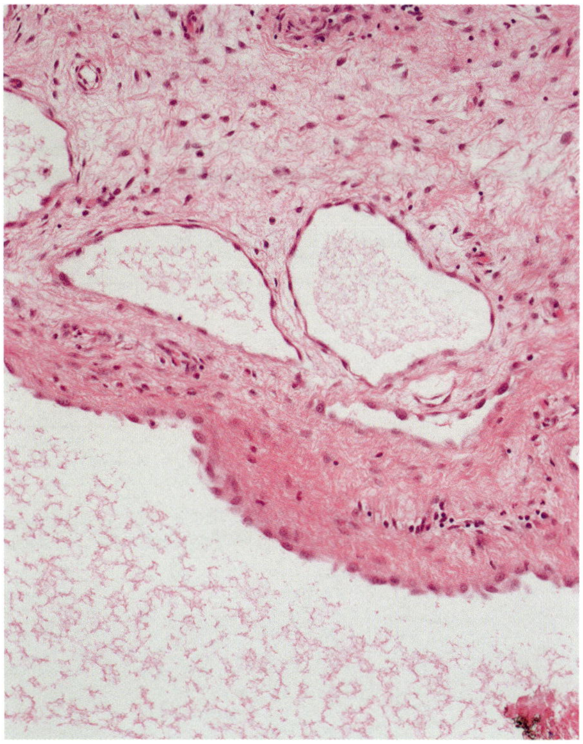

Figure 4-9

MULTICYSTIC MESOTHELIOMA

The mesothelial lining is flat and bland.

sequencing with t(7;17)(p12;q23) and t(8;11)(q23;p13) (55). There appears to be no loss of BAP1 staining (55). Many authors view multicystic mesothelioma as a reactive disorder (56), although others describe morphologic overlap with WDPMT and adenomatoid tumors (9,12), so it is possible that the pathogenesis of multicystic mesothelioma may be multifactorial.

Differential Diagnosis. The differential diagnosis includes lymphangioma (57) and cystic lymphangioleiomyomatosis (58), but these can be excluded by morphology and mesothelial-targeting immunohistochemistry. Cystic adenomatoid tumors can have overlapping features, and in fact, some combination lesions of multicystic mesothelioma and adenomatoid tumors suggest a common relationship (fig. 4-10) (9). On occasion, diffuse mesothelioma can have cystic spaces, but mesothelial proliferations in such tumors are more atypical and infiltrative.

Treatment and Prognosis. Multicystic mesothelioma typically follows a benign course, but

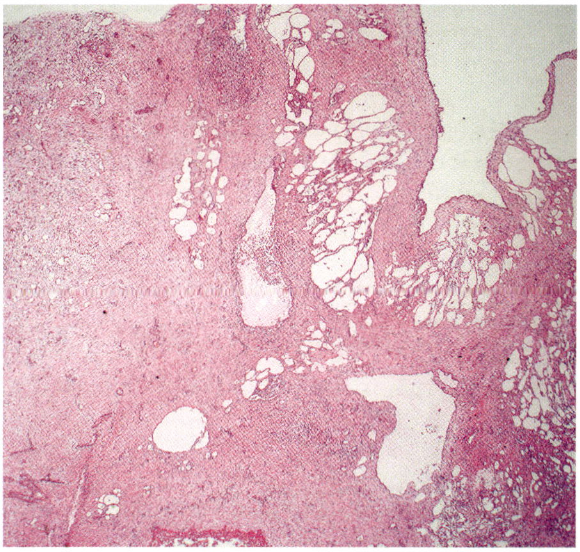

Figure 4-10

MULTICYSTIC MESOTHELIOMA

This multicystic lesion has a more solid area on the left which represents morphologic overlap with an adenomatoid tumor.

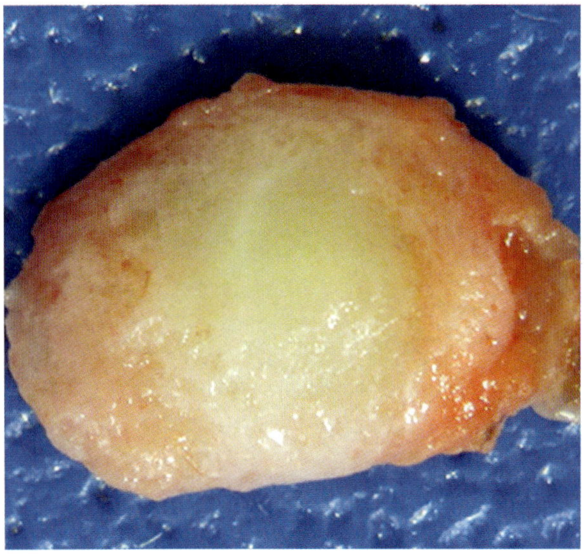

Figure 4-11

ADENOMATOID TUMOR

The lesion in the epididymis is circumscribed with a glistening solid cut surface.

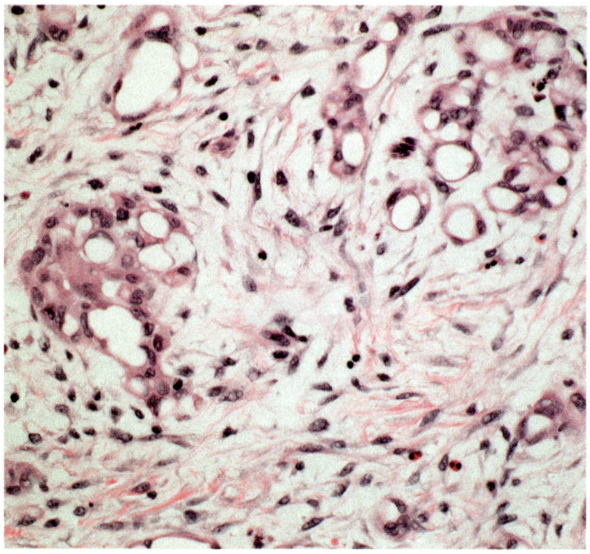

Figure 4-12

ADENOMATOID TUMOR

Bland clusters of mesothelial cells show cystic change and frequent vacuolization. Stromal reaction can mimic invasion.

recurrences are possible. Surgery is the treatment of choice (59). Some patients are treated with a combination of cytoreductive surgery and hyperthermic intraperitoneal chemotherapy (HIPEC) (60,61). Recurrences can result in death, but this is unusual. Rare examples are associated with malignant transformation (62,63), but this mostly occurs after recurrence and is usually not a consideration in the first presentation. As a progressive lesion, it is assumed to be neoplastic, although an associated history of prior surgery, endometriosis, or inflammation suggests the possibility of a reactive lesion. It is of note that reported molecular alterations are limited to one case report (55).

ADENOMATOID TUMOR

Definition. *Adenomatoid tumors* are benign localized mesothelial neoplasms. Studies of gynecologic cases by X-inactivation patterns confirm a neoplastic clonal proliferation (64), and identification of *TRAF7* mutations support clonality as well as suggests a relationship with WDPMTs (65).

Clinical Features. Tumors are generally found in the genitourinary tract (66), most of which are in the epididymis. In women, they are found in relation to the uterus, fallopian tube, and broad ligament, although not confined to these areas (67,68). Rare examples are seen in the omentum, adrenal gland, liver, and pleura (69,70).

Gross Findings. These circumscribed tumors span a few millimeters to several centimeters, with a cut surface that is gray, tan, or yellow (fig. 4-11). Rare cases are reported as cystic and multiple (70).

Microscopic Findings. Histologically, these tumors are composed of plump cells with a microcystic appearance, with cells variably arranged in trabeculae and tubules (70–72). The cells have eosinophilic and vacuolated cytoplasm (fig. 4-12). The stroma is fibrous, with reactive myofibroblastic cells as well as smooth muscle cells. While they can be infiltrative (73), adenomatoid tumors are not malignant.

Immunohistochemical Findings. The cells are mesothelial by immunohistochemistry (positive for cytokeratin, calretinin, D2-40, and WT-1) (72,74,75) and electron microscopy (76), showing classic long slender microvilli.

Differential Diagnosis. Diffuse mesotheliomas may show an adenomatoid/microcystic morphology but they are diffuse by definition and show greater nuclear atypia, as opposed to adenomatoid tumors which are localized and cytologically bland. Adenomatoid tumors do not show BAP1 loss, which may also help in distinction from diffuse mesothelioma with prominent adenomatoid/microcystic growth (77).

Treatment and Prognosis. Adenomatoid tumors are benign and not associated with recurrence or metastasis. They are typically removed as incidental lesions during gynecologic surgery but can be the target lesion in men with scrotal masses. The proper identification of this lesion can lead to testis-sparing surgery.

REFERENCES

1. WHO Classification of Tumours Editorial Board. Thoracic tumours, 5th ed. Lyon: IARC; 2021.
2. Foyle A, Al-Jabi M, McCaughey WT. Papillary peritoneal tumors in women. Am J Surg Pathol 1981;5:241-9.
3. Daya D, McCaughey WT. Well-differentiated papillary mesothelioma of the peritoneum. A clinicopathologic study of 22 cases. Cancer 1990;65:292-6.
4. Gan AM, Plantinga P, Punjani N, Hussey A, Power N. Images—well-differentiated papillary mesothelioma of the tunica vaginalis. Can Urol Assoc J 2018;12:E425-7.
5. Ko KW, Tse KS, Shek KW, Hau MN, Ting SH. Well-differentiated papillary mesothelioma of tunica vaginalis testis of unknown malignant potential: sonographic appearance. J Clin Ultrasound 2018;46:364-7.
6. Chetty R. Well differentiated (benign) papillary mesothelioma of the tunica vaginalis. J Clin Pathol 1992;45:1029-30.
7. Galateau-Salle F, Vignaud JM, Burke L, et al. Well-differentiated papillary mesothelioma of the pleura: a series of 24 cases. Am J Surg Pathol 2004;28:534-40.
8. Sane AC, Roggli VL. Curative resection of a well-differentiated papillary mesothelioma of the pericardium. Arch Pathol Lab Med 1995;119:266-7.
9. Sun M, Zhao L, Weng Lao I, Yu L, Wang J. Well-differentiated papillary mesothelioma: a 17-year single institution experience with a series of 75 cases. Ann Diagn Pathol 2019;38:43-50.
10. Butnor KJ, Sporn TA, Hammar SP, Roggli VL. Well-differentiated papillary mesothelioma. Am J Surg Pathol 2001;25:1304-9.
11. Lee YK, Jun HJ, Nahm JH, et al. Therapeutic strategies for well-differentiated papillary mesothelioma of the peritoneum. Jpn J Clin Oncol 2013;43:996-1003.
12. Chen X, Sheng W, Wang J. Well-differentiated papillary mesothelioma: a clinicopathological and immunohistochemical study of 18 cases with additional observation. Histopathology 2013;62:805-13.
13. Hatano Y, Hirose Y, Matsunaga K, et al. Combined adenomatoid tumor and well differentiated papillary mesothelioma of the omentum. Pathol Int 2011;61:681-5.
14. Churg A, Allen T, Borczuk AC, et al. Well-differentiated papillary mesothelioma with invasive foci. Am J Surg Pathol 2014;38:990-8.
15. Stevers M, Rabban JT, Garg K, et al. Well-differentiated papillary mesothelioma of the peritoneum is genetically defined by mutually exclusive mutations in TRAF7 and CDC42. Mod Pathol 2019;32:88-99.
16. Trpkov K, Barr R, Kulaga A, Yilmaz A. Mesothelioma of tunica vaginalis of "uncertain malignant potential"—an evolving concept: case report and review of the literature. Diagn Pathol 2011;6:78.
17. Brimo F, Illei PB, Epstein JI. Mesothelioma of the tunica vaginalis: a series of eight cases with uncertain malignant potential. Mod Pathol 2010; 23:1165-72.
18. Lee M, Alexander HR, Burke A. Diffuse mesothelioma of the peritoneum: a pathological study of 64 tumours treated with cytoreductive therapy. Pathology 2013;45:464-73.
19. Shinozaki-Ushiku A, Ushiku T, Morita S, Anraku M, Nakajima J, Fukayama M. Diagnostic utility of BAP1 and EZH2 expression in malignant mesothelioma. Histopathology 2017;70:722-33.
20. Lee HE, Molina JR, Sukov WR, Roden AC, Yi ES. BAP1 loss is unusual in well-differentiated papillary mesothelioma and may predict development of malignant mesothelioma. Hum Pathol 2018;79:18-76.
21. Tandon RT, Jimenez-Cortez Y, Taub R, Borczuk AC. Immunohistochemistry in peritoneal mesothelioma: a single-center experience of 244 cases. Arch Pathol Lab Med 2018;142:236-42.
22. Xing D, Banet N, Sharma R, Vang R, Ronnett BM, Illei PB. Aberrant Pax-8 expression in well-differentiated papillary mesothelioma and malignant mesothelioma of the peritoneum: a clinicopathologic study. Hum Pathol 2018;72:160-6.

23. Chapel DB, Husain AN, Krausz T, McGregor SM. PAX8 expression in a subset of malignant peritoneal mesotheliomas and benign mesothelium has diagnostic implications in the differential diagnosis of ovarian serous carcinoma. Am J Surg Pathol 2017;41:1675-82.
24. Nasit JG, Dhruva G. Well-differentiated papillary mesothelioma of the peritoneum: a diagnostic dilemma on fine-needle aspiration cytology. Am J Clin Pathol 2014;142:233-42.
25. Ikeda K, Suzuki T, Tate G, Mitsuya T. Cytomorphologic features of well-differentiated papillary mesothelioma in peritoneal effusion: a case report. Diagn Cytopathol 2008;36:512-5.
26. Nemoto H, Tate G, Kishimoto K, et al. Heterozygous loss of NF2 is an early molecular alteration in well-differentiated papillary mesothelioma of the peritoneum. Cancer Genet 2012;205:594-8.
27. Yu W, Chan-On W, Teo M, et al. First somatic mutation of E2F1 in a critical DNA binding residue discovered in well-differentiated papillary mesothelioma of the peritoneum. Genome Biol 2011;12:R96.
28. Shrestha R, Nabavi N, Volik S, et al. Well-differentiated papillary mesothelioma of the peritoneum is genetically distinct from malignant mesothelioma. Cancers (Basel) 2020;12:1568.
29. Ribeiro C, Campelos S, Moura CS, Machado JC, Justino A, Parente B. Well-differentiated papillary mesothelioma: clustering in a Portuguese family with a germline BAP1 mutation. Ann Oncol 2013;24:2147-50.
30. Kim M, Kim HS. Clinicopathological characteristics of well-differentiated papillary mesothelioma of the peritoneum: a single-institutional experience of 12 cases. In Vivo 2019;33:633-42.
31. Vogin G, Hettal L, Vignaud JM, et al. Well-differentiated papillary mesothelioma of the peritoneum: a retrospective study from the RENAPE Observational Registry. Ann Surg Oncol 2019;26:852-60.
32. Deraco M, Nizri E, Glehen O, et al. Well differentiated papillary peritoneal mesothelioma treated by cytoreduction and hyperthermic intraperitoneal chemotherapy-the experience of the PSOGI registry. Eur J Surg Oncol 2019;45:371-5.
33. Washimi K, Yokose T, Amitani Y, et al. Well-differentiated papillary mesothelioma, possibly giving rise to diffuse malignant mesothelioma: a case report. Pathol Int 2013;63:220-5.
34. Costanzo L, Scarlata S, Perrone G, et al. Malignant transformation of well-differentiated papillary mesothelioma 13 years after the diagnosis: a case report. Clin Respir J 2014;8:124-9.
35. Marchevsky AM, Khoor A, Walts AE, et al. Localized malignant mesothelioma, an unusual and poorly characterized neoplasm of serosal origin: best current evidence from the literature and the International Mesothelioma Panel. Mod Pathol 2020;33:281-96.
36. Crotty TB, Myers JL, Katzenstein AL, Tazelaar HD, Swensen SJ, Churg A. Localized malignant mesothelioma. A clinicopathologic and flow cytometric study. Am J Surg Pathol 1994;18:357-63.
37. Espinal-Witter R, Servais EL, Klimstra DS, Lieberman MD, Yantiss RK. Localized intrapancreatic malignant mesothelioma: a rare entity that may be confused with other pancreatic neoplasms. Virchows Arch 2010;456:455-61.
38. Giansanti M, Bellezza G, Guerriero A, Pireddu A, Sidoni A. Localized intrasplenic mesothelioma: a case report. Int J Surg Pathol 2014;22:451-5.
39. Hung YP, Dong F, Dubuc AM, Dal Cin P, Bueno R, Chirieac LR. Molecular characterization of localized pleural mesothelioma. Mod Pathol 2020;33:271-80.
40. Hmeljak J, Sanchez-Vega F, Hoadley KA, et al. Integrative molecular characterization of malignant pleural mesothelioma. Cancer Discov 2018;8:1548-65.
41. Bueno R, Stawiski EW, Goldstein LD, et al. Comprehensive genomic analysis of malignant pleural mesothelioma identifies recurrent mutations, gene fusions and splicing alterations. Nat Genet 2016;48:407-16.
42. Gelvez-Zapata SM, Gaffney D, Scarci M, Coonar AS. What is the survival after surgery for localized malignant pleural mesothelioma? Interact Cardiovasc Thorac Surg 2013;16:533-7.
43. Ross MJ, Welch WR, Scully RE. Multilocular peritoneal inclusion cysts (so-called cystic mesotheliomas). Cancer 1989;64:1336-46.
44. Weiss SW, Tavassoli FA. Multicystic mesothelioma. An analysis of pathologic findings and biologic behavior in 37 cases. Am J Surg Pathol 1988;12:737-46.
45. Shakya VC, Agrawal CS, Karki S, Sah PL, Poudel P, Adhikary S. Benign cystic mesothelioma of the peritoneum in a child-case report and review of the literature. J Pediatr Surg 2011;46:e23-6.
46. Durell J, Dagash H, Eradi B, Nour S. Pediatric benign cystic peritoneal mesothelioma. J Pediatr Adolesc Gynecol 2016;29:e33-4.
47. Morita S, Goto A, Sakatani T, et al. Multicystic mesothelioma of the pericardium. Pathol Int 2011;61:319-21.
48. Drut R, Quijano G. Multilocular mesothelial inclusion cysts (so-called benign multicystic mesothelioma) of pericardium. Histopathology 1999;34:472-4.
49. Tobioka H, Manabe K, Matsuoka S, Sano F, Mori M. Multicystic mesothelioma of the spermatic cord. Histopathology 1995;27:479-81.
50. Sawh RN, Malpica A, Deavers MT, Liu J, Silva EG. Benign cystic mesothelioma of the peritoneum: a clinicopathologic study of 17 cases and immunohistochemical analysis of estrogen and progesterone receptor status. Hum Pathol 2003;34:369-74.

51. Sugarbaker P, Yan T, Zappa L, Haveric N, Brun E. Thin-walled cysts as a pathognomonic CT finding in cystic mesothelioma. Tumori 2008;94:14-8.
52. Holtzman RN, Heymann AD, Bordone F, Marinoni G, Barillari P, Wahl SJ. Carbohydrate antigen 19-9 and carcinoembryonic antigen immunostaining in benign multicystic mesothelioma of the peritoneum. Arch Pathol Lab Med 2001;125:944-7.
53. Assaly M, Bongiovanni M, Kumar N, et al. Cytology of benign multicystic peritoneal mesothelioma in peritoneal washings. Cytopathology 2008;19:224-8.
54. Devaney K, Kragel PJ, Devaney EJ. Fine-needle aspiration cytology of multicystic mesothelioma. Diagn Cytopathol 1992;8:68-72.
55. Panagopoulos I, Gorunova L, Davidson B, Heim S. Novel TNS3-MAP3K3 and ZFPM2-ELF5 fusion genes identified by RNA sequencing in multicystic mesothelioma with t(7;17)(p12;q23) and t(8;11)(q23;p13). Cancer Lett 2015;357:502-9.
56. Rapisarda AM, Cianci A, Caruso S, et al. Benign multicystic mesothelioma and peritoneal inclusion cysts: are they the same clinical and histopathological entities? A systematic review to find an evidence-based management. Arch Gynecol Obstet 2018;297:1353-75.
57. Fernandez Eire P, Lemos Bouzas MX, Ortiz Rey JA, Herreros Villaraviz M, Prada Arias M. Benign cystic mesothelioma in a boy: an uncommon case mimicking a lymphangioma. J Paediatr Child Health 2015;51:841-2.
58. Slodkowska J, Patera J, Breborowicz J, et al. Extrapulmonary lymphangioleiomyomatosis presented as the asymptomatic retroperitoneal tumours—two cases report. Pol J Pathol 2006;57:205-7.
59. Wang TB, Dai WG, Liu DW, Shi HP, Dong WG. Diagnosis and treatment of benign multicystic peritoneal mesothelioma. World J Gastroenterol 2013;19:6689-92.
60. Chua TC, Yan TD, Deraco M, et al. Multi institutional experience of diffuse intra-abdominal multicystic peritoneal mesothelioma. Br J Surg 2011;98:60-4.
61. Nizri E, Baratti D, Guaglio M, et al. Multicystic mesothelioma: operative and long-term outcomes with cytoreductive surgery and hyperthermic intra peritoneal chemotherapy. Eur J Surg Oncol 2018;44:1100-4.
62. Mino JS, Monteiro R, Pigalarga R, Varghese S, Guisto L, Rezac C. Diffuse malignant epithelioid mesothelioma in a background of benign multicystic peritoneal mesothelioma: a case report and review of the literature. BMJ Case Report 2014;2014:bcr2013200212.
63. Gonzalez-Moreno S, Yan H, Alcorn KW, et al. Malignant transformation of "benign" cystic mesothelioma of the peritoneum. J Surg Oncol 2002;79:243–51.
64. Wang W, Zhu H, Wang J, et al. Clonality assessment of adenomatoid tumor supports its neoplastic nature. Hum Pathol 2016;48:88-94.
65. Goode B, Joseph NM, Stevers M, et al. Adenomatoid tumors of the male and female genital tract are defined by TRAF7 mutations that drive aberrant NF-kB pathway activation. Mod Pathol 2018;31:660-73.
66. Golden A, Ash JE. Adenomatoid tumors of the genital tract. Am J Pathol 1945;21:63-79.
67. Young RH, Silva EG, Scully RE. Ovarian and juxtaovarian adenomatoid tumors: a report of six cases. Int J Gynecol Pathol 1991;10:364-71.
68. Marcussen N, Donna A. Adenomatoid tumour of the uterus. Histopathology 1988;13:582-3.
69. Minato H, Nojima T, Kurose N, Kinoshita E. Adenomatoid tumor of the pleura. Pathol Int 2009;59:567-71.
70. Karpathiou G, Hiroshima K, Peoc'h M. Adenomatoid tumor: a review of pathology with focus on unusual presentations and sites, histogenesis, differential diagnosis, and molecular and clinical aspects with a historic overview of its description. Adv Anat Pathol 2020;27:394-407.
71. Quigley JC, Hart WR. Adenomatoid tumors of the uterus. Am J Clin Pathol 1981;76:627-35.
72. Wachter DL, Wunsch PH, Hartmann A, Agaimy A. Adenomatoid tumors of the female and male genital tract. A comparative clinicopathologic and immunohistochemical analysis of 47 cases emphasizing their site-specific morphologic diversity. Virchows Arch 2011;458:593-602.
73. Miller F, Lieberman MK. Local invasion in adenomatoid tumors. Cancer 1968;21:933-9.
74. Terada T. An immunohistochemical study of adenomatoid tumors of the uterus and fallopian tube. Appl Immunohistochem Mol Morphol 2012;20:173-6.
75. Schwartz EJ, Longacre TA. Adenomatoid tumors of the female and male genital tracts express WT1. Int J Gynecol Pathol 2004;23:123-8.
76. Mackay B, Bennington JL, Skoglund RW. The adenomatoid tumor: fine structural evidence for a mesothelial origin. Cancer 1971;27:109-15.
77. Erber R, Warth A, Muley T, Hartmann A, Herpel E, Agaimy A. BAP1 loss is a useful adjunct to distinguish malignant mesothelioma including the adenomatoid-like variant from benign adenomatoid tumors. Appl Immunohistochem Mol Morphol 2020;28:67-73.

5 MESENCHYMAL NEOPLASMS

As in the mediastinum, the serosal membranes and their surrounding tissues contain nearly all mesenchymal elements as part of their normal histology, and so most types of soft tissue tumors have been described as arising from and around the serosal membranes, in particular, the pleura. Most literature relates to tumors arising in the pleura, rather than the peritoneum, pericardium, or tunica vaginalis. Their histopathologic features are the same as for mesenchymal tumors arising within the soft tissues, and readers are referred to the World Health Organization (WHO) Classification of Tumours of Soft Tissue and Bone (1), rather than reviewing them all herein.

A few mesenchymal neoplasms, however, occur with greater frequency or present specific diagnostic dilemmas for the reporting pathologist when presenting as primary tumors involving the serosal membranes, and these are discussed in more detail in this chapter. In particular, the field of soft tissue tumors is advancing rapidly in relation to the identification of tumor-related or tumor-specific molecular abnormalities, especially gene fusions, and the reporting pathologist should be aware of these new potential diagnostic findings and the best methodologies for their identification.

SOLITARY FIBROUS TUMOR

Definition. *Solitary fibrous tumor* (SFT) is typically a localized tumor made of spindle cells of likely fibroblastic derivation that may show both benign and malignant behaviors. SFTs harbor a specific *NAB2-STAT6* gene fusion. Historically, these tumors were classified as "submesothelial fibroma," "localized mesothelioma," or "fibrous mesothelioma," but these terms should be avoided so as not to confuse SFT with mesothelioma as currently defined. Localized fibrous tumor is sometimes used, although the 2021 WHO Classification of Thoracic Tumours recommends using the term solitary fibrous tumor (2), despite the fact they are occasionally multiple (3).

Clinical Features. There is no gender preference and the age range is wide, averaging in the 50s. The most common symptoms are cough, chest pain, and dyspnea, although some patients are asymptomatic at diagnosis. Both hypertrophic osteoarthropathy and hypoglycemia are reported as associations, the latter due to production of an insulin-like growth factor (Doege-Potter syndrome) (4). There is no known etiologic factor, and in particular, there is no link to exposure to asbestos. SFTs are quite common in the pleura and rarer at other serosal sites, although reported in the peritoneum (5), pericardium (6,7), and testis (8).

Imaging studies characteristically show soft tissue masses with well-defined margins (fig. 5-1). Pleural effusion may be present although tumor cells are rarely seen in the fluid. There is considerable variation in size and, when large, computerized tomography (CT) or magnetic resonance imaging (MRI) may be needed to identify a pleural rather than pulmonary origin. Tumors often show heterogeneous contrast enhancement and tend to compress rather than invade adjacent lung (9). Rarely, a pedicle is seen. Large size, lobulated borders, the presence of calcification, and ipsilateral pleural effusion are CT features reported as predictive of malignancy, although none are definitive (10).

Gross Findings. Most pleural tumors arise from the visceral rather than parietal pleura, and they are often pedunculated (fig. 5-2, left) (11). The cut surface is usually firm due to their fibrous nature and white/gray in color, sometimes with a whorled or lobulated appearance (fig. 5-2, right). Areas with a softer consistency or hemorrhage suggest malignancy and these areas should be sampled widely, as should any areas including necrosis.

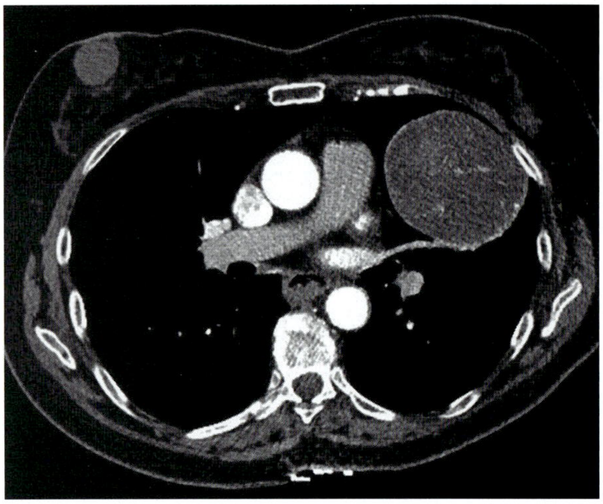

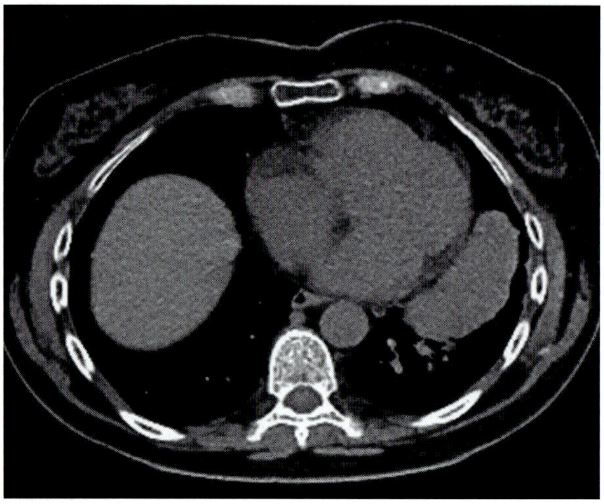

Figure 5-1

SOLITARY FIBROUS TUMOR

Left: High-resolution computerized tomography (HRCT) with contrast shows a localized mass within the pleural space. An incidental superficial benign lesion is present in the right breast.

Right: A rare site for solitary fibrous tumors is the pericardium. HRCT shows a lobulated pericardial solid mass.

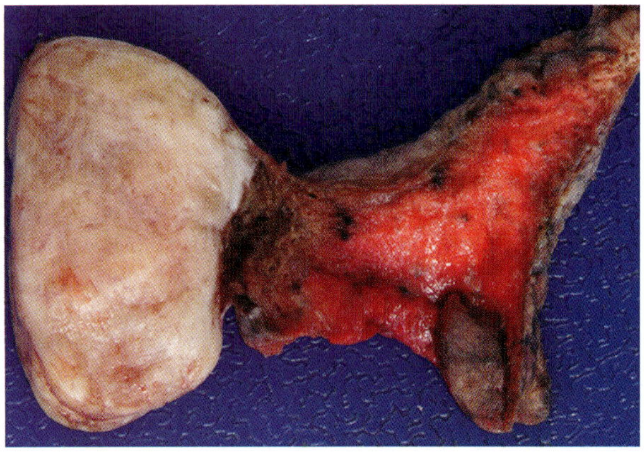

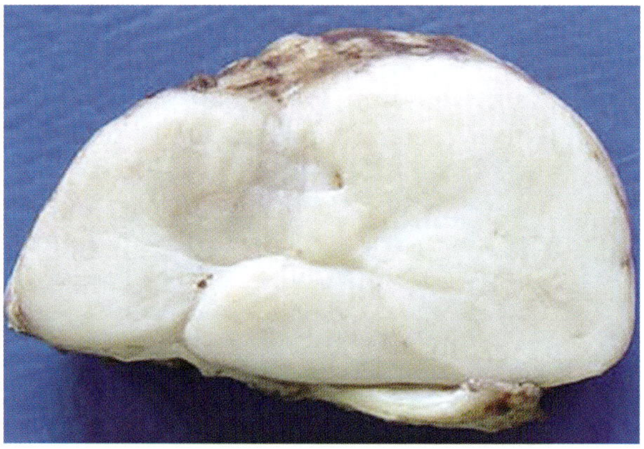

Figure 5-2

SOLITARY FIBROUS TUMOR

Left: A well-circumscribed mass with a cream-colored cut surface is attached to the lung by a broad pedicle.

Right: A well-circumscribed, encapsulated and lobulated mass is attached to parietal pleura.

Microscopic Findings. SFTs usually have sharply demarcated borders with a fibrous capsule. In the pleura, they typically lie in the pleural space, although they can infiltrate peripheral lung, even in benign variants. Architecturally, they usually show what is called a patternless architecture (not obviously storiform or fascicular), with areas of variable cellularity and background fibrous stroma within which there may be prominent hemangiopericytoma-like branching blood vessels (fig. 5-3). Occasionally, tumors are strikingly myxoid.

In benign variants, the spindle cells are small and cytologically bland, and mitoses are absent or sparse. The term *malignant solitary fibrous tumor* is assigned when there is a mitotic rate of more than 4 mitoses per 10 high-power fields, moderate to marked cellular atypia, necrosis, and increased cellularity (11), although the latter three usually accompany a raised mitotic

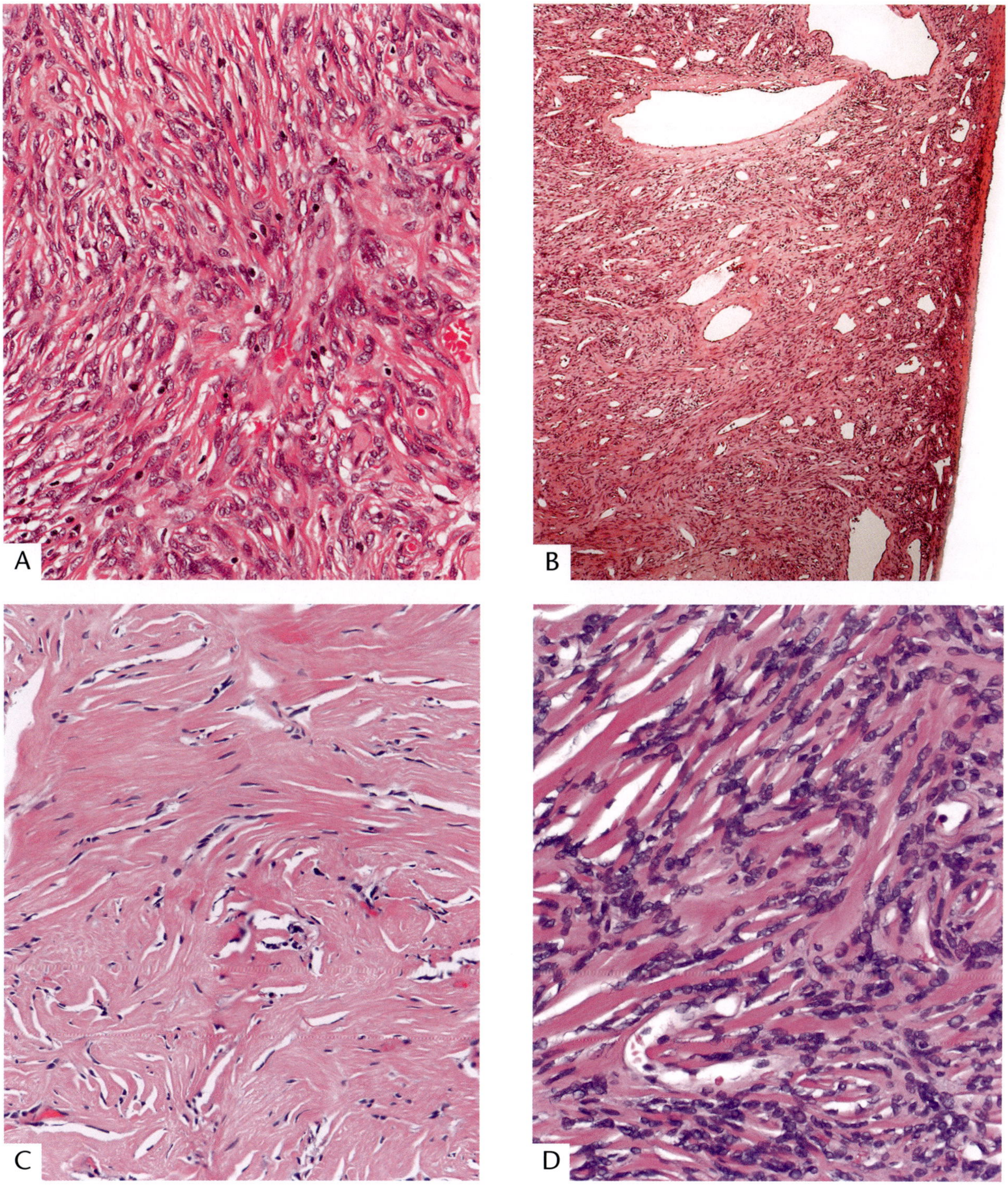

Figure 5-3

BENIGN SOLITARY FIBROUS TUMOR

A,B: The spindle cells may have a patternless arrangement (A), and there may be prominent hemangiopericytoma-like branching blood vessels (B).

C,D: Benign solitary fibrous tumors may be densely fibrotic (C) or moderately cellular (D) but lack pleomorphism, necrosis, or mitotic activity.

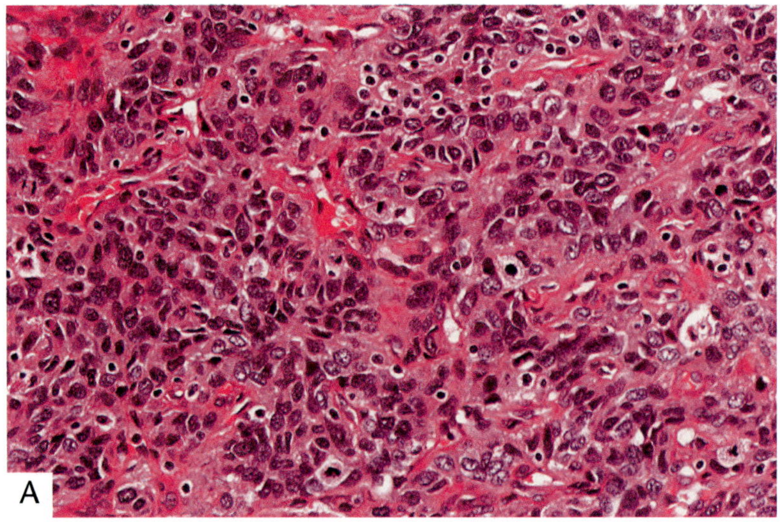

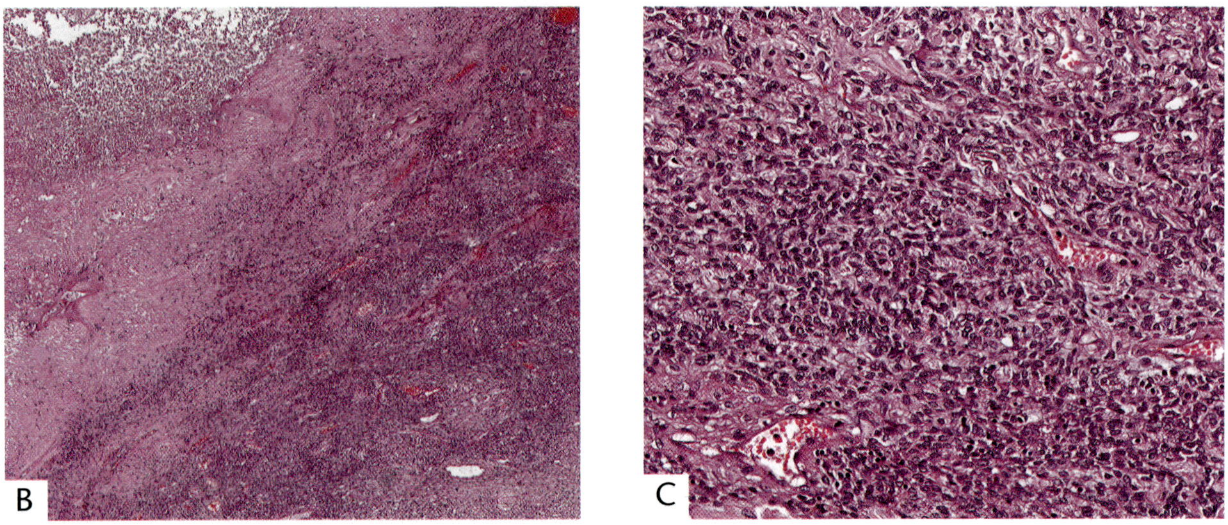

Figure 5-4

MALIGNANT SOLITARY FIBROUS TUMOR

Malignant features are increased pleomorphism and mitotic activity (A), necrosis (B), and marked cellularity (C).

rate (fig. 5-4). The presence of necrosis alone may reflect the site of a previous biopsy, and this possibility should be considered when assessing for features of malignancy. Some tumors show malignant transformation within an otherwise benign tumor (fig. 5-5).

Cytologic Findings. Cytomorphologic features include low to moderate cellularity of mixed oval, elongated, round, and stellate cells; pink collagenous stroma; and hypercellular clusters with infrequent atypia. Immunohistochemistry on a cell block is recommended to help confirm diagnosis (12,13).

Immunohistochemical Findings. Benign SFTs are nearly always positive for CD34 (fig. 5-5B), BCL2 protein (fig. 5-5C), and CD99, and negative for cytokeratins and other mesenchymal markers (14–16). Malignant tumors either partially or completely lose expression of CD34 (fig. 5-5B) (17). Nearly all cases, both malignant and benign, show strong and diffuse nuclear expression of STAT6 (fig. 5-6; see fig. 3-122), an immunohistochemical marker of the *NAB2-STAT6* gene fusion. Still, care is required in its assessment since other mesenchymal tumors show nuclear positivity, although usually with a patchy distribution (16,18–21). ALDH1 is another possible marker with reasonable specificity for SFTs (22).

Molecular Genetic Findings. The presence of a *NAB2-STAT6* gene fusion in SFT was first reported in 2013 and confirmed in subsequent papers (23–27). This fusion is thought to

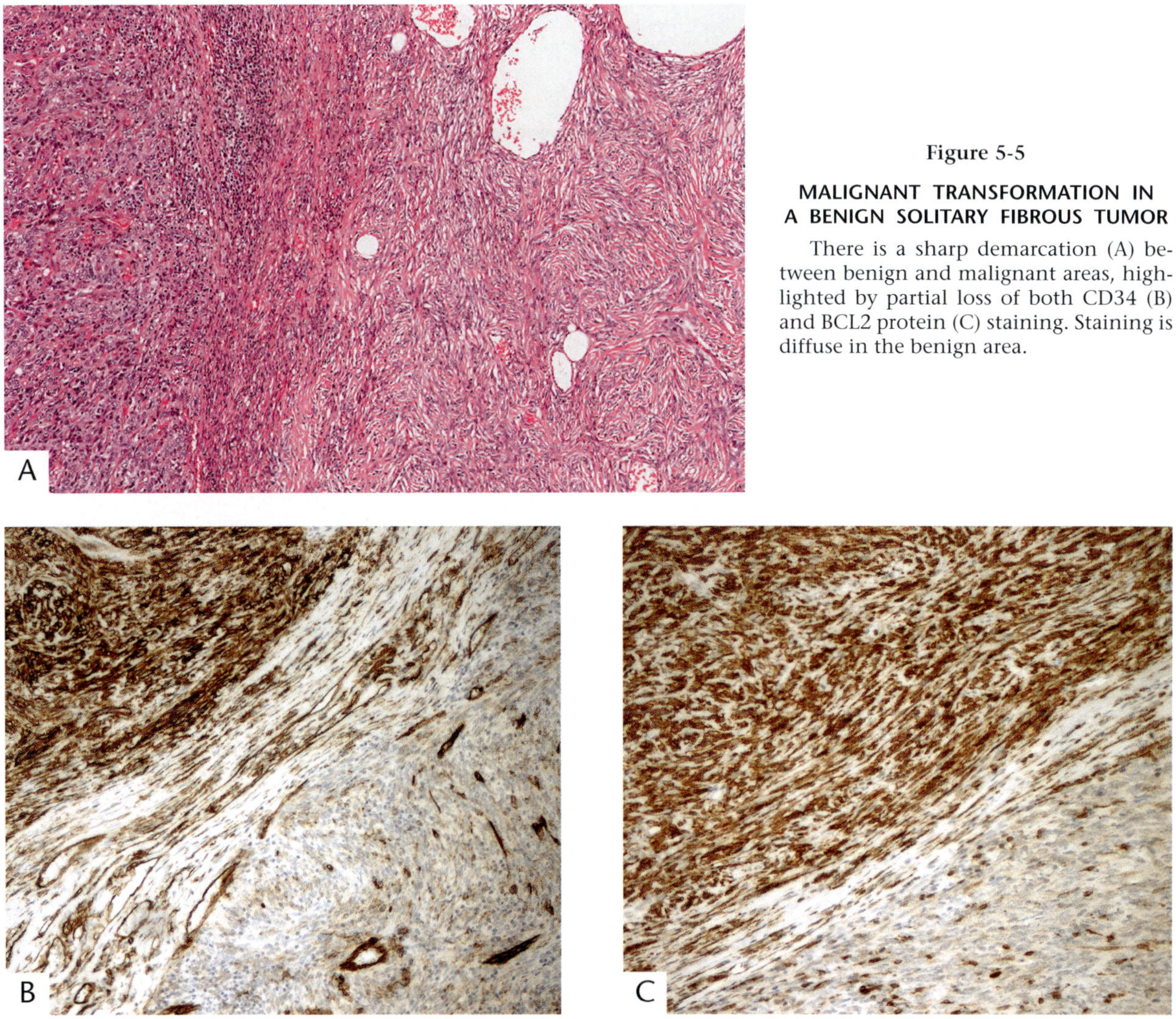

Figure 5-5

MALIGNANT TRANSFORMATION IN A BENIGN SOLITARY FIBROUS TUMOR

There is a sharp demarcation (A) between benign and malignant areas, highlighted by partial loss of both CD34 (B) and BCL2 protein (C) staining. Staining is diffuse in the benign area.

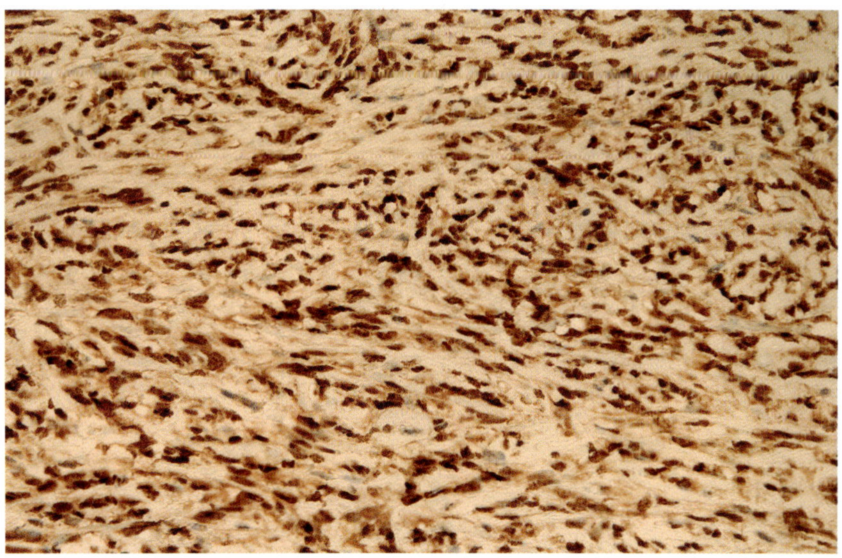

Figure 5-6

BENIGN SOLITARY FIBROUS TUMOR

Tumor cells show diffuse nuclear positivity for STAT6. Weak and patchy nuclear or cytoplasmic staining should be regarded as a negative result.

activate the expression of early growth response 1 (EGR1) transcription factor (25). Because the *NAB2* and *STAT6* genes are located in close proximity on 12q13, fluorescence in situ hybridization (FISH) has limited sensitivity, and currently, immunohistochemistry for STAT6 protein is thought to be a better diagnostic test. Molecular analysis may also be used (23,24) but immunohistochemistry is usually sufficient.

Cytogenetic studies report recurrent chromosomal abnormalities, including losses on chromosome arms 13q (33 percent), 4q, and 21q (17 percent each) and rare gains at chromosome 8 and 15q, although numbers are small (25,28). A rearrangement at 12q13-15, similar to that described in a subset of hemangiopericytomas of soft tissue and meninges, has also been identified (29). Comparative genomic hybridization (CGH) of 12 SFTs of pleura showed IGF2 overexpression (30), which may explain the hypoglycemia seen in some patients.

Differential Diagnosis. Benign variants are readily distinguished from other mesenchymal neoplasms by the immunophenotype described above and a lack of staining for smooth muscle actin, desmin, CD31, and factor VIII. Malignant variants may be more problematic, especially on biopsy, as CD34 and BCL2 staining may be partly or completely lost (fig. 5-5). Identification of the *NAB2-STAT6* gene fusion (23,24), particularly by using immunohistochemistry to assess nuclear positivity for STAT6 protein, is a useful marker in this situation (fig. 5-6; see fig. 3-122) (16,18,21–23). In rare cases, immunohistochemistry either is not available or not definitive, and microscopic features cannot advance further than a differential diagnosis of malignant SFT (or other sarcoma) versus cytokeratin-negative sarcomatoid mesothelioma or sarcomatoid carcinoma, although this situation is thankfully rare. Multidisciplinary review, particularly in relation to whether a malignant spindle cell tumor is localized or diffuse, may then be of value in coming to a favored diagnosis. Of note, rare cases of malignant SFT showing focal cytokeratin positivity have been reported (43).

Treatment and Prognosis. Treatment is primarily surgical and, if excision is complete, the prognosis is good, even in malignant variants. Completeness of resection is reported as the most important prognostic factor (11,31). For benign tumors, recurrence is rare and survival is excellent.

A staging system and prognostic algorithm were proposed by de Perrot in 2002 (32), which was shown to have prognostic significance in a series of 88 patients, along with the presence of necrosis and high p53 expression (33). Malignant histology, origin from the parietal pleura, and the de Perrot system have been shown to be significant adverse predictors of disease-free survival in other series (34,35). More recently proposed prognostic models have shown even better stratification using a variety of clinical and histologic parameters such as parietal/visceral pleural origin, sessile/pedunculated morphology, size, presence of hypercellularity, necrosis, and mitotic activity (36–40). Proliferation rates are also prognostically significant (41,42).

EPITHELIOID HEMANGIOENDOTHELIOMA

Definition. *Epithelioid hemangioendothelioma* (EHE) is a low- to intermediate-grade vascular tumor of endothelial origin made up of cords and nests of epithelioid endothelial cells that lie within fibrous and/or myxoid stroma. Most cases are characterized by *WWTR1-CAMTA1* gene fusions.

Clinical Features. The mean age at presentation is about 50 years, with a male predominance. Patients usually present with diffuse pleural thickening, pleural effusion, and pleuritic chest pain. Some patients have both pulmonary as well as pleural involvement. Imaging characteristically demonstrates pleural thickening, and pleural effusions may represent the primary manifestation (44), sometimes accompanied by pulmonary nodules (45–47).

Gross Findings. EHE may involve the pleura diffusely and mimic the gross appearance of diffuse malignant mesothelioma (fig. 5-7) (45–47).

Microscopic Findings. The tumors usually consist of cords or small nests of epithelioid endothelial cells, often with glassy eosinophilic cytoplasm. A tubulopapillary architecture may sometimes be seen. Nuclei are relatively uniform, ovoid, and vesicular. Cytoplasmic vacuoles, which sometimes contain fragmented erythrocytes, may be seen within the tumor cells, sometimes with vacuolation and with or without septation (fig. 5-8). The tumor cells lie in variably dense myxoid or hyaline stroma. Occasionally,

EHEs show higher-grade nuclear morphology, with a greater degree of pleomorphism and mitotic activity, and these tumors are viewed as intermediate grade rather than low grade (45).

Immunohistochemical Findings. Most tumors stain with one or more endothelial markers including CD31, CD34 (fig. 5-8B), ERG, factor VIII (von Willebrand factor), and FLI1 (45,48,49). Cytokeratin is expressed in up to 40 percent of cases, causing some difficulty in differentiating EHE from carcinoma (45,49); the staining, however, is usually weak to moderate (49,50). Nuclear staining for CAMTA1, in relation to the *WWTR1-CAMTA1* gene fusion, has been reported as a useful marker (50–52).

Ultrastructural Findings. Electron microscopy reveals abundant intermediate filaments, micropinocytosis, and Weibel-Palade bodies. The basal lamina surrounding the tumor cells is uninterrupted and cytoplasmic lumens may be seen (49).

Molecular Genetic Findings. Most examples of EHE show a specific translocation, t(1:3)(p36:q2325), which results in a *WWTR1-CAMTA1* gene fusion (45,50,53,54). Occasionally, other gene fusions, such as *YAP1-TFE3,* have been reported, sometimes related to a more vasoformative architecture (55).

Differential Diagnosis. There is a wide differential diagnosis that includes chronic fibrous pleuritis, malignant mesothelioma, metastatic carcinoma, and melanoma. All these, however, can be discounted by identification of the relevant gene fusion. If molecular testing cannot be undertaken, or in cases where a gene fusion is not identified, the characteristic appearance of the tumor, together with its endothelial phenotype, should allow distinction from the majority of differential diagnoses.

Distinction from mesothelioma may be problematic, especially when presentation is diffuse. Some cases of EHE are positive focally for cytokeratins, but mesotheliomas are negative for endothelial markers, and vascular dedifferentiation within mesothelioma is rare (57). Occasionally, chronic fibrous pleuritis has florid granulation tissue on the surface, but the cytologic features of the endothelial cells in the capillaries are generally not epithelioid and therefore differ from those in EHE. EHE also usually shows a more disorganized growth pattern. The *WWTR1-CAMTA1* gene fusion is absent in epithelioid angiosarcomas, allowing distinction between this tumor type and EHE (45).

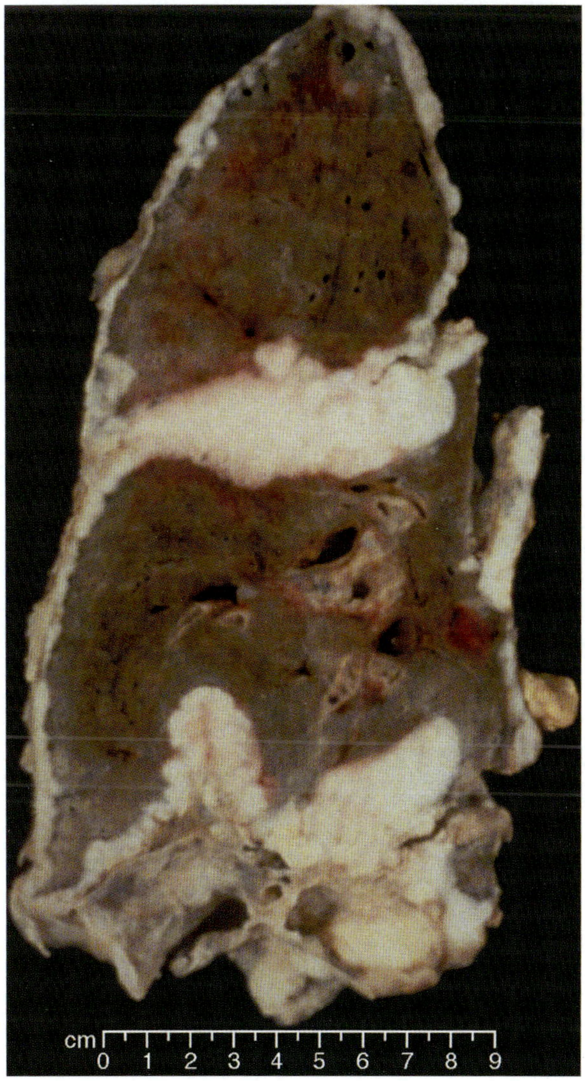

Figure 5-7

EPITHELIOID HEMANGIOENDOTHELIOMA (EHE)

EHE diffusely involves the pleura, mimicking a mesothelioma.

Treatment and Prognosis. Unlike EHE presenting in the soft tissues where the behavior is typically indolent, most cases arising in the pleura are aggressive, with survival generally less than 1 year, usually due to either extensive local spread or metastatic disease. This may reflect late presentation at such a "deep" site rather than a different biology. A poorer prognosis is

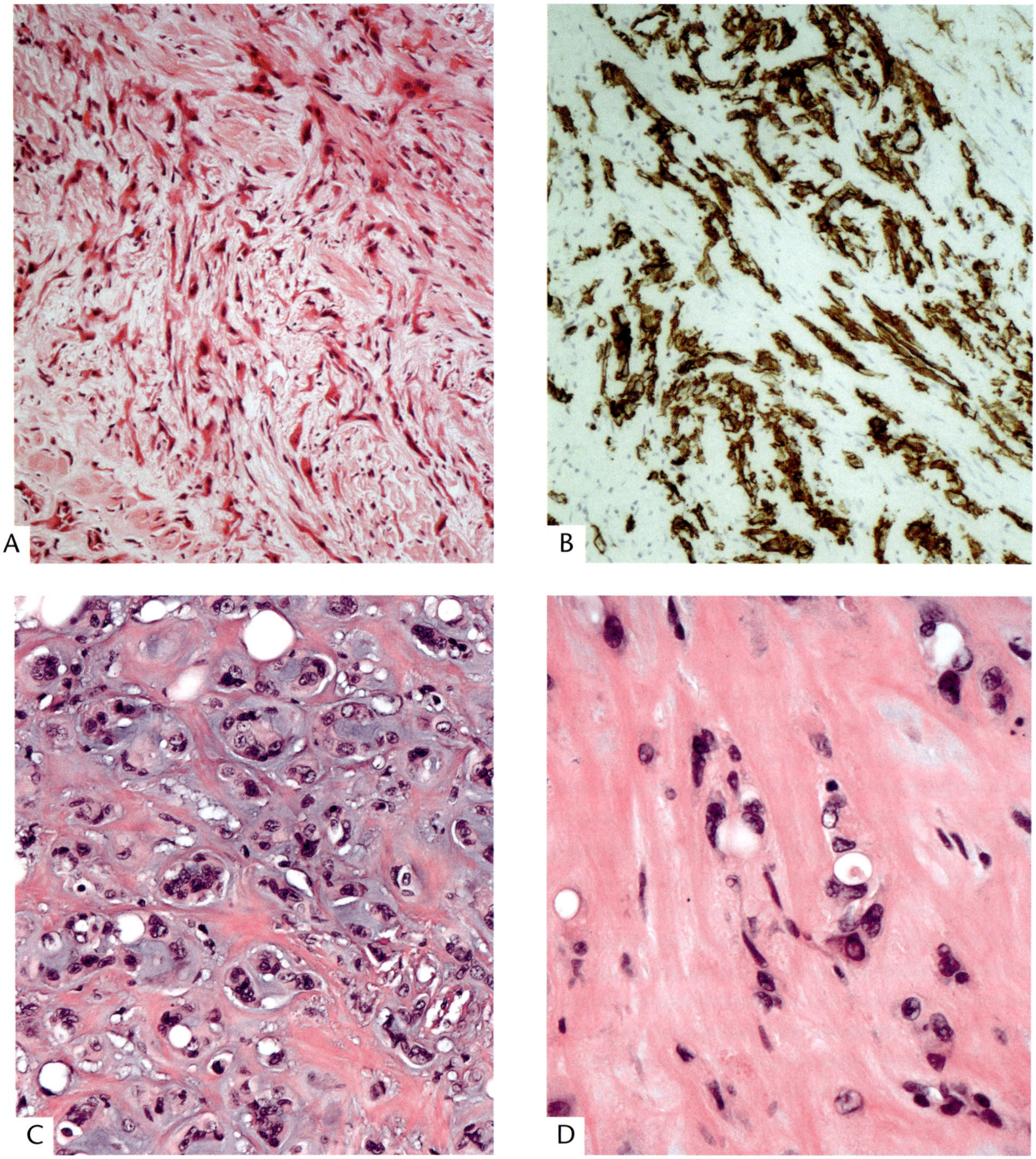

Figure 5-8

EPITHELIOID HEMANGIOENDOTHELIOMA (EHE)

A: EHE shows trabeculae of mildly atypical cells lying within fibrous stroma.

B: Tumor cells stain for CD34.

C: In this example, the tumor cells are more pleomorphic and form nests of ovoid and polygonal cells with abundant, pale, eosinophilic cytoplasm, vesicular nuclei, and inconspicuous nucleoli, within a prominent fibromyxoid or sclerotic stroma.

D: Tumor cells sometimes contain intracytoplasmic lumens due to cytoplasmic vacuolization, thought to represent primitive vascular differentiation. Red blood cells may be seen within the lumens (center of field).

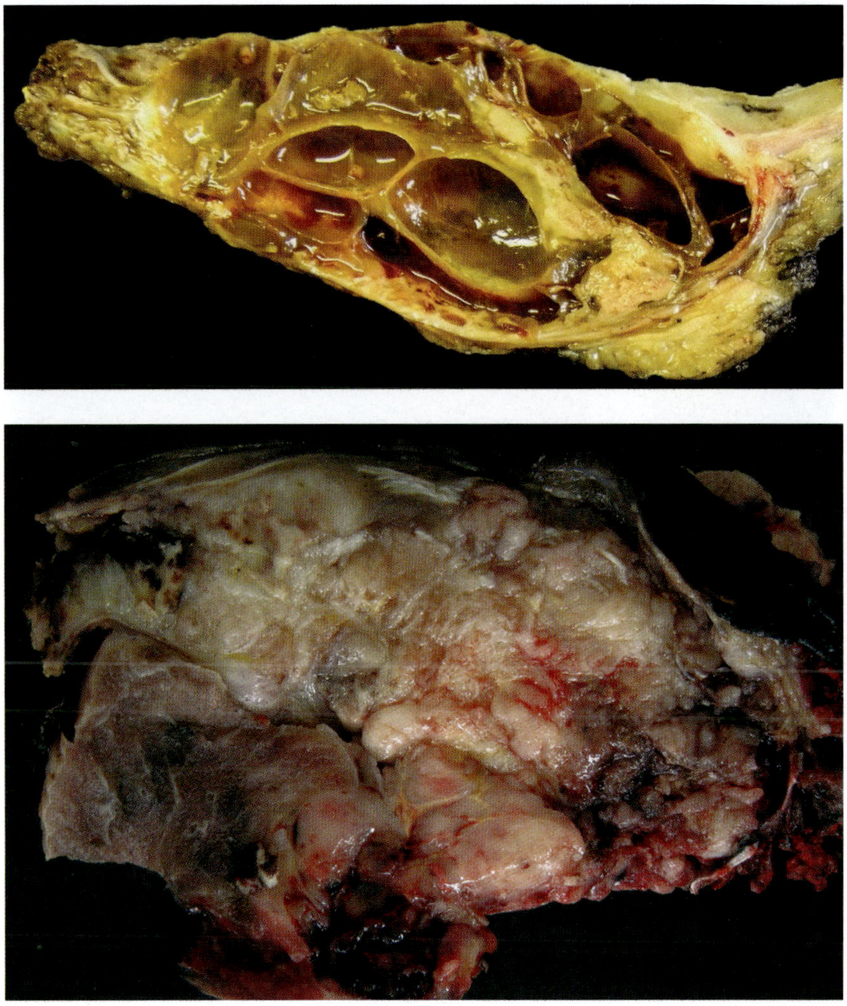

Figure 5-9

PRIMARY SYNOVIAL SARCOMA OF THE PLEURA

Top: A part solid, part cystic mass occupies the pleural space.

Bottom: A separate case shows a focally hemorrhagic solid mass filling the pleural space.

associated with intermediate- compared to low-grade EHEs (45). There is no known effective chemotherapy for these patients. Response to immune checkpoint inhibitors has occasionally been reported (56).

SYNOVIAL SARCOMA

Definition. *Synovial sarcoma* is a soft tissue sarcoma showing variable epithelial and mesenchymal differentiation, characterized by a specific chromosomal translocation: t(X;18)(p11.2;q11.2). Synonyms are now obsolete and should not be used (synovial cell sarcoma, malignant synovioma, synovioblastic sarcoma).

Clinical Features. Patients typically present at a younger age than patients with mesothelioma (mean, 25 years; range, 9 to 50 years), especially those with biphasic tumors (mean age, 25 versus 47 years). There is no gender predilection. Chest pain is the most common presenting symptom, and both dyspnea and dysphagia also occur. Patients also present with pleural effusions and pneumothorax (58–62).

Gross Findings. Pleural synovial sarcomas are usually localized, but can present with diffuse pleural thickening mimicking mesothelioma. They are usually solid, although occasionally are cystic (fig. 5-9), even within the pleura (58,61, 63–65). There is considerable variation in size and some have a fibrous capsule or grow on a pedicle. Necrosis may be a feature.

Microscopic Findings. The spindle cell component consists of relatively small and elongated, uniform spindle cells forming sheets or fascicles (fig. 5-10). In biphasic tumors, there are variable proportions of an epithelial component that contains gland-like or slit-like elements. Mucin may be present. The epithelial

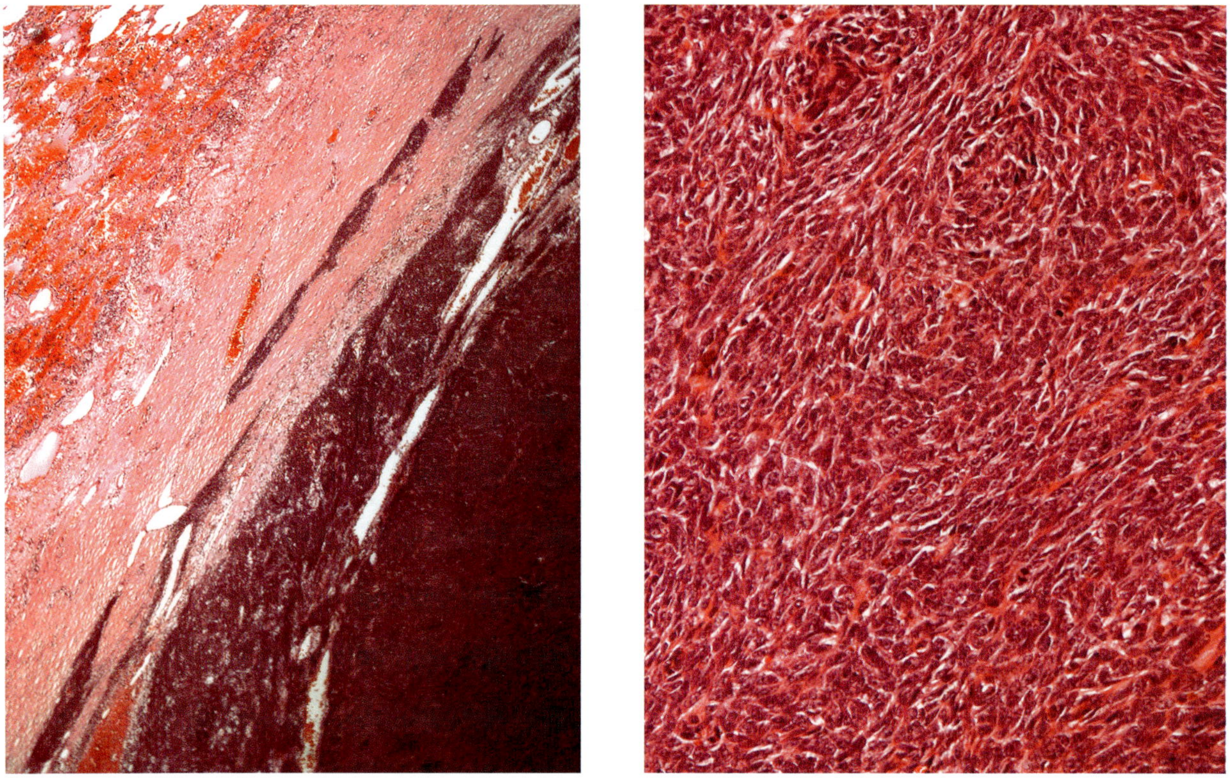

Figure 5-10

PRIMARY SYNOVIAL SARCOMA OF THE PLEURA

Left: At low power, a cellular hyperchromatic tumor expands the pleural space. There is a fibrous pseudocapsule.

Right: At high power, there are sheets of small spindle cells with relatively uniform morphology and a mixture of fascicular and storiform architecture.

cells are typically cuboidal. Tumors may show hemangiopericytoma-like vasculature, hyalinized stroma, or myxoid changes. The mitotic rate is variable (58,60–62,66,67).

Immunohistochemical Findings. At least one epithelial marker should be positive, either keratin or epithelial membrane antigen (EMA). Synovial sarcomas are also positive for BCL2 protein, BerEP4, CD99, and vimentin. Positive staining for S-100 protein, smooth muscle actin (SMA), CD56, calponin, and calretinin may also be seen (58,67–69). TLE-1 is reported as having high specificity for synovial sarcoma (70), although in pleural tumors, TLE-1 staining of mesothelioma may occur, so positive staining should not be regarded as 100 percent specific (71). More recently antibodies targeting the SS18-SSX fusion protein, SSX-C, and SS18-SSX, have been shown to have high sensitivities and specificities (72).

Molecular Genetic Findings. Regardless of site of origin, synovial sarcomas contain a translocation fusing the gene *SS18*, also called *SYT*, on chromosome 18 to either *SSX1* or *SSX2*, both located at Xp11.2. Rarer variants are described in synovial sarcomas arising at other body sites (e.g., t(X:20) translocation). *SS18-SSX1* variants are more likely to show glandular epithelial differentiation than *SS18-SSX2* variants (58,66).

Differential Diagnosis. Before a diagnosis of primary pleural synovial sarcoma is made, metastasis from a soft tissue primary must be excluded. Once excluded, both biphasic and monophasic types can be readily confused with diffuse/localized mesothelioma or pulmonary sarcomatoid carcinoma. Identification of the translocation characteristic of synovial sarcoma cells resolves most diagnostic dilemmas (74), although it may fail to be identified in very rare cases (75).

Compared to mesothelioma, synovial sarcomas typically show more uniform cytology, more focal staining for cytokeratins and EMA, and also usually stain for BerEP4 (58,61,66,74). TLE-1 appears to be of little value as it stains both tumors (74). Pulmonary sarcomatoid carcinomas may stain for TTF1 (76) and are generally much more pleomorphic than synovial sarcomas (77), with pulmonary blastomas showing blastomatous stroma and primitive fetal-type glandular structures (78).

Treatment and Prognosis. Intrathoracic synovial sarcomas are aggressive. In one series (66), the median survival period was 1 to 2 years, and in another (58), 50 months with a 5-year survival of 32 percent. If localized, surgical resection may lead to prolonged survival, although recurrence is common (58). Diffuse variants have a particularly poor outcome. Synovial sarcomas are reported as being sensitive to chemotherapy in a large series not limited to the thorax (73).

DESMOPLASTIC SMALL ROUND CELL TUMOR

Definition. *Desmoplastic small round cell tumor* is a malignant mesenchymal neoplasm with round cell morphology, and polyphenotypic differentiation; it is characterized by an *EWSR1-WT1* gene fusion.

Clinical Features. Primary pleural desmoplastic small round cell tumors are rare. They usually present in teenagers and young adults, more commonly in males. Chest pain and pleural effusion are presenting manifestations (79,80).

Gross Findings. Desmoplastic small round cell tumors consist typically of multiple tumor nodules, sometimes encasing the lung and spreading into the mediastinum. Intrathoracic metastases are common. The cut surface is typically firm and gray-white. Areas of hemorrhage and necrosis may be seen (79,80).

Microscopic Findings. Variably sized and well-demarcated islands of small, round, uniform cells are seen within dense fibrous stroma (fig. 5-11). The stroma may be highly vascular. Tumor cells have hyperchromatic nuclei, with small amounts of cytoplasm, although some tumors contain eosinophilic rhabdoid inclusions. Mitoses are readily apparent.

Immunohistochemical Findings. The tumors are typically polyphenotypic, with positive expression of cytokeratins, EMA, desmin, WT-1 (C-terminus), and neuron-specific enolase. Desmin positivity typically shows a perinuclear dot-like pattern (fig. 5-11) (79–82).

Molecular Genetic Findings. Cases are characterized by an *EWSR1-WT1* gene fusion (82–84).

Differential Diagnosis. Entities are readily excluded by identification of the relevant gene fusion. If molecular analysis cannot be undertaken, other small round cell tumors should be excluded by routine immunohistochemistry for lymphoid markers to exclude lymphoma and neuroendocrine markers to exclude small cell carcinoma of the lung involving the pleura. Desmoplastic small round cell tumors are negative for myogenin and MyoD1, which excludes embryonal rhabdomyosarcoma (79,80,83,84).

Treatment and Prognosis. Desmoplastic small round cell tumors may initially be chemosensitive, but most patients die of disease within 2 years (79,80). One recent report describes response to the drug pazopanib (85).

EWING SARCOMA

Definition. *Ewing sarcoma* is a malignant small round cell sarcoma that is part of the Ewing sarcoma family of pediatric tumors. The sarcoma is defined by the presence of specific chromosomal translocations involving, in nearly all cases, the *EWSR1* gene on chromosome 22 and a member of the ETS family of transcription factors. *Askin tumor* is a term used for Ewing sarcomas that arise in the chest wall.

Clinical Features. Most tumors occur in children or young adults. They usually present with chest pain and pleural effusion (86). Rare cases are reported arising in the pericardium (87).

Gross Findings. The tumors present either as a single mass or as multiple nodules involving the chest wall and other thoracic structures (fig. 5-12) (86,88,89).

Microscopic Findings. Most cases consist of uniform small round cells with round nuclei containing fine chromatin and scanty clear or eosinophilic cytoplasm. There are indistinct cell membranes (fig. 5-13A) (90). Cases often are reported after neoadjuvant therapy, and viable tumor may be scarce (fig. 5-13B). The extent of viable tumor should be reported in this situation.

Immunohistochemical Findings. Nearly all cases stain for CD99 (fig. 5-13C), with

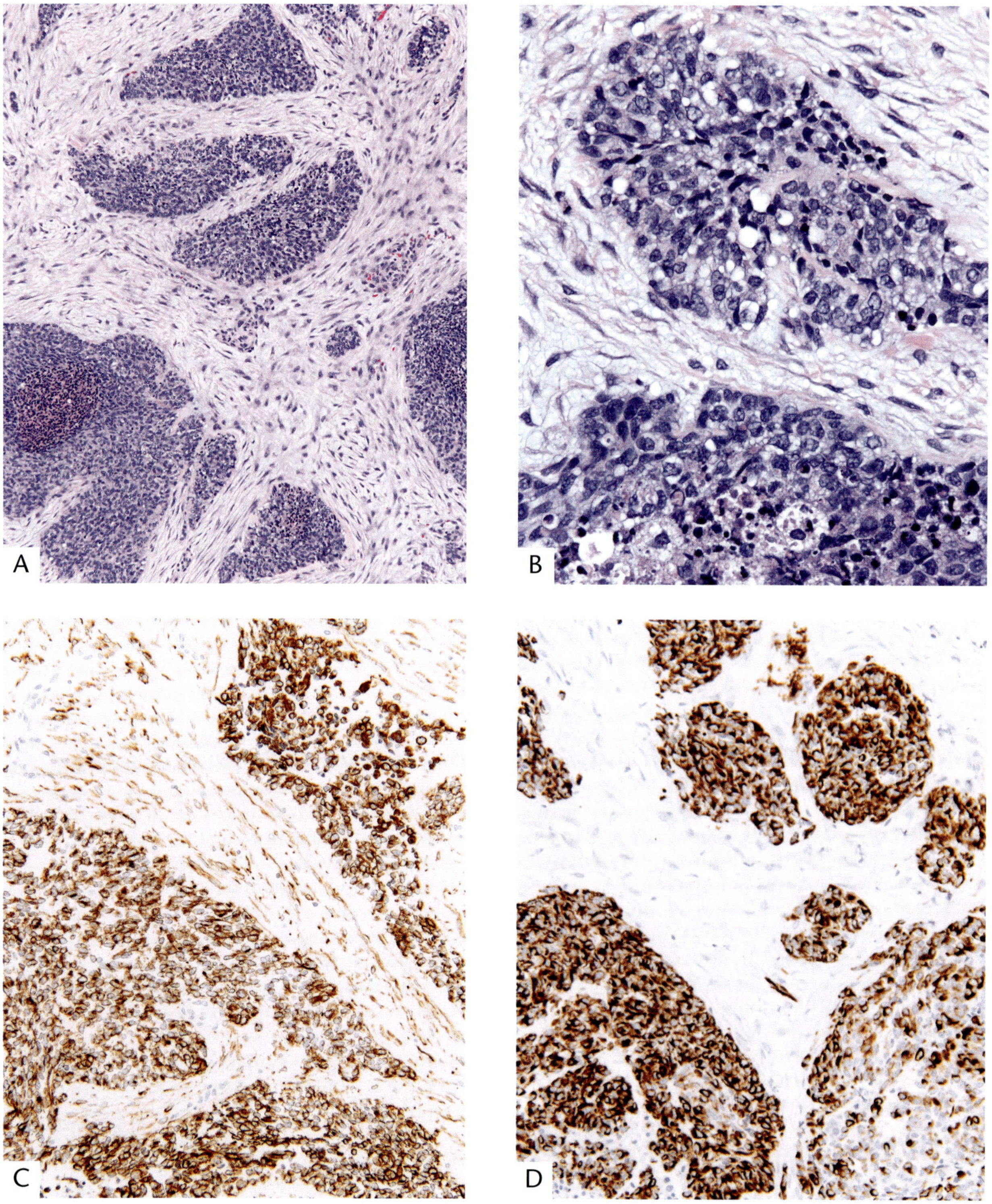

Figure 5-11

DESMOPLASTIC SMALL ROUND CELL TUMOR

A,B: Islands of small round hyperchromatic cells lie in dense fibrous stroma.
C,D: Tumor cells stain for both cytokeratins (C) and desmin (D).

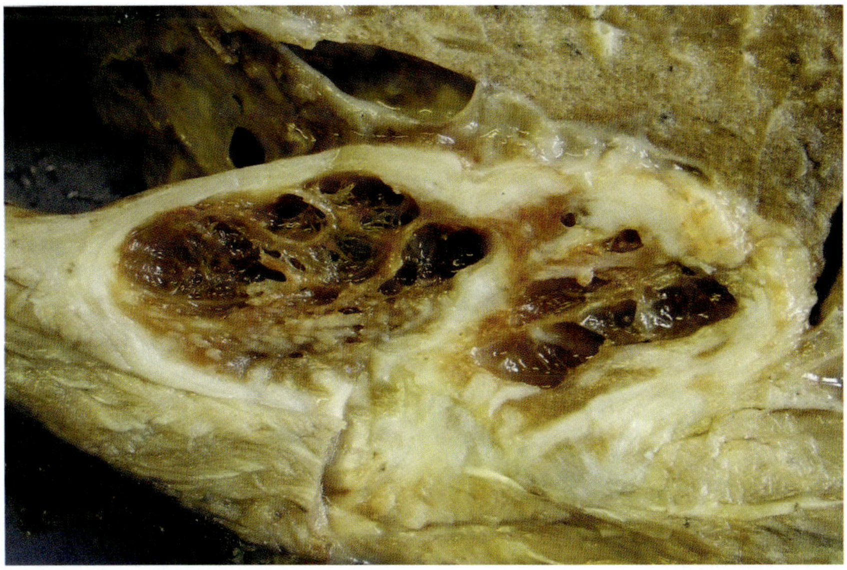

Figure 5-12

EWING SARCOMA

Resection of residual Ewing sarcoma post-therapy shows a partial response to therapy, with fibrous scarring and cystic degeneration, although areas of viable tumor remain.

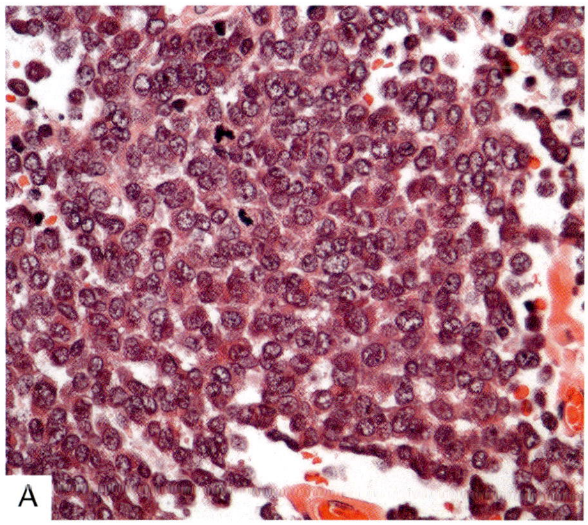

Figure 5-13

EWING SARCOMA

A: The tumor consists of sheets of mitotically active "small round blue cells."

B: Most of the tumor (seen in fig. 5-12), has responded to therapy, with cystic change and dense fibrosis containing scattered hemosiderin-laden macrophages.

C: CD99 staining highlights the focal residual viable tumor.

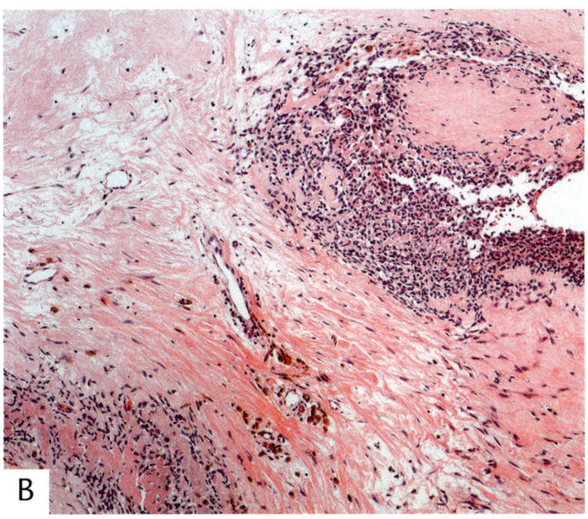

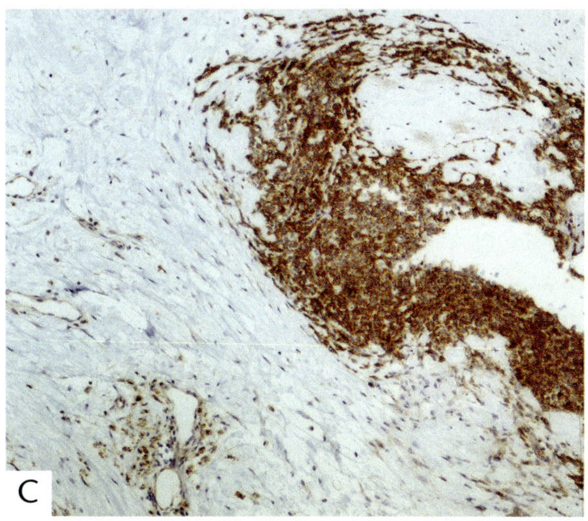

staining for S-100 protein, cytokeratins, and synaptophysin also reported. Although CD99 lacks specificity, strong membranous staining is a characteristic feature. NKX2.2 is another relatively specific marker when considering malignant small round cell tumors (91). FLI1 also commonly stains Ewing sarcoma, although it is not very specific (92).

Molecular Genetic Findings. Approximately 85 percent of cases contain a somatic reciprocal chromosomal translocation, t(11:22)(q24;q12), that fuses *EWSR1* to *FLI1* with the formation of the EWSR1-FLI1 oncoprotein. The remainder have translocations fusing *EWSR1* with other ETS family members, or more rarely, *FUS* (93–95).

Differential Diagnosis. Other small round cell tumors are readily excluded by identification of the relevant gene fusion. The identification of *EWSR1* breakapart using FISH testing is extremely helpful in the correct morphologic setting. If molecular analysis cannot be undertaken, other small round cell tumors are excluded by routine immunohistochemistry for lymphoid markers to exclude lymphoma, neuroendocrine and epithelial markers to exclude metastatic small cell carcinoma, and a combination of epithelial and mesenchymal markers to exclude desmoplastic small round cell tumor. Ewing sarcoma is negative for myogenin and MyoD1, which helps exclude embryonal rhabdomyosarcoma.

Treatment and Prognosis. Prognosis has improved considerably for patients with Ewing sarcomas, with two thirds being cured. However, disseminated disease and early relapse are poor prognostic indicators. Often, the pathologist receives a resection specimen after induction chemotherapy, and histologic assessment of the extensive tumor necrosis/response is of prognostic value (86,96).

OTHER PRIMARY MESENCHYMAL NEOPLASMS OF SEROSAL MEMBRANES

Desmoid-Type Fibromatosis. This tumor arises, rarely, as a primary pleural proliferation. This is now viewed as a locally aggressive but nonmetastasizing myofibroblastic neoplasm exhibiting a *CTNNB1* gene mutation in about 95 percent of cases (97). This tumor usually presents in adults with no gender predominance. Some cases are associated with previous trauma, and there is an association with familial adenomatous polyposis.

Presentation is typically with chest pain and dyspnea. Lesions are often diffusely infiltrative, although are occasionally polypoid, with a firm and sometimes gritty cut surface (fig. 5-14A). Margins are often ill-defined.

At microscopy, there is a typically dense collagenous fibrous stroma, within which are uniform, cytologically bland, fibroblastic, or myofibroblastic cells, typically with a fascicular architecture (fig. 5-14B). Stromal vessels are prominent. The infiltrate may involve lung (fig. 5-14C).

The spindle cells stain variably for SMA, with desmin and CD34 negative in most cases; the latter allows distinction from solitary fibrous tumor. Around three quarters of cases show nuclear positivity for beta-catenin, although this does not help discriminate fibromatosis from a solitary fibrous tumor (98,99).

Most cases are treated by surgical resection, with around 25 percent recurring locally. Recurrence cannot be predicted on the basis of histologic data, although the progression arrest rate seen in a phase 2 studies with imatinib therapy may be related to *CTNNB1* mutation status (100). Sorafenib has also been shown to benefit those with progressive, refractory or symptomatic fibromatosis (101).

Calcifying Fibrous Tumor. This tumor, sometimes referred to as *calcifying fibrous pseudotumor,* is a rare benign tumor occurring in the visceral pleura. It consists of collagenous fibrous stroma associated with chronic inflammation, psammomatous/dystrophic calcification, and a poorly cellular, cytologically bland, spindle cell infiltrate. The histogenesis is unknown, although these tumors may reflect a "burning out" of other neoplasms or reactive proliferations, such as an inflammatory myofibroblastic tumor (102,103) or IgG4-related disease, which lack the requisite diagnostic features by the time of histologic assessment. Some calcifying tumors have recently been shown to carry identifying novel deleterious mutations in *ZN717*, *FRG1*, and *CDC27* genes as well as significantly recurrent copy number losses, most notably involving chromosome 6 (104).

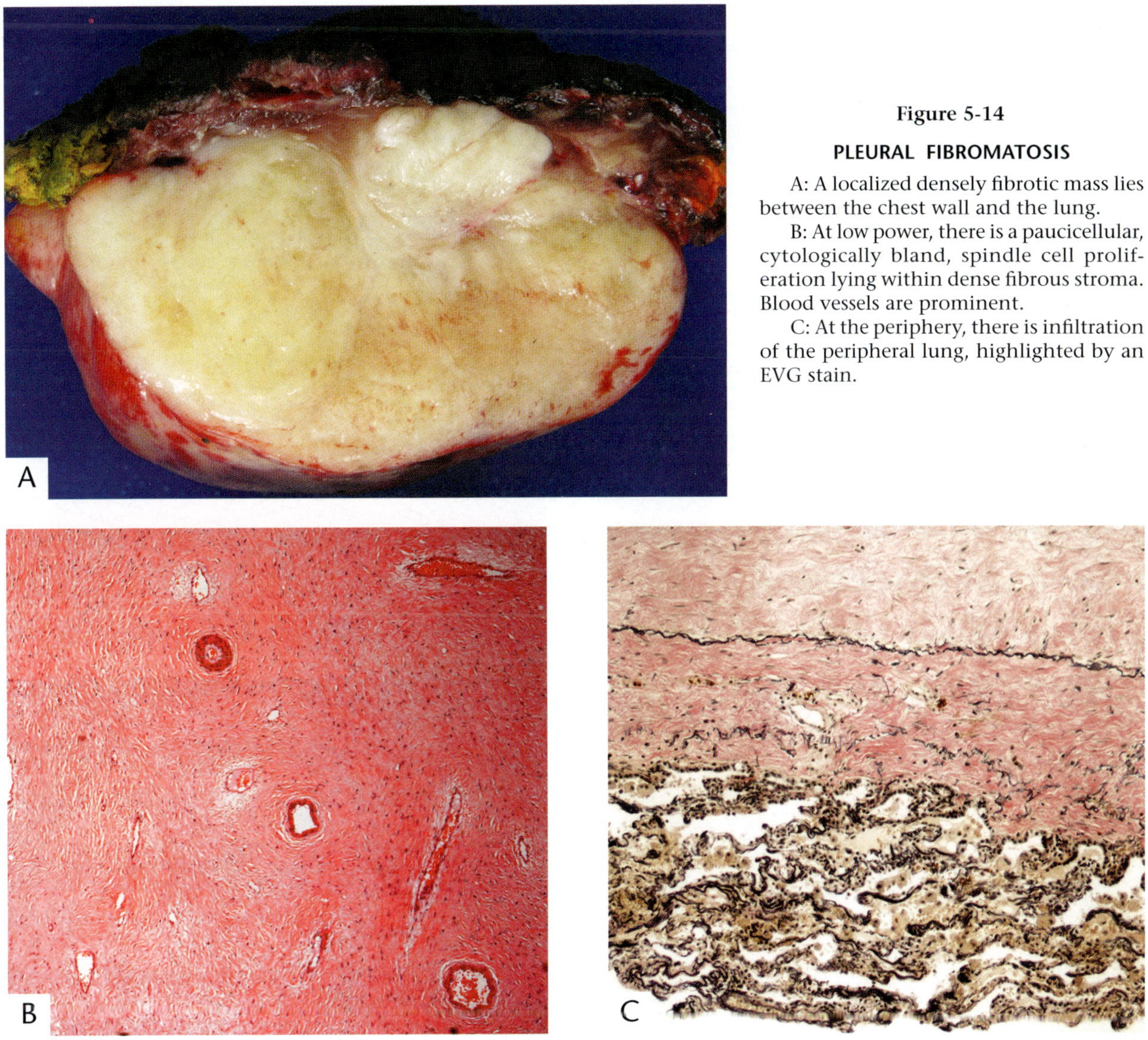

Figure 5-14

PLEURAL FIBROMATOSIS

A: A localized densely fibrotic mass lies between the chest wall and the lung.

B: At low power, there is a paucicellular, cytologically bland, spindle cell proliferation lying within dense fibrous stroma. Blood vessels are prominent.

C: At the periphery, there is infiltration of the peripheral lung, highlighted by an EVG stain.

Calcifying fibrous tumors of the visceral pleura typically present in females, of a median age of 39 years, different than those at other sites where the age at presentation is younger. Rarely, these tumors are multiple (105). Presentation may be asymptomatic, or with chest pain and cough. Imaging shows local or multiple nodular masses, sometimes with calcification (fig. 5-15).

Microscopy shows hypocellular, densely collagenous fibrosis, within which there is a mild, patchy, chronic inflammatory cell infiltrate (fig. 5-15), often at the periphery. Calcification can be prominent. The spindle cells may be positive for CD34, with negative staining for beta-catenin and ALK1. Local recurrence is reported in 10 to 15 percent of patients. Metastases have not been described (106,107).

Schwannoma and Other Neurogenic Tumors. These are frequently seen arising within the pleural space, often arising from an intercostal nerve (fig. 5-16). Variants (cellular, melanotic) may be encountered at this site (108).

Lipomas. Lipomas arising in the chest wall fat are common. Variants of adipose tumors (*angiolipoma, lipoblastoma*) may also be found.

Elastofibroma. This may also cause clinical concern regarding malignancy. It usually presents as an irregular mass in the subscapular region of the chest wall, although diagnosis

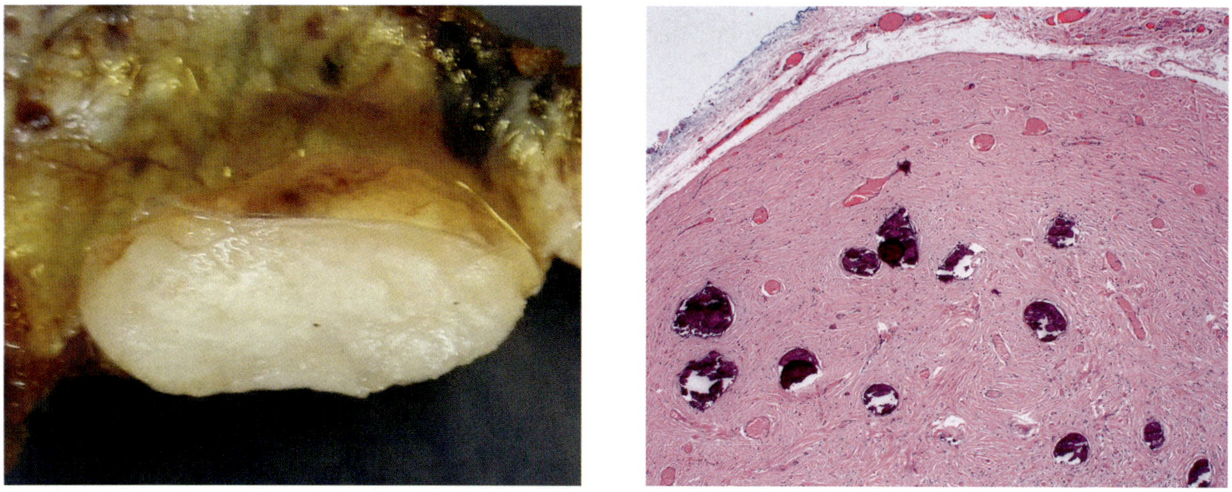

Figure 5-15

CALCIFYING FIBROUS TUMOR

Left: A localized fibrous nodule lies within the parietal pleura.

Right: The tumor is well circumscribed and consists of thick collagen bundles with minimal chronic inflammation and foci of dystrophic calcification within the stroma.

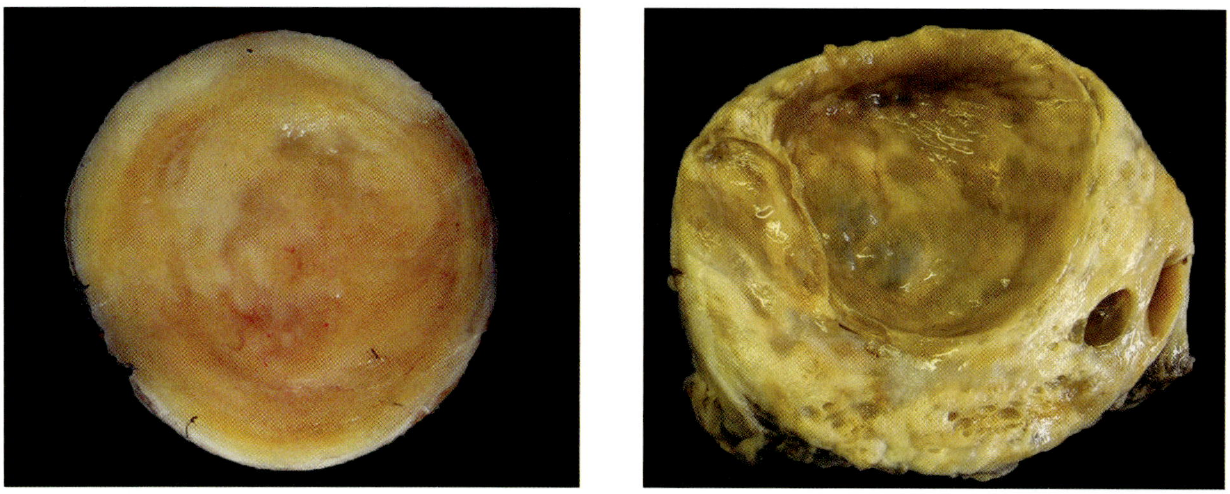

Figure 5-16

SCHWANNOMA

Left: A well-circumscribed mass has a solid, whorled, and yellowish cut surface.

Right: Schwannomas may show degenerative cystic change.

is usually made with relative ease with EVG staining (fig. 5-17).

Sarcomas. In relation to primary sarcomas of the serosal membranes, small series and case reports abound, with variably sized cohorts of *angiosarcomas* (fig. 5-18) (45,49,109–111), *leiomyosarcomas* (112,113), *liposarcomas* (114–117), *malignant peripheral nerve sheath tumors* (118), and even *clear cell sarcomas* (119). Most reports describe poor prognosis unless tumors are localized and therefore resectable in the first instance (fig. 5-19A,B).

Metastatic disease from an occult site or a previous tumor is more common than primary

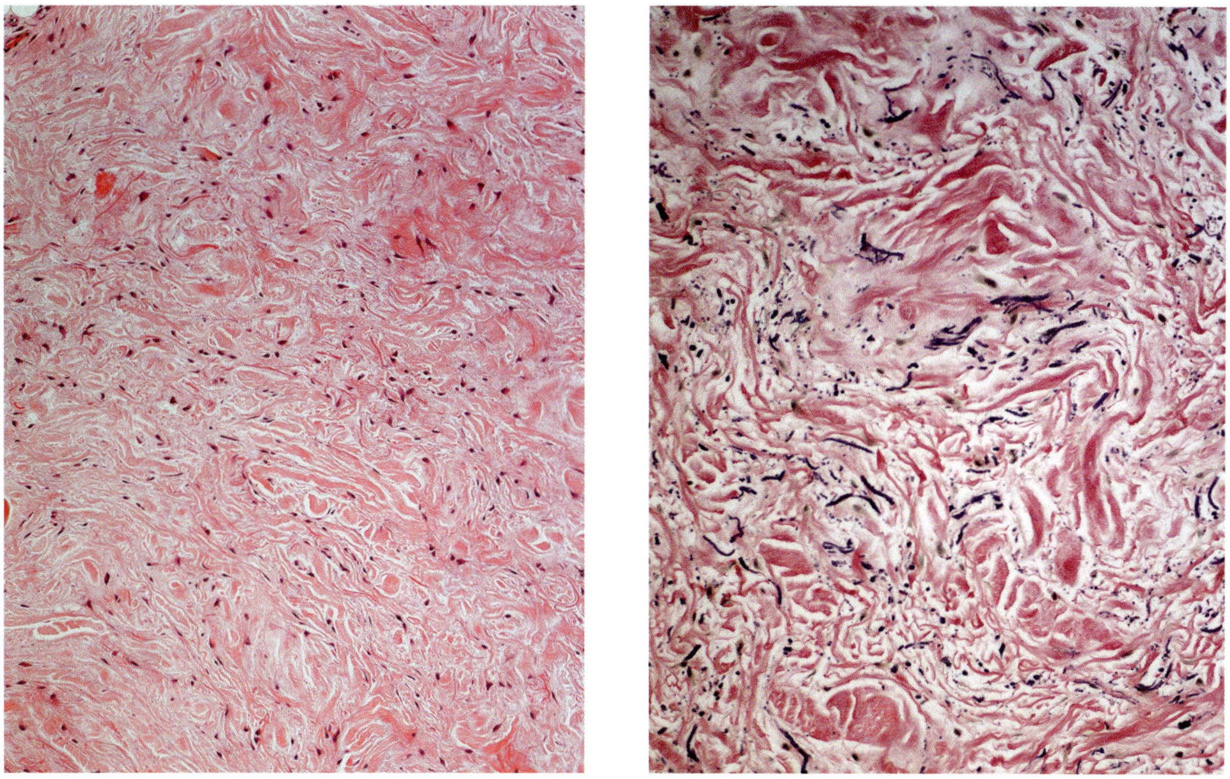

Figure 5-17

ELASTOFIBROMA

An elastofibroma (left) consists of dense collagenous fibrosis within which are fragmented elastin fibres, highlighted with an EVG stain (right).

sarcoma, and this must be excluded before assigning origin from the serosal membranes. Somatic malignancy within a mediastinal teratoma also requires exclusion if a lesion involves either pericardium or mediastinal pleura (120). In some instances, it remains uncertain whether these tumors are arising from pleura, mediastinum, chest wall, or even lung, and several series group these tumors as "intrathoracic," not least because management is based more on the type of tumor than the exact site of origin (45,108,114).

Cases of primary *osteosarcoma* (121) and *chondrosarcoma* (122) have also been reported but, if there is no evidence of tumor origin from bone (fig. 5-19C), specifically, direct spread from rib to involve the pleura, then osteosarcomatous or chondrosarcomatous dedifferentiation of a mesothelioma cannot be excluded (123). In undifferentiated spindle cell neoplasms, any positive staining for keratins argues in favor of sarcomatoid mesothelioma over a sarcoma, not otherwise specified (NOS). Conversely, staining for SMA does not exclude a diagnosis of sarcomatoid mesothelioma, since up 60 percent of cases are positive. These issues are discussed in more detail in relation to the differential diagnosis of mesothelioma (see chapter 3). Pathologists need to be aware of the rapidly increasing number tumor-related and tumor-specific molecular abnormalities, specifically gene fusions, so that relevant investigations can be undertaken to confirm or refute diagnoses.

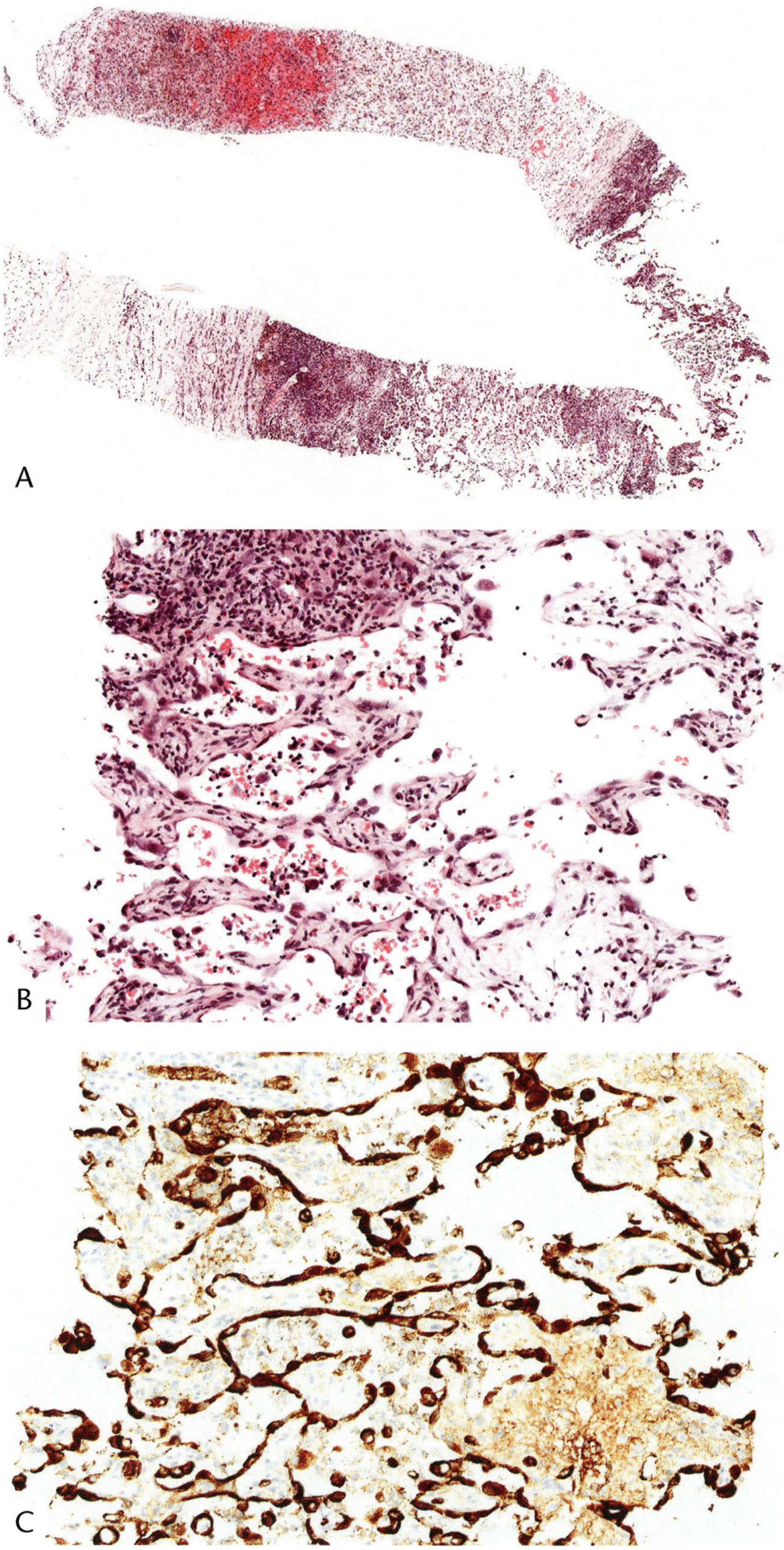

Figure 5-18

PRIMARY PLEURAL ANGIOSARCOMA

A,B: An angiosarcoma recapitulates vascular endothelium, with atypical cells lining irregular vascular spaces containing blood. Areas of fibrosis and hemorrhage are also present.

C: The tumor cells stain for factor VIII.

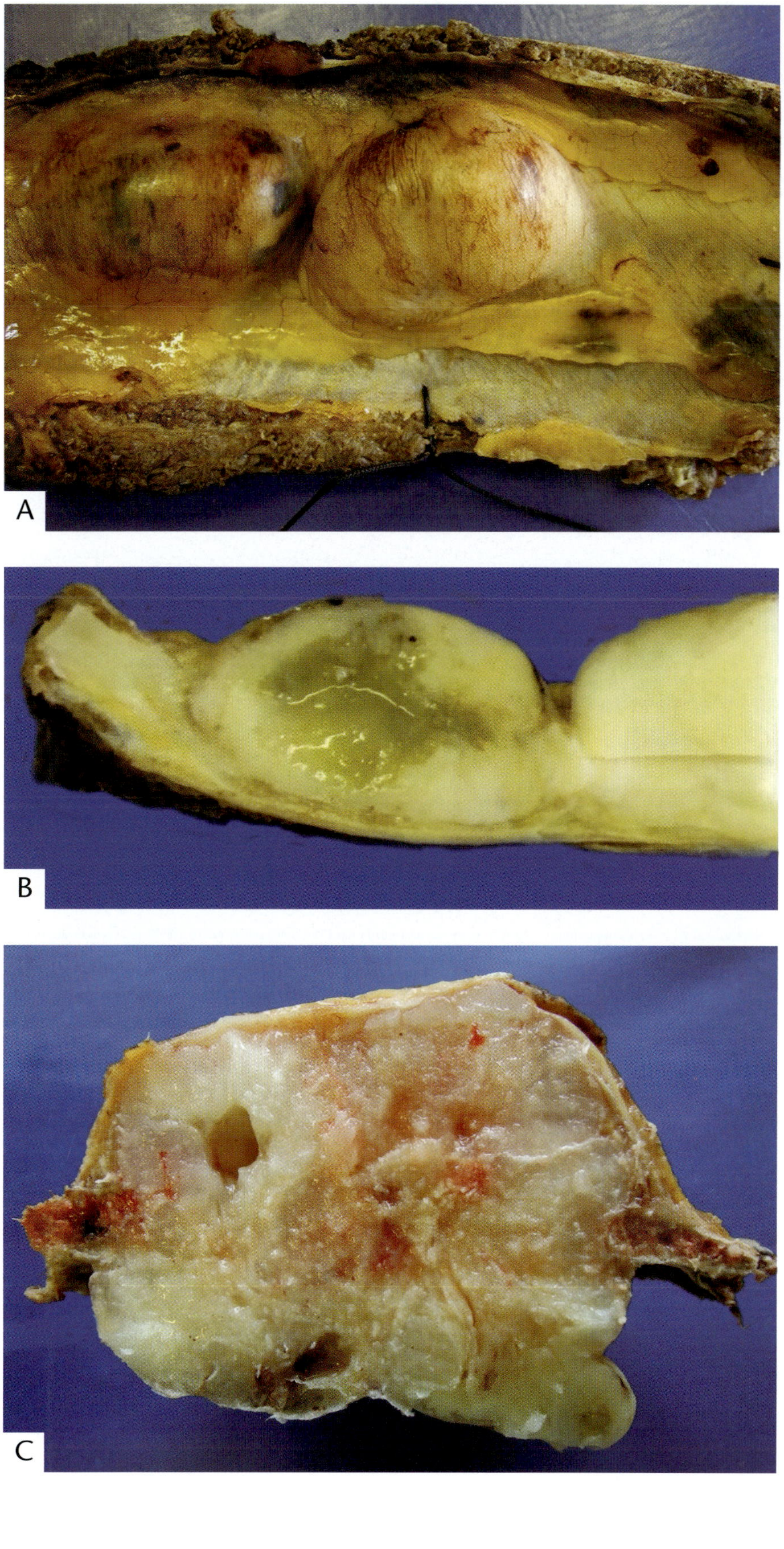

Figure 5-19

PRIMARY PLEURAL SARCOMA

A,B: A primary malignant peripheral nerve sheath tumor presents as a nodular mass between two ribs.

C: A primary chondrosarcoma arises from a rib and extends to involve the pleural space. If no origin from bone is seen in tumors containing bone or chondroid stroma, and especially if diffuse disease, then a mesothelioma with chondrosarcomatous or osteosarcomatous heterologous elements should be considered.

REFERENCES

1. WHO Classification of Tumours Editorial Board. Soft tissue and bone tumours, 5th ed. Lyon: IARC; 2020.
2. WHO Classification of Tumours Editorial Board. Thoracic tumours, 5th ed. Lyon: IARC; 2021.
3. Urschel JD, Brooks JS, Werness BA, Antkowiak JG, Takita H. Metachronous benign solitary fibrous tumours of the pleura (localized "mesotheliomas"): a case report. Can J Surg 1998;41:467-9.
4. Reuvers JR, van Dorp M, Van Schil PE. Solitary fibrous tumor of the pleura with associated Doege-Potter syndrome. Acta Chir Belg 2016; 116:386-7.
5. Goodlad JR, Fletcher CD. Solitary fibrous tumour arising at unusual sites: analysis of a series. Histopathology 1991;19:515-22.
6. Corgnati G, Drago S, Bonamini R, Trevi GP, Carra R, Di Summa M. Solitary fibrous tumor of the pericardium presenting itself as a pericardial effusion and right ventricular obstruction. J Cardiovasc Surg (Torino) 2004;45:393-4.
7. Andreani SM, Tavecchio L, Giardini R, Bedini AV. Extrapericardial solitary fibrous tumour of the pericardium. Eur J Cardiothorac Surg 1998;14:98-100.
8. Kouba E, Simper NB, Chen S, et al. Solitary fibrous tumour of the genitourinary tract: a clinicopathological study of 11 cases and their association with the NAB2-STAT6 fusion gene. J Clin Pathol 2017;70:508-14.
9. Cardinale L. CT imaging of solitary fibrous tumour of the pleura (SFTP): typical patterns and pitfalls. J Thorac Dis 2012;4:112-3.
10. Gupta A, Souza CA, Sekhon HS, et al. Solitary fibrous tumour of pleura: CT differentiation of benign and malignant types. Clin Radiol 2017;72:796.e9-796.e17.
11. England DM, Hochholzer L, McCarthy MJ. Localized benign and malignant fibrous tumors of the pleura. A clinicopathologic review of 223 cases. Am J Surg Pathol 1989;13:640-58.
12. Tani E, Wejde J, Astrom K, Wingmo IL, Larsson O, Haglund F. FNA cytology of solitary fibrous tumors and the diagnostic value of STAT6 immunocytochemistry. Cancer Cytopathol 2018;126:36-43.
13. Weynand B, Collard P, Galant C. Cytopathological features of solitary fibrous tumor of the pleura: a study of 5 cases. Diagn Cytopathol 1998;18:118-24.
14. Flint A, Weiss SW. CD-34 and keratin expression distinguishes solitary fibrous tumor (fibrous mesothelioma) of pleura from desmoplastic mesothelioma. Hum Pathol 1995;26:428-31.
15. Ordonez NG. Localized (solitary) fibrous tumor of the pleura. Adv Anat Pathol 2000;7:327-40.
16. Han Y, Zhang Q, Yu X, et al. Immunohistochemical detection of STAT6, CD34, CD99 and BCL-2 for diagnosing solitary fibrous tumors/hemangiopericytomas. Int J Clin Exp Pathol 2015;8:13166-75.
17. Yokoi T, Tsuzuki T, Yatabe Y, et al. Solitary fibrous tumour: significance of p53 and CD34 immunoreactivity in its malignant transformation. Histopathology 1998;32:423-32.
18. Cheah AL, Billings SD, Goldblum JR, Carver P, Tanas MZ, Rubin BP. STAT6 rabbit monoclonal antibody is a robust diagnostic tool for the distinction of solitary fibrous tumour from its mimics. Pathology 2014;46:389-5.
19. Koelsche C, Schweizer L, Renner M, et al. Nuclear relocation of STAT6 reliably predicts NAB2-STAT6 fusion for the diagnosis of solitary fibrous tumour. Histopathology 2014;65:613-22.
20. Yoshida A, Tsuta K, Ohno M, et al. STAT6 immunohistochemistry is helpful in the diagnosis of solitary fibrous tumors. Am J Surg Pathol 2014;38:552-9.
21. Doyle LA, Vivero M, Fletcher CD, Mertens F, Hornick JL. Nuclear expression of STAT6 distinguishes solitary fibrous tumor from histologic mimics. Mod Pathol 2014;27:390-5.
22. Guner G, Bishop JA, Bezerra SM, et al. The utility of STAT6 and ALDH1 expression in the differential diagnosis of solitary fibrous tumor versus prostate-specific stromal neoplasms. Hum Pathol 2016;54:184-8.
23. Rekhi B, Shetty O, Tripathi P, et al. Molecular characterization of a series of solitary fibrous tumors, including immunohistochemical expression of STAT6 and NATB2-STAT6 fusion transcripts, using Reverse Transcriptase(RT)-Polymerase chain reaction(PCR) technique: an Indian experience. Pathol Res Pract 2017;213:1404-11.
24. Guseva NV, Tanas MR, Stence AA, et al. The NAB2-STAT6 gene fusion in solitary fibrous tumor can be reliably detected by anchored multiplexed PCR for targeted next-generation sequencing. Cancer Genet 2016;209:303-12.
25. Mohajeri A, Tayebwa J, Collin A, et al. Comprehensive genetic analysis identifies a pathognomonic NAB2/STAT6 fusion gene, nonrandom secondary genomic imbalances, and a characteristic gene expression profile in solitary fibrous tumor. Genes Chromosomes Cancer 2013; 52:873-86.

26. Robinson DR, Wu YM, Kalyana-Sundaram S, et al. Identification of recurrent NAB2-STAT6 gene fusions in solitary fibrous tumor by integrative sequencing. Nat Genet 2013;45:180-5.
27. Chmielecki J, Crago AM, Rosenberg M, et al. Whole-exome sequencing identifies a recurrent NAB2-STAT6 fusion in solitary fibrous tumors. Nat Genet 2013;45:131-2.
28. Krismann M, Adams H, Jaworska M, Muller KM, Johnen G. Patterns of chromosomal imbalances in benign solitary fibrous tumours of the pleura. Virchows Arch 2000;437:248-55.
29. Donner LR, Silva MT, Dobin SM. Solitary fibrous tumor of the pleura: a cytogenetic study. Cancer Genet Cytogenet 1999;111:169-71.
30. Hajdu M, Singer S, Maki RG, Schwartz GK, Keohan ML, Antonescu CR. IGF2 over-expression in solitary fibrous tumours is independent of anatomical location and is related to loss of imprinting. J Pathol 2010;221:300-7.
31. Ge W, Yu DC, Chen G, Ding YT. Clinical analysis of 47 cases of solitary fibrous tumor. Oncol Lett 2016;12:2475-80.
32. de Perrot M, Fischer S, Brundler MA, Sekine Y, Keshavjee S. Solitary fibrous tumors of the pleura. Ann Thorac Surg 2002;74:285-93.
33. Schirosi L, Lantuejoul S, Cavazza A, et al. Pleuro-pulmonary solitary fibrous tumors: a clinicopathologic, immunohistochemical, and molecular study of 88 cases confirming the prognostic value of de Perrot staging system and p53 expression, and evaluating the role of c-kit, BRAF, PDGFRs (alpha/beta), c-met, and EGFR. Am J Surg Pathol 2008;32:1627-42.
34. Bellini A, Marulli G, Breda C, et al. Predictors of behaviour in solitary fibrous tumours of the pleura surgically resected: analysis of 107 patients. J Surg Oncol 2019;120:761-7.
35. Bylicki O, Rouviere D, Cassier P, et al. Assessing the multimodal management of advanced solitary fibrous tumors of the pleura in a routine practice setting. J Thorac Oncol 2015;10:309-15.
36. Tapias LF, Mercier O, Ghigna MR, et al. Validation of a scoring system to predict recurrence of resected solitary fibrous tumors of the pleura. Chest 2015;147:216-23.
37. Tapias LF, Mino-Kenudson M, Lee H, et al. Risk factor analysis for the recurrence of resected solitary fibrous tumours of the pleura: a 33-year experience and proposal for a scoring system. Eur J Cardiothorac Surg 2013;44:111-7.
38. Demicco EG, Park MS, Araujo DM, et al. Solitary fibrous tumor: a clinicopathological study of 110 cases and proposed risk assessment model. Mod Pathol 2012;25:1298-306.
39. Demicco EG, Wagner MJ, Maki RG, et al. Risk assessment in solitary fibrous tumors: validation and refinement of a risk stratification model. Mod Pathol 2017;30:1433-42.
40. Demicco EG, Griffin AM, Gladdy RA, et al. Comparison of published risk models for prediction of outcome in patients with extrameningeal solitary fibrous tumour. Histopathology 2019;75:723-37.
41. Diebold M, Soltermann A, Hottinger S, et al. Prognostic value of MIB-1 proliferation index in solitary fibrous tumors of the pleura implemented in a new score—a multicenter study. Respir Res 2017;18:210.
42. Franzen D, Diebold M, Soltermann A, et al. Determinants of outcome of solitary fibrous tumors of the pleura: an observational cohort study. BMC Pulm Med 2014;14:138.
43. Barak S, Wang Z, Miettinen M. Immunoreactivity for calretinin and keratins in desmoid fibromatosis and other myofibroblastic tumors: a diagnostic pitfall. Am J Surg Pathol 2012;36:1404-9.
44. Jiang J, Li Y, Hu C, et al. Primary pleural intermediate hemangioendothelioma with pleural effusion as the only manifestation. J Thorac Dis 2016;8:E369-73.
45. Anderson T, Zhang L, Hameed M, Rusch V, Travis WD, Antonescu CR. Thoracic epithelioid malignant vascular tumors: a clinicopathologic study of 52 cases with emphasis on pathologic grading and molecular studies of WWTR1-CAMTA1 fusions. Am J Surg Pathol 2015;39:132-9.
46. Crotty EJ, McAdams HP, Erasmus JJ, Sporn TA, Roggli VL. Epithelioid hemangioendothelioma of the pleura: clinical and radiologic features. AJR Am J Roentgenol 2000;175:1545-9.
47. Fan Y, Wang F, Li S, Ye C, Ying Y, Mao H. Pleural epithelioid hemangioendothelioma: a case report and literature review. J Natl Med Assoc 2016;108:124-9.
48. Folpe AL, Chand EM, Goldblum JR, Weiss SW. Expression of Fli-1, a nuclear transcription factor, distinguishes vascular neoplasms from potential mimics. Am J Surg Pathol 2001;25:1061-6.
49. Lin BT, Colby T, Gown AM, et al. Malignant vascular tumors of the serous membranes mimicking mesothelioma. A report of 14 cases. Am J Surg Pathol 1996;20:1431-9.
50. Flucke U, Vogels RJ, de Saint Aubain Somerhausen N, et al. Epithelioid hemangioendothelioma: clinicopathologic, immunhistochemical, and molecular genetic analysis of 39 cases. Diagn Pathol 2014;9:131.
51. Doyle LA, Fletcher CD, Hornick JL. Nuclear expression of CAMTA1 distinguishes epithelioid hemangioendothelioma from histologic mimics. Am J Surg Pathol 2016;40:94-102.

52. Shibuya R, Matsuyama A, Shiba E, Harada H, Yabuki K, Hisaoka M. CAMTA1 is a useful immunohistochemical marker for diagnosing epithelioid haemangioendothelioma. Histopathology 2015;67:827-35.
53. Ha SY, Choi IH, Han J, et al. Pleural epithelioid hemangioendothelioma harboring CAMTA1 rearrangement. Lung Cancer 2014;83:411-5.
54. Errani C, Zhang L, Sung YS, et al. A novel WWTR1-CAMTA1 gene fusion is a consistent abnormality in epithelioid hemangioendothelioma of different anatomic sites. Genes Chromosomes Cancer 2011;50:644-53.
55. Antonescu CR, Le Loarer F, Mosquera JM, et al. Novel YAP1-TFE3 fusion defines a distinct subset of epithelioid hemangioendothelioma. Genes Chromosomes Cancer 2013;52:775-84.
56. Kanemura S, Kuribayashi K, Moriya Y, Shimizu S, Tsujimura T, Nakano T. Pemetrexed for epithelioid haemangioendothelioma of the pleura. Respirol Case Rep 2016;4:e00191.
57. Klabatsa A, Nicholson AG, Dulay K, Rudd RM, Sheaff MT. Diffuse pleural mesothelioma with epithelioid and angiosarcomatous components—a hitherto undescribed pattern of differentiation. Histopathology 2012;60:1164-6.
58. Begueret H, Galateau-Salle F, Guillou L, et al. Primary intrathoracic synovial sarcoma: a clinicopathologic study of 40 t(X;18)-positive cases from the French Sarcoma Group and the Mesopath Group. Am J Surg Pathol 2005;29:339-46.
59. Colwell AS, D'Cunha J, Vargas SO, Parker B, Dal Cin P, Maddaus MA. Synovial sarcoma of the pleura: a clinical and pathologic study of three cases. J Thorac Cardiovasc Surg 2002;124:828-32.
60. Aubry MC, Bridge JA, Wickert R, Tazelaar HD. Primary monophasic synovial sarcoma of the pleura: five cases confirmed by the presence of SYT-SSX fusion transcript. Am J Surg Pathol 2001;25:776-81.
61. Nicholson AG, Goldstraw P, Fisher C. Synovial sarcoma of the pleura and its differentiation from other primary pleural tumours: a clinicopathological and immunohistochemical review of three cases. Histopathology 1998;33:508-13.
62. Gaertner E, Zeren EH, Fleming MV, Colby TV, Travis WD. Biphasic synovial sarcomas arising in the pleural cavity. A clinicopathologic study of five cases. Am J Surg Pathol 1996;20:36-45.
63. Sharif A, Akhtar T, Akhtar M, Zia N. Cystic pleural synovial sarcoma. J Coll Physicians Surg Pak 2016;26:127-9.
64. Tajima S, Takahashi T, Itaya T, Koda K, Neyatani H. Cystic synovial sarcoma of the pleura mimicking a cystic thymoma: a case report illustrating the role of decreased INI-1 expression in differential diagnosis. Int J Clin Exp Pathol 2015;8:3262-9.
65. Cummings NM, Desai S, Thway K, et al. Cystic primary pulmonary synovial sarcoma presenting as recurrent pneumothorax: report of 4 cases. Am J Surg Pathol 2010;34:1176-9.
66. Lan T, Chen H, Xiong B, et al. Primary pleuropulmonary and mediastinal synovial sarcoma: a clinicopathologic and molecular study of 26 genetically confirmed cases in the largest institution of southwest China. Diagn Pathol 2016;11:62.
67. Miettinen M, Limon J, Niezabitowski A, Lasota J. Calretinin and other mesothelioma markers in synovial sarcoma: analysis of antigenic similarities and differences with malignant mesothelioma. Am J Surg Pathol 2001;25:610-7.
68. Fisher C, Montgomery E, Healy V. Calponin and h-caldesmon expression in synovial sarcoma; the use of calponin in diagnosis. Histopathology 2003;42:588-93.
69. Fisher C, Schofield JB. S-100 protein positive synovial sarcoma. Histopathology 1991;19:375-7.
70. Lino-Silva LS, Flores-Gutierrez JP, Vilches-Cisneros N, Dominguez-Malagon HR. TLE1 is expressed in the majority of primary pleuropulmonary synovial sarcomas. Virchows Arch 2011;459:615-21.
71. Matsuyama A, Hisaoka M, Iwasaki M, Iwashita M, Hisanaga S, Hashimoto H. TLE1 expression in malignant mesothelioma. Virchows Arch 2010; 457:577-83.
72. Baranov E, McBride MJ, Bellizzi AM, et al. A novel SS18-SSX Fusion-specific antibody for the diagnosis of synovial sarcoma. Am J Surg Pathol 2020;44:922-33.
73. Spurrell EL, Fisher C, Thomas JM, Judson IR. Prognostic factors in advanced synovial sarcoma: an analysis of 104 patients treated at the Royal Marsden Hospital. Ann Oncol 2005;16:437-44.
74. Klebe S, Prabhakaran S, Hocking A, et al. Pleural malignant mesothelioma versus pleuropulmonary synovial sarcoma: a clinicopathological study of 22 cases with molecular analysis and survival data. Pathology 2018;50:629-34.
75. Weinbreck N, Vignaud JM, Begueret H, et al. SYT-SSX fusion is absent in sarcomatoid mesothelioma allowing its distinction from synovial sarcoma of the pleura. Mod Pathol 2007;20:617-21.
76. Terra SB, Aubry MC, Yi ES, Boland JM. Immunohistochemical study of 36 cases of pulmonary sarcomatoid carcinoma—ensitivity of TTF-1 is superior to napsin. Hum Pathol 2014;45:294-302.
77. Mochizuki T, Ishii G, Nagai K, et al. Pleomorphic carcinoma of the lung: clinicopathologic characteristics of 70 cases. Am J Surg Pathol 2008;32:1727-35.
78. Yousem SA, Wick MR, Randhawa P, Manivel JC. Pulmonary blastoma. An immunohistochemical analysis with comparison with fetal lung in its pseudoglandular stage. Am J Clin Pathol 1990; 93:167-75.

79. Fois AG, Pirina P, Arcadu A, et al. Desmoplastic small round cell tumors of the pleura: a review of the clinical literature. Multidiscip Respir Med 2017;12:22.
80. Parkash V, Gerald WL, Parma A, Miettinen M, Rosai J. Desmoplastic small round cell tumor of the pleura. Am J Surg Pathol 1995;19:659-65.
81. Karavitakis EM, Moschovi M, Stefanaki K, et al. Desmoplastic small round cell tumor of the pleura. Pediatr Blood Cancer 2007;49:335-8.
82. Syed S, Haque AK, Hawkins HK, Sorensen PH, Cowan DF. Desmoplastic small round cell tumor of the lung. Arch Pathol Lab Med 2002;126:1226-8.
83. Gerald WL, Ladanyi M, de Alava E, et al. Clinical, pathologic, and molecular spectrum of tumors associated with t(11;22)(p13;q12): desmoplastic small round-cell tumor and its variants. J Clin Oncol 1998;16:3028-36.
84. Ladanyi M, Gerald WL. Specificity of the EWS/WT1 gene fusion for desmoplastic small round cell tumour. J Pathol 1996;180:462.
85. Ikeue T, Ohi I, Noguchi S, et al. Desmoplastic small round cell tumor of the pleura successfully treated with a lower dose of Pazopanib. Intern Med 2016;55:2463-7.
86. Laskar S, Nair C, Mallik S, et al. Prognostic factors and outcome in Askin-Rosai tumor: a review of 104 patients. Int J Radiat Oncol Biol Phys 2011; 79:202-7.
87. Mohandas KM, Chinoy RF, Merchant NH, Lotliker RG, Desai PB. Malignant small cell tumour (Askin-Rosai) of the pericardium. Postgrad Med J 1992;68:140-2.
88. Basharkhah A, Pansy J, Urban C, Hollwarth ME. Outcomes after interdisciplinary management of 7 patients with Askin tumor. Pediatr Surg Int 2013;29:431-6.
89. Sallustio G, Pirronti T, Lasorella A, Natale L, Bray A, Marano P. Diagnostic imaging of primitive neuroectodermal tumour of the chest wall (Askin tumour). Pediatr Radiol 1998;28:697-702.
90. Travis WD, Brambilla E, Burke AP, Marx A, Nicholson AG. WHO classification of tumours of the lung, pleura, thymus, and heart, 4th ed. Lyons: IARC; 2015.
91. Shibuya R, Matsuyama A, Nakamoto M, Shiba E, Kasai T, Hisaoka M. The combination of CD99 and NKX2.2, a transcriptional target of EWSR1-FLI1, is highly specific for the diagnosis of Ewing sarcoma. Virchows Arch 2014;465:599-605.
92. Tomlins SA, Palanisamy N, Brenner JC, et al. Usefulness of a monoclonal ERG/FLI1 antibody for immunohistochemical discrimination of Ewing family tumors. Am J Clin Pathol 2013;139:771-9.
93. Delattre O, Zucman J, Melot T, et al. The Ewing family of tumors—a subgroup of small-round-cell tumors defined by specific chimeric transcripts. N Engl J Med 1994;331:294-9.
94. Delattre O, Zucman J, Plougastel B, et al. Gene fusion with an ETS DNA-binding domain caused by chromosome translocation in human tumours. Nature 1992;359:162-5.
95. Lessnick SL, Ladanyi M. Molecular pathogenesis of Ewing sarcoma: new therapeutic and transcriptional targets. Annu Rev Pathol 2012;7:145-59.
96. Zhang KE, Lu R, Zhang P, Shen S, Li X. Askin's tumor: 11 cases and a review of the literature. Oncol Lett 2016;11:253-6.
97. Crago AM, Chmielecki J, Rosenberg M, et al. Near universal detection of alterations in CTNNB1 and Wnt pathway regulators in desmoid-type fibromatosis by whole-exome sequencing and genomic analysis. Genes Chromosomes Cancer 2015;54:606-15.
98. Tokarek T, Szpor J, Pankowski J, Okon K. Desmoid tumor of lung with pleural involvement—the case of unique location of aggressive fibromatosis. Folia Med Cracov 2015;55:53-9.
99. Andino L, Cagle PT, Murer B, et al. Pleuropulmonary desmoid tumors: immunohistochemical comparison with solitary fibrous tumors and assessment of beta-catenin and cyclin D1 expression. Arch Pathol Lab Med 2006;130:1503-9.
100. Kasper B, Gruenwald V, Reichardt P, Bauer S, Hohenberger P, Haller F. Correlation of ctnnb1 mutation status with progression arrest rate in RECIST progressive desmoid-type fibromatosis treated with imatinib: translational research results from a phase 2 study of the German Interdisciplinary Sarcoma Group (GISG-01). Ann Surg Oncol 2016;23:1924-7.
101. Gounder MM, Mahoney MR, Van Tine BA, et al. Sorafenib for advanced and refractory desmoid tumors. N Engl J Med 2018;379:2417-28.
102. Lee D, Suh YL, Lee SK. Calcifying fibrous pseudotumour arising in a gastric inflammatory myofibroblastic tumour. Pathology 2006;38:588-91.
103. Pomplun S, Goldstraw P, Davies SE, Burke MM, Nicholson AG. Calcifying fibrous pseudotumour arising within an inflammatory pseudotumour: evidence of progression from one lesion to the other? Histopathology 2000;37:380-2.
104. Mehrad M, LaFramboise WA, Lyons MA, Trejo Bittar HE, Yousem SA. Whole-exome sequencing identifies unique mutations and copy number losses in calcifying fibrous tumor of the pleura: report of 3 cases and review of the literature. Hum Pathol 2018;78:36-43.
105. Jiang K, Nie J, Wang J, Li J. Multiple calcifying fibrous pseudotumor of the bilateral pleura. Jpn J Clin Oncol 2011;41:130-3.

106. Erasmus JJ, McAdams HP, Patz EF Jr, Murray JG, Pinkard NB. Calcifying fibrous pseudotumor of pleura: radiologic features in three cases. J Comput Assist Tomogr 1996;20:763-5.
107. Pinkard NB, Wilson RW, Lawless N, et al. Calcifying fibrous pseudotumor of pleura. A report of three cases of a newly described entity involving the pleura. Am J Clin Pathol 1996;105:189-94.
108. Boland JM, Colby TV, Folpe AL. Intrathoracic peripheral nerve sheath tumors-a clinicopathological study of 75 cases. Hum Pathol 2015; 46:419-25.
109. Zhang S, Zheng Y, Liu W, Yu X. Primary epithelioid angiosarcoma of the pleura: a case report and review of literature. Int J Clin Exp Pathol 2015;8:2153-8.
110. Falconieri G, Bussani R, Mirra M, Zanella M. Pseudomesotheliomatous angiosarcoma: a pleuropulmonary lesion simulating malignant pleural mesothelioma. Histopathology 1997; 30:419-24.
111. McCaughey WT, Dardick I, Barr JR. Angiosarcoma of serous membranes. Arch Pathol Lab Med 1983;107:304-7.
112. Rais G, Raissouni S, Mouzount H, et al. Primary pleural leiomyosarcoma with rapid progression and fatal outcome: a case report. J Med Case Rep 2012;6:101.
113. Gibbs AR. Smooth muscle tumours of the pleura. Histopathology 1995;27:295-6.
114. Chen M, Yang J, Zhu L, Zhou C, Zhao H. Primary intrathoracic liposarcoma: a clinicopathologic study and prognostic analysis of 23 cases. J Cardiothorac Surg 2014;9:119.
115. Benchetritt M, Hofman V, Venissac N, et al. Dedifferentiated liposarcoma of the pleura mimicking a malignant solitary fibrous tumor and associated with dedifferentiated liposarcoma of the mediastinum: usefulness of cytogenetic and molecular genetic analyses. Cancer Genet Cytogenet 2007;179:150-5.
116. Goldsmith P, Papagiannopoulos K. Pleural myxoid liposarcoma: features of 2 cases and associated literature review. J Cardiothorac Surg 2007;2:48.
117. Okby NT, Travis WD. Liposarcoma of the pleural cavity: clinical and pathologic features of 4 cases with a review of the literature. Arch Pathol Lab Med 2000;124:699-703.
118. Ordonez NG, Tornos C. Malignant peripheral nerve sheath tumor of the pleura with epithelial and rhabdomyoblastic differentiation: report of a case clinically simulating mesothelioma. Am J Surg Pathol 1997;21:1515-21.
119. Joshi A, McAndrew N, Birdsall S, Attanoos RL. Clear cell sarcoma mimicking malignant pleural mesothelioma. Histopathology 2008;53:359-61.
120. Colecchia M, Necchi A, Paolini B, Nicolai N, Salvioni R. Teratoma with somatic-type malignant components in germ cell tumors of the testis: a clinicopathologic analysis of 40 cases with outcome correlation. Int J Surg Pathol 2011;19:321-7.
121. Rapicetta C, Lococo F, Carlinfante G, Ricchetti T, Paci M. Extraskeletal osteosarcoma arising from the pleura. Lung 2017;195:523-5.
122. Jain A, Safaya R, Jagan C, Sharma SK. Extraskeletal mesenchymal chondrosarcoma of the pleura: report of a rare case. Indian J Pathol Microbiol 2011;54:144-6.
123. Klebe S, Mahar A, Henderson DW, Roggli VL. Malignant mesothelioma with heterologous elements: clinicopathological correlation of 27 cases and literature review. Mod Pathol 2008;21: 1084-94.

6 LYMPHOPROLIFERATIVE NEOPLASMS

Primary lymphoproliferative neoplasms involving the serosal membranes are very rare. There are two main categories, *primary effusion lymphoma* and *diffuse large B-cell lymphoma associated with chronic inflammation.* Primary presentation of other lymphoma types, such as extranodal marginal zone lymphoma and follicular lymphoma, is exceptional (1,2), although secondary involvement of serosal membranes, often with accumulation of fluid, is much more common. Erdheim-Chester disease, a neoplastic systemic histiocytic disorder (3,4), is also briefly discussed.

PRIMARY EFFUSION LYMPHOMA

Definition. *Primary effusion lymphoma* (PEL) is a tumor consisting of large, atypical B-lymphoid cells that are nearly always positive for human herpesvirus 8 (HHV8), also called Kaposi sarcoma-associated herpesvirus (KSHV). PEL presents as an effusion in serous body cavities, although mass lesions may develop during the course of the disease. The term body cavity-based lymphoma has been used to classify cases with mass lesions in the body cavities, with little or no dissemination, although such cases are now included under the term primary effusion lymphoma. The term *primary effusion lymphoma-like lymphoma* is used for cases where there is an Epstein-Barr virus (EBV)-positive/HHV8-negative effusion lymphoma.

Clinical Features. Patients typically present with signs and symptoms of effusion. Symptoms are also related to the immunocompromised state, the latter being the norm and generally severe (with less than 200 CD4-positive T cells per μL) (5–10). Kaposi sarcoma, or more rarely, HHV8-associated multicentric Castleman disease, may coexist (8,11).

Most cases of PEL occur in human immunodeficiency virus (HIV)-positive patients. The majority of patients are HIV-positive men in their fourth to sixth decades who have sex with men. PEL accounts for 3 to 4 percent of HIV-associated lymphomas (10,12). It occasionally occurs in immunosuppressed organ transplant recipients (13,14), the elderly (15), and in areas of high HHV8 seroprevalence. Cases have also been reported in children with congenital immunodeficiency (16) and after therapy with tyrosine kinase inhibitors (17). Immunosenescence and possibly chronic liver disease are contributing factors (18–20).

Gross Findings. PELs are most common in the pleural cavity, followed by peritoneal and pericardial cavities. Involvement of multiple cavities may occur (21–23). Local or distant lymph node involvement or extranodal solid tissue involvement is seen in 30 to 40 percent of patients at presentation (10,21,22,24). Rarely, an effusion is absent; in these situations, the term *extracavitary primary effusion lymphoma* has been used (9,25–27).

Microscopic Findings. The tumor cells are pleomorphic, with large nuclei, one or more prominent nucleoli, and abundant variably staining cytoplasm (fig. 6-1). Features vary from immunoblastic, with more rounded and very large nuclei, to cells that are more plasmacytoid (fig. 6-2), with basophilic cytoplasm, eccentric nuclei, and a prominent Golgi zone. Features may also be mixed. Reed-Sternberg-like cells may be present. Mitoses are often seen. The features seen in tissue sections are generally similar to those in cytologic preparations, but the cells may appear less pleomorphic (21,28,29).

In solid tissue, which may take the form of cell blocks from effusions, biopsies from serosal tissues, or autopsy material, the appearances are similar, with the cells infiltrating diffusely within the native stroma. Necrosis may be present.

Immunohistochemical Findings. PELs almost always express CD45, and CD138 is usually positive, reflecting a plasmablastic phenotype, but surface and cytoplasmic immunoglobulins

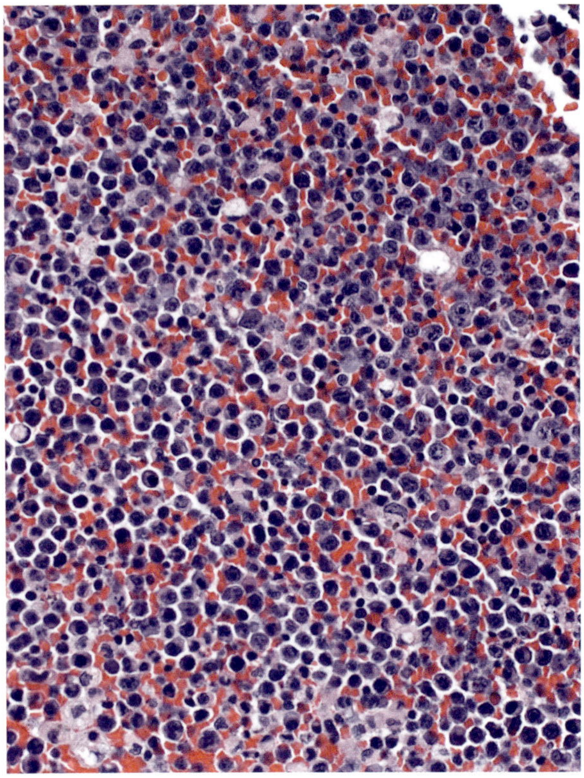

Figure 6-1

PRIMARY EFFUSION LYMPHOMA

Sheets of monotonous large blastic lymphoid cells are present within a cell block from a pleural effusion.

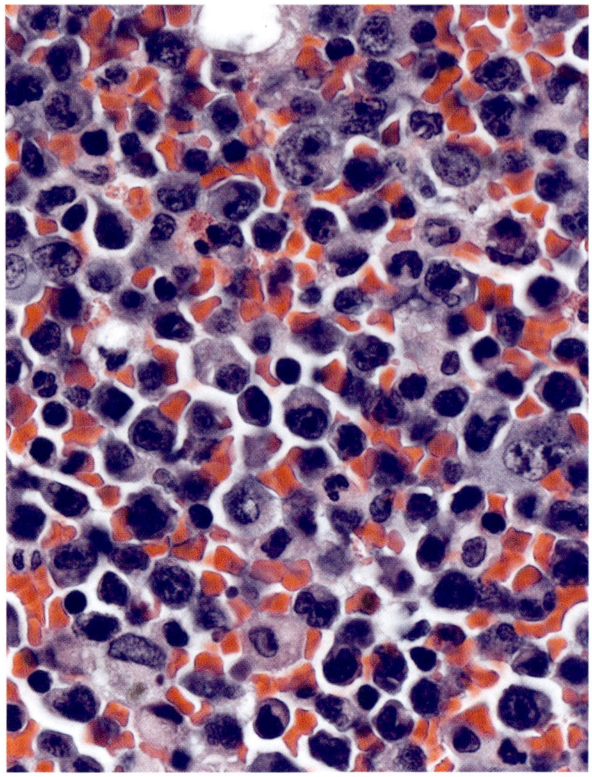

Figure 6-2

PRIMARY EFFUSION LYMPHOMA

The cells have prominent single nucleoli and focally show plasmacytoid features.

are rarely present. CD30, CD38, and epithelial membrane antigen (EMA) are expressed in most cases. Pan-B-cell antigens such as CD19, CD20, CD79a, and PAX5 are usually negative, as is BCL6. IRF4/MUM1 and PRDM1/BLIMP1 may be expressed. Pan-T-cell antigens, either with or without immunophenotypic or molecular evidence of B-cell lineage, are expressed in a small number of cases (9,12,30,31). Rare cases express cytokeratins (32,33). Independent of lineage, nuclear expression of latency-associated nuclear protein (LANA) is present in all cases (fig. 6-3) (34). EBV may be detected by in situ hybridization for EBV-encoded small RNA, but LMP1 is immunohistochemically undetectable.

Molecular Genetic Findings. By definition, cases of PEL are HHV8 positive, with most tumor cells containing 40 to 80 monoclonal/oligoclonal episomal viral genomes and displaying a latency-associated gene expression pattern (9,35–37). LANA and viral microRNAs play diverse roles in lymphomagenesis (38,39). A few tumor cells express lytic-phase genes such as *vIL6* and *vGPCR*, which may also have important oncogenic roles, perhaps partly in a paracrine manner (9,40,41).

Approximately 80 percent of HIV-associated cases display monoclonal coinfection with EBV, although EBV is usually lacking in PELs that occur in HIV-negative patients (9,23,28). EBV exhibits a restricted type I latency-like gene expression pattern, with expression of EBNA1 and EBV-encoded small RNA, but little or no expression of LMP1 or LMP2A (42). EBV and HHV8 may cooperate in the infection and initial transformation of B cells (43,44).

Immunoglobulin genes are nearly always clonally rearranged (9,28). Occasional cases show both immunoglobulin and T-cell receptor gene rearrangements (9,30,31), and rare HHV8-

positive cases with a T-cell immunoprofile and only T-cell receptor gene rearrangement have been reported (45–47). The somatic mutation pattern suggests derivation from antigen-selected postgerminal center B cells, and gene expression profiling suggests plasmablastic/incomplete plasma cell differentiation (48,49).

PELs typically show clonal cytogenetic heterogeneity, with numerical abnormalities that include trisomies of chromosomes 7, 8, and 12 (50). Array-based analyses have revealed numerous regions and several discrete genes showing recurrent genomic copy number gain or loss, including frequent loss of fragile site candidate tumor suppressor genes in PEL cell lines (51,52). EBV-positive PEL cell lines show fewer copy number changes than EBV-negative lines, consistent with a cooperative role for EBV in lymphomagenesis (51).

c-MET is highly activated in PEL although mutations are not found (53). MALT1 activation may also play a role in tumor development (54).

Differential Diagnosis. PELs must be distinguished from effusions secondary to other diffuse large B-cell lymphomas. The presence of lymphomatous cells within an effusion should prompt investigation for a previous or undiagnosed lymphoma. PELs are readily distinguished from most other lymphoid neoplasms by HHV8 positivity. Other types of lymphoma originating in serosal membranes are rare, although diffuse large B-cell lymphomas associated with chronic inflammation, specifically pyothorax-associated lymphoma, may present as masses in the pleura or peripheral lung. These are typically EBV positive but are HHV8 negative (55). They also rarely present with effusions. Burkitt lymphoma may present with effusions but again is HHV8 negative.

HHV8-negative effusion-based diffuse large B-cell lymphoma typically expresses pan-B-cell antigens and occurs mostly in elderly HIV-negative individuals. These tumors may be associated with fluid overload states (23,56,57).

HHV8-associated multicentric Castleman disease and the Kaposi sarcoma-associated herpesvirus inflammatory cytokine syndrome associated with severe Kaposi sarcoma and high viral loads may cause effusions. However, there is a lack of atypical lymphoid cells seen in PEL (58).

Treatment and Prognosis. Patients may be treated with anthracycline-containing chemotherapy, and if HIV positive, with highly active antiretroviral therapy (HAART). Some patients achieve durable remission (10,22,24,56). Occasional responses to HAART alone, or in HIV-negative patients, to chemical pleurodesis or intracavitary cidofovir have been reported (5–8,18,59,60). Complete response rates of 25 to 50 percent and 1- to 2-year overall survival rates of 20 to 40 percent are reported; the median overall survival period is about 6 months. In a recent review of Surveillance, Epidemiology, and End Results (SEER) data, the 1-year overall survival rate was 25 percent in advanced stages compared with 52 percent in early stages (6). In this review, Ann Arbor staging showed that 65 percent of the patients had advanced stages (stages III and IV) at the time of diagnosis. Adverse prognostic factors include absence of HAART before PEL diagnosis in HIV-positive patients, disease involving more than one cavity, and

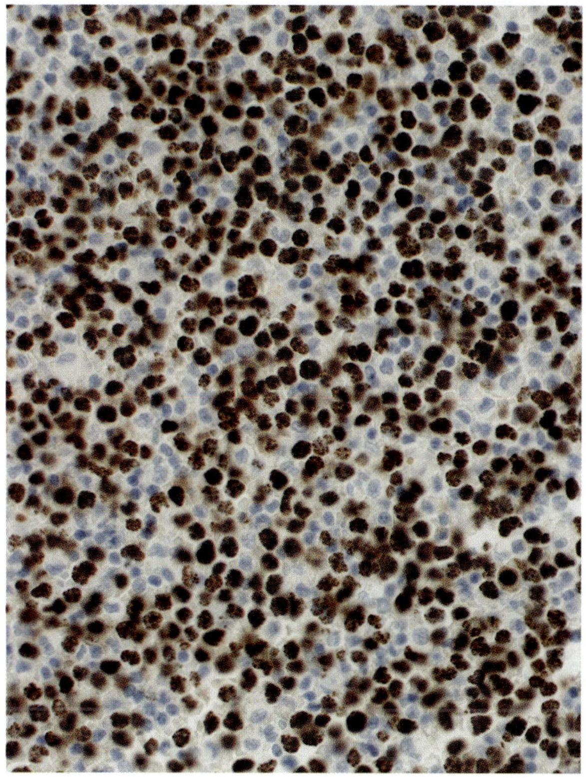

Figure 6-3

PRIMARY EFFUSION LYMPHOMA

Tumor cells are positive for HHV8 latency-associated nuclear protein (LANA).

poor performance status (22,24). In PEL-like lymphomas, old age and low levels of lactate dehydrogenase are favorable features (15).

DIFFUSE LARGE B-CELL LYMPHOMA ASSOCIATED WITH CHRONIC INFLAMMATION

Definition. *Diffuse large B-cell lymphoma associated with chronic inflammation* is an EBV-associated B-cell neoplasm that usually presents in body cavities, most commonly the pleural cavity. It occurs in the context of longstanding chronic inflammation. The term *pyothorax-associated lymphoma* may be used.

Clinical Features. Most patients have pyothorax for several decades, often the result of treatment of pulmonary tuberculosis or tuberculous pleurisy, but other causes such as trauma have been reported. Most cases have been reported in the Japanese literature. Presentation usually relates to the presence of a mass, with symptoms including chest pain and cough in relation to, for example, the pleural cavity (61). Effusions may be present, although tumor cells may not be identifiable. Imaging shows mass lesions, sometimes with infiltration of adjacent structures such as lung and chest wall in relation to the pleural cavity (61). Disease tends to be more localized than other forms of diffuse large B-cell lymphoma, and lymph node involvement is uncommon.

Gross Findings. The tumors are mass lesions that are generally solid and firm. Necrosis, intratumoral hemorrhage, and fibrous pseudocapsules are all reported.

Microscopic Findings. Morphologically, the tumor cells are similar to those seen in other diffuse large B-cell lymphomas, ranging from central blasts to immunoblasts (fig. 6-4A). Necrosis and individual cell apoptosis are prominent. There may be marked associated fibrosis (fig. 6-4B), likely reflecting the longstanding associated chronic inflammatory process.

Immunohistochemical Findings. The cells express a mature B-cell phenotype, with positivity for CD20, CD19, and CD79a (fig. 6-4C). CD30 is often expressed, but CD15 is negative. Some cases show evidence of plasmacytic differentiation with positive staining for CD138 and loss of CD20. IRF4/MUM1 is positive in nearly all cases. Aberrant expression of T-cell-associated antigens may be seen (62), although a complete T-cell phenotype is absent. In the presence of EBV (fig. 6-4D), tumor cells express a type III latency phenotype with positivity for LMP1, EBNA1, and EBNA2 (63).

Molecular Genetic Findings. EBV sequences are detected by polymerase chain reaction (PCR); in situ hybridization for EBV-encoded small RNA is positive. The immunoglobulin genes show clonal rearrangement, and gene expression profiling shows a postgerminal center pattern (64–66).

Differential Diagnosis. PELs also show EBV positivity, but they are also HHV8 positive. Also, PELs generally lack a mass lesion, and do not express pan-B-cell antigens. Secondary involvement of serosal membranes by diffuse large B-cell lymphoma is usually identifiable from the clinical history, and most are negative for EBV. In cases involving the pleura, the differential diagnosis includes primary pulmonary diffuse large B-cell lymphoma, which again is usually negative for EBV. Lymphomatoid granulomatosis is positive for EBV, but is predominantly intrapulmonary and shows striking angiocentricity (67–69).

Treatment and Prognosis. The 5-year survival rate is 50 percent for those who achieve complete remission with chemotherapy and/or radiotherapy. Response to rituximab has been reported (70). Complete surgical resection has been reported to provide prolonged remission. Nevertheless, the 5-year overall survival rate ranges from 20 to 35 percent. Adverse prognostic factors include poor performance status, high serum levels of lactate dehydrogenase, and advanced clinical stage (based on the Ann Arbor staging system employed for other forms of lymphomas) (61).

ERDHEIM-CHESTER DISEASE

Definition. *Erdheim-Chester disease* is a systemic histiocytic disorder typically involving the skeleton and viscera.

Clinical Features. Erdheim-Chester disease involves the lungs and pleura in 20 to 30 percent of patients, and is slightly more common in males, with a peak age of 50 to 70 years. In patients with thoracic involvement, pleural thickening and effusions are commonly seen (71–74).

Gross Findings. The pleural surface is firm and variably thickened. There is nearly always coexistent lung involvement.

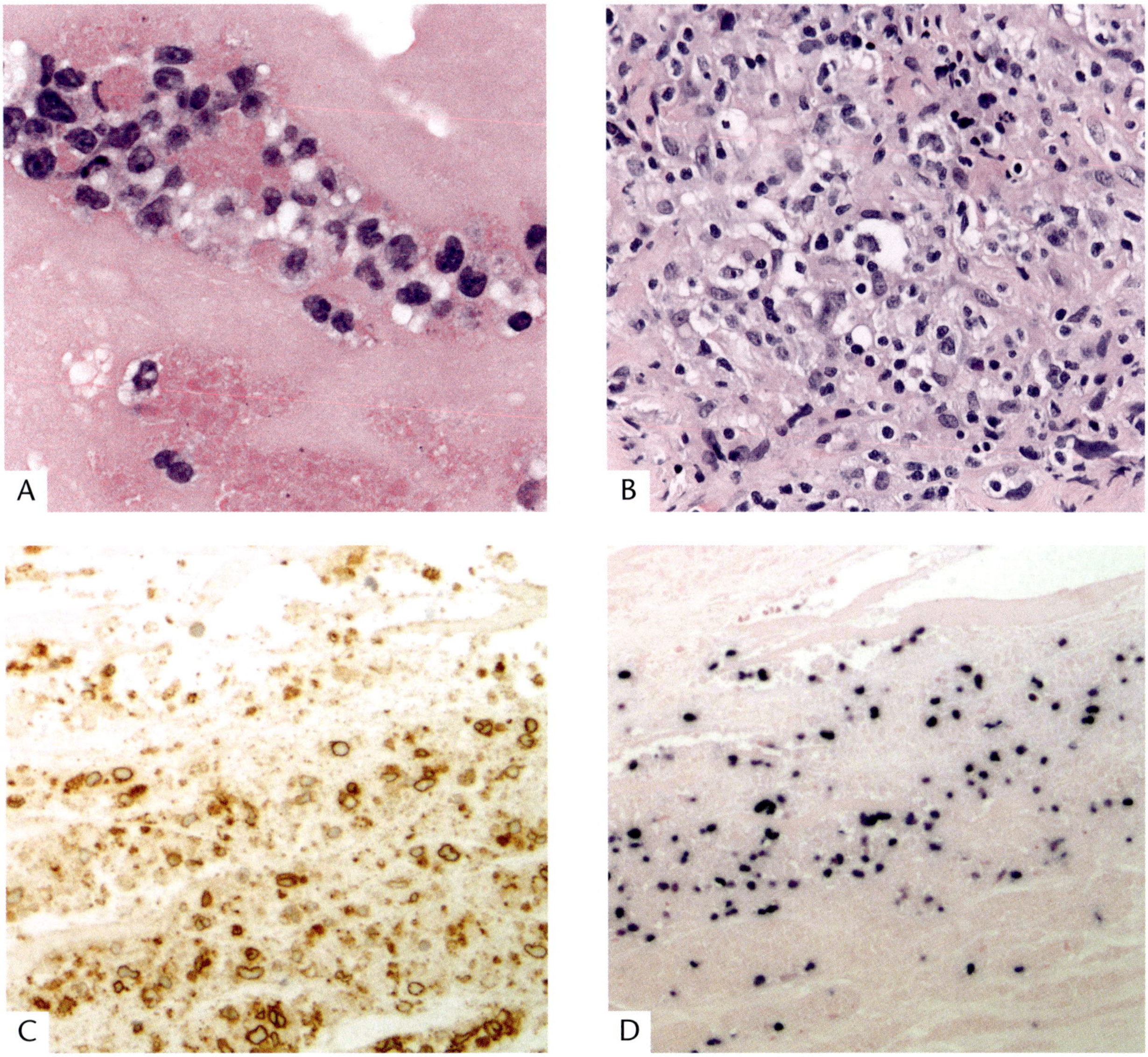

Figure 6-4

DIFFUSE LARGE B-CELL LYMPHOMA ASSOCIATED WITH CHRONIC INFLAMMATION

A: Clusters of large dyscohesive cells are present within a pleural biopsy. Necrosis is seen.
B: A separate case shows large dyscohesive cells within fibrous pleura.
C,D: The tumor cells are positive for CD79a (C), and are also positive for Epstein-Barr virus (EBV) (EBER stain) (D).

Microscopic Findings. Histiocytic infiltration and fibrosis predominate along the distribution of the pulmonary lymphatics (visceral pleura, bronchovascular bundles, and interlobular septa) (fig. 6-5A,B). Histiocytes are often foamy, with Touton giant cells often seen (fig. 6-5C), and associated with variably dense fibrosis, lymphocytes, plasma cells, and eosinophils.

Immunohistochemical Findings. The lesional cells stain for histiocytic markers such as CD68. Staining for S-100 protein is variable (fig. 6-5D). Cells may also stain for factor XIIIa (75). They do not stain for CD1a. BRAF immunohistochemistry may be positive in cases with *BRAF* mutations (3).

Molecular Genetic Findings. *BRAF* mutations are seen in 50 to 60 percent of cases

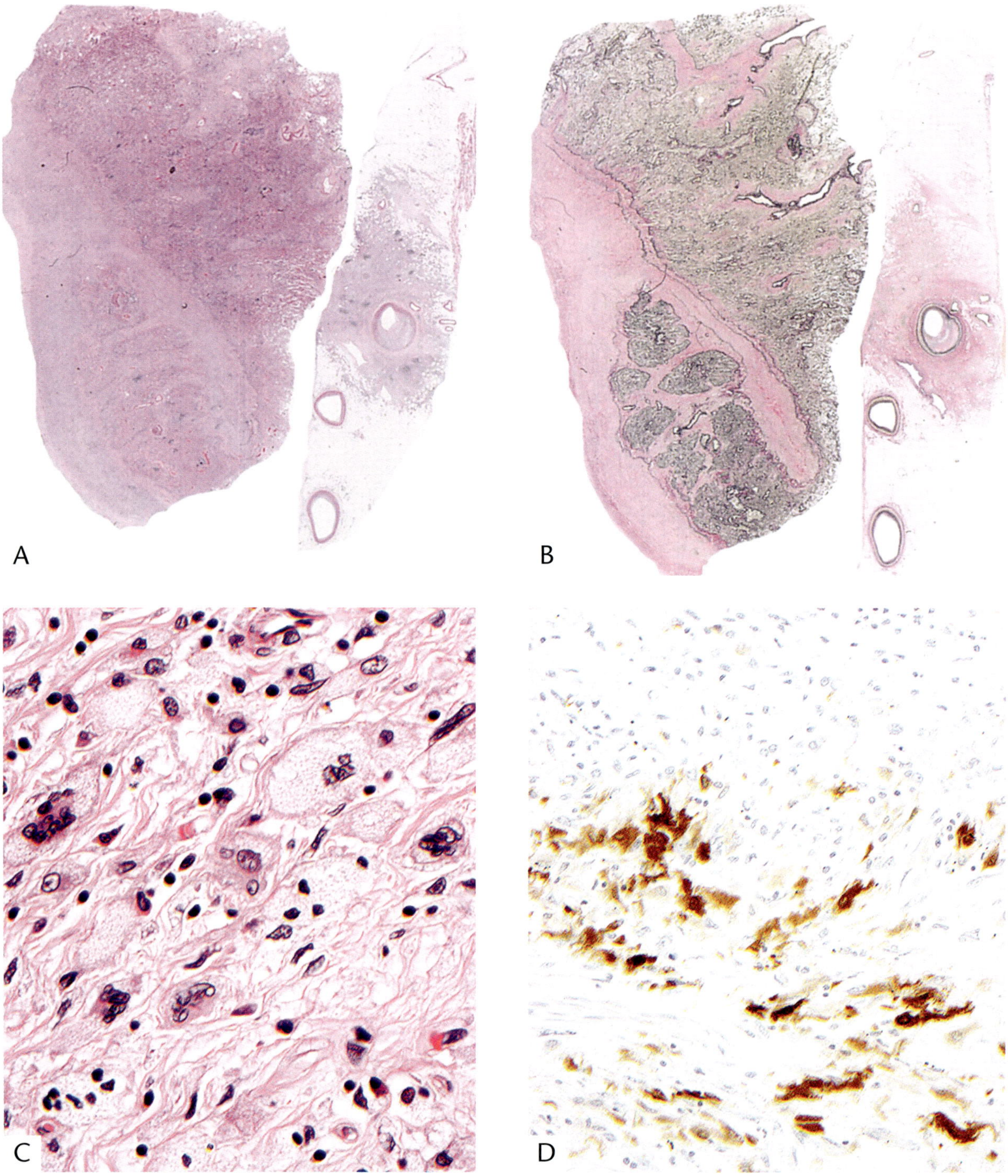

Figure 6-5

ERDHEIM-CHESTER DISEASE INVOLVING SEROSAL MEMBRANES

A,B: At autopsy, there is marked fibrous thickening of the pleura, as well as involvement of the pericardium around a coronary artery (hematoxylin and eosin [H&E] [A] and EVG [B] staining).

C: High power shows foamy histiocytic cells within fibrous stroma.

D: These cells stain focally for S-100 protein.

(3,76,77) and clonality has also been reported (78,79). Somatic mutations involving *MAP2K1* (15 percent), *ARAF* (5 percent), *MAP2K2* (2.5 percent), *KRAS* (2.5 percent), and *NRAS* (2.5 percent) have been detected in one series, along with 65 percent showing *BRAF* mutations (3).

Differential Diagnosis. In the thorax, Erdheim-Chester disease may mimic other neoplasms, in particular, other lymphoproliferative diseases, lymphangitis carcinomatosis (80), and even mesothelioma due to its prominent pleural involvement (76,81). On microscopy, the main differential diagnosis is Langerhans cell histiocytosis, although the distribution is considerably different, being perilymphatic in Erdheim-Chester disease rather than the parenchymal disease that is bronchiolocentric, nodular, or cystic in Langerhans cell histiocytosis. In addition, the lesional cells do not stain for CD1a or contain Birbeck granules on electron microscopy.

Treatment and Prognosis. Targeted therapy with vemurafenib has shown rapid clinical and biologic response (82–84). Tocizilumab, an interleukin-6 inhibitor, is also effective (85). Patients should be followed according to consensus guidelines (74).

REFERENCES

1. Motta G, Conticello C, Amato G, et al. Pleuric presentation of extranodal marginal zone lymphoma of mucosa-associated lymphoid tissue: a case report and a review of the literature. Int J Hematol 2010;92:369-73.
2. Steiropoulos P, Kouliatsis G, Karpathiou G, Popidou M, Froudarakis ME. Rare cases of primary pleural Hodgkin and non-Hodgkin lymphomas. Respiration 2009;77:459-63.
3. Ozkaya N, Rosenblum MK, Durham BH, et al. The histopathology of Erdheim-Chester disease: a comprehensive review of a molecularly characterized cohort. Mod Pathol 2018;31:581-97.
4. Goyal G, Young JR, Koster MJ, et al. The Mayo Clinic Histiocytosis Working Group consensus statement for the diagnosis and evaluation of adult patients with histiocytic neoplasms: Erdheim-Chester disease, Langerhans cell histiocytosis, and Rosai-Dorfman disease. Mayo Clin Proc 2019;94:2054-71.
5. Arora N, Gupta A, Sadeghi N. Primary effusion lymphoma: current concepts and management. Curr Opin Pulm Med 2017;23:365-70.
6. El-Fattah MA. Clinical characteristics and survival outcome of primary effusion lymphoma: a review of 105 patients. Hematol Oncol 2017;35:878-83.
7. Gupta A, Sen S, Marley E, Chen W, Naina HV. Management and outcomes of HIV-associated primary effusion lymphoma: a single center experience. Clin Lymphoma Myeloma Leuk 2016;16(Suppl):S175-80.
8. Guillet S, Gerard L, Meignin V, et al. Classic and extracavitary primary effusion lymphoma in 51 HIV-infected patients from a single institution. Am J Hematol 2016;91:233-7.
9. Chadburn A, Hyjek E, Mathew S, Cesarman E, Said J, Knowles DM. KSHV-positive solid lymphomas represent an extra-cavitary variant of primary effusion lymphoma. Am J Surg Pathol 2004;28:1401-16.
10. Simonelli C, Spina M, Cinelli R, et al. Clinical features and outcome of primary effusion lymphoma in HIV-infected patients: a single-institution study. J Clin Oncol 2003;21:3948-54.
11. Chadburn A, Hyjek EM, Tam W, et al. Immunophenotypic analysis of the Kaposi sarcoma herpesvirus (KSHV; HHV-8)-infected B cells in HIV+ multicentric Castleman disease (MCD). Histopathology 2008;53:513-24.
12. Boulanger E, Agbalika F, Maarek O, et al. A clinical, molecular and cytogenetic study of 12 cases of human herpesvirus 8 associated primary effusion lymphoma in HIV-infected patients. Hematol J 2001;2:172-9.
13. Ariza-Heredia EJ, Razonable RR. Human herpes virus 8 in solid organ transplantation. Transplantation 2011;92:837-44.
14. Cain O, Yoong A, Lipkin G, et al. Rapidly progressive intravascular primary effusion lymphoma in an HIV-positive renal transplant recipient. Histopathology 2018;72:339-41.
15. Zaimoku Y, Takahashi W, Iwaki N, et al. Human herpesvirus-8-unrelated primary effusion lymphoma of the elderly not associated with an increased serum lactate dehydrogenase level: a benign sub-group of effusion lymphoma without chemotherapy. Leuk Lymphoma 2016;57:1625-32.
16. Lam GK, Abdelhaleem M, Somers GR, Roifman C, Read S, Abla O. Primary effusion lymphoma (PEL)-like lymphoma in a child with congenital immunodeficiency. Pediatr Blood Cancer 2016; 63:1674-6.

17. Kojima M, Nakamura N, Amaki J, et al. Human herpesvirus 8-unrelated primary effusion lymphoma-like lymphoma following tyrosine kinase inhibitor treatment for chronic myelogenous leukemia. J Clin Exp Hematop 2017;57:69-73.
18. Yiakoumis X, Pangalis GA, Kyrtsonis MC, et al. Primary effusion lymphoma in two HIV-negative patients successfully treated with pleurodesis as first-line therapy. Anticancer Res 2010;30:271-6.
19. Boulanger E, Hermine O, Fermand JP, et al. Human herpesvirus 8 (HHV-8)-associated peritoneal primary effusion lymphoma (PEL) in two HIV-negative elderly patients. Am J Hematol 2004;76:88-91.
20. Ascoli V, Scalzo CC, Danese C, Vacca K, Pistilli A, Lo Coco F. Human herpes virus-8 associated primary effusion lymphoma of the pleural cavity in HIV-negative elderly men. Eur Respir J 1999; 14:1231-4.
21. Ammari ZA, Mollberg NM, Abdelhady K, Mansueto MD, Massad MG. Diagnosis and management of primary effusion lymphoma in the immunocompetent and immunocompromised hosts. Thorac Cardiovasc Surg 2013;61:343-9.
22. Castillo JJ, Shum H, Lahijani M, Winer ES, Butera JN. Prognosis in primary effusion lymphoma is associated with the number of body cavities involved. Leuk Lymphoma 2012;53:2378-82.
23. Kobayashi Y, Kamitsuji Y, Kuroda J, et al. Comparison of human herpes virus 8 related primary effusion lymphoma with human herpes virus 8 unrelated primary effusion lymphoma-like lymphoma on the basis of HIV: report of 2 cases and review of 212 cases in the literature. Acta Haematol 2007;117:132-44.
24. Boulanger E, Gerard L, Gabarre J, et al. Prognostic factors and outcome of human herpesvirus 8-associated primary effusion lymphoma in patients with AIDS. J Clin Oncol 2005;23:4372-80.
25. Kim Y, Leventaki V, Bhaijee F, Jackson CC, Medeiros LJ, Vega F. Extracavitary/solid variant of primary effusion lymphoma. Ann Diagn Pathol 2012;16:441-6.
26. Carbone A, Gloghini A, Vaccher E, et al. Kaposi's sarcoma-associated herpesvirus/human herpesvirus type 8-positive solid lymphomas: a tissue-based variant of primary effusion lymphoma. J Mol Diagn 2005;7:17-27.
27. DePond W, Said JW, Tasaka T, et al. Kaposi's sarcoma-associated herpesvirus and human herpesvirus 8 (KSHV/HHV8)-associated lymphoma of the bowel. Report of two cases in HIV-positive men with secondary effusion lymphomas. Am J Surg Pathol 1997;21:719-24.
28. Nador RG, Cesarman E, Chadburn A, et al. Primary effusion lymphoma: a distinct clinicopathologic entity associated with the Kaposi's sarcoma-associated herpes virus. Blood 1996;88:645-56.
29. Carbone A, Gloghini A, Vaccher E, et al. Kaposi's sarcoma-associated herpesvirus DNA sequences in AIDS-related and AIDS-unrelated lymphomatous effusions. Br J Haematol 1996;94:533-43.
30. Said JW, Shintaku IP, Asou H, et al. Herpesvirus 8 inclusions in primary effusion lymphoma: report of a unique case with T-cell phenotype. Arch Pathol Lab Med 1999;123:257-60.
31. Hollingsworth HC, Stetler-Stevenson M, Gagneten D, Kingma DW, Raffeld M, Jaffe ES. Immunodeficiency-associated malignant lymphoma. Three cases showing genotypic evidence of both T- and B-cell lineages. Am J Surg Pathol 1994;18:1092-101.
32. Yamani F, Chen W. Cytokeratin-positive primary effusion lymphoma: a diagnostic challenge. Br J Haematol 2018;180:9.
33. Wang HY, Wang L. Diagnostic pitfall: primary effusion lymphoma with rare cytokeratin immunoreactivity. Blood 2016;127:3102.
34. Dupin N, Fisher C, Kellam P, et al. Distribution of human herpesvirus-8 latently infected cells in Kaposi's sarcoma, multicentric Castleman's disease, and primary effusion lymphoma. Proc Natl Acad Sci U S A 1999;96:4546-51.
35. Parravicini C, Chandran B, Corbellino M, et al. Differential viral protein expression in Kaposi's sarcoma-associated herpesvirus-infected diseases: Kaposi's sarcoma, primary effusion lymphoma, and multicentric Castleman's disease. Am J Pathol 2000;156:743-9.
36. Katano H, Sato Y, Kurata T, Mori S, Sata T. Expression and localization of human herpesvirus 8-encoded proteins in primary effusion lymphoma, Kaposi's sarcoma, and multicentric Castleman's disease. Virology 2000;269:335-44.
37. Cesarman E, Chang Y, Moore PS, Said JW, Knowles DM. Kaposi's sarcoma-associated herpesvirus-like DNA sequences in AIDS-related body-cavity-based lymphomas. N Engl J Med 1995;332:1186-91.
38. Mesri EA, Cesarman E, Boshoff C. Kaposi's sarcoma and its associated herpesvirus. Nat Rev Cancer 2010;10:707-19.
39. Giffin L, Damania B. KSHV: pathways to tumorigenesis and persistent infection. Adv Virus Res 2014;88:111-59.
40. Sakakibara S, Tosato G. Viral interleukin-6: role in Kaposi's sarcoma-associated herpesvirus: associated malignancies. J Interferon Cytokine Res 2011;31:791-801.
41. Nador RG, Milligan LL, Flore O, et al. Expression of Kaposi's sarcoma-associated herpesvirus G protein-coupled receptor monocistronic and bicistronic transcripts in primary effusion lymphomas. Virology 2001;287:62-70.

42. Horenstein MG, Nador RG, Chadburn A, et al. Epstein-Barr virus latent gene expression in primary effusion lymphomas containing Kaposi's sarcoma-associated herpesvirus/human herpesvirus-8. Blood 1997;90:1186-91.
43. Mack AA, Sugden B. EBV is necessary for proliferation of dually infected primary effusion lymphoma cells. Cancer Res 2008;68:6963-8.
44. Kliche S, Kremmer E, Hammerschmidt W, Koszinowski U, Haas J. Persistent infection of Epstein-Barr virus-positive B lymphocytes by human herpesvirus 8. J Virol 1998;72:8143-9.
45. Coupland SE, Charlotte F, Mansour G, Maloum K, Hummel M, Stein H. HHV-8-associated T-cell lymphoma in a lymph node with concurrent peritoneal effusion in an HIV-positive man. Am J Surg Pathol 2005;29:647-52.
46. Lechapt-Zalcman E, Challine D, Delfau-Larue MH, Haioun C, Desvaux D, Gaulard P. Association of primary pleural effusion lymphoma of T-cell origin and human herpesvirus 8 in a human immunodeficiency virus-seronegative man. Arch Pathol Lab Med 2001;125:1246-8.
47. Nakayama-Ichiyama S, Yokote T, Kobayashi K, et al. Primary effusion lymphoma of T-cell origin with t(7;8)(q32;q13) in an HIV-negative patient with HCV-related liver cirrhosis and hepatocellular carcinoma positive for HHV6 and HHV8. Ann Hematol 2011;90:1229-31.
48. Jenner RG, Maillard K, Cattini N, et al. Kaposi's sarcoma-associated herpesvirus-infected primary effusion lymphoma has a plasma cell gene expression profile. Proc Natl Acad Sci U S A 2003; 100:10399-404.
49. Klein U, Gloghini A, Gaidano G, et al. Gene expression profile analysis of AIDS-related primary effusion lymphoma (PEL) suggests a plasmablastic derivation and identifies PEL-specific transcripts. Blood 2003;101:4115-21.
50. Wilson KS, McKenna RW, Kroft SH, Dawson DB, Ansari Q, Schneider NR. Primary effusion lymphomas exhibit complex and recurrent cytogenetic abnormalities. Br J Haematol 2002;116:113-21.
51. Roy D, Sin SH, Damania B, Dittmer DP. Tumor suppressor genes FHIT and WWOX are deleted in primary effusion lymphoma (PEL) cell lines. Blood 2011;118:e32-9.
52. Luan SL, Boulanger E, Ye H, et al. Primary effusion lymphoma: genomic profiling revealed amplification of SELPLG and CORO1C encoding for proteins important for cell migration. J Pathol 2010;222:166-79.
53. Lam BQ, Dai L, Li L, Qiao J, Lin Z, Qin Z. Molecular mechanisms of activating c-MET in KSHV+ primary effusion lymphoma. Oncotarget 2017;8:18373-80.
54. Bonsignore L, Passelli K, Pelzer C, et al. A role for MALT1 activity in Kaposi's sarcoma-associated herpes virus latency and growth of primary effusion lymphoma. Leukemia 2017;31:614-24.
55. Aozasa K, Takakuwa T, Nakatsuka S. Pyothorax-associated lymphoma: a lymphoma developing in chronic inflammation. Adv Anat Pathol 2005;12:324-31.
56. Alexanian S, Said J, Lones M, Pullarkat ST. KSHV/HHV8-negative effusion-based lymphoma, a distinct entity associated with fluid overload states. Am J Surg Pathol 2013;37:241-9.
57. Wu W, Youm W, Rezk SA, Zhao X. Human herpesvirus 8-unrelated primary effusion lymphoma-like lymphoma: report of a rare case and review of 54 cases in the literature. Am J Clin Pathol 2013;140:258-73.
58. Bower M, Pria AD, Coyle C, Nelson M, Naresh K. Diagnostic criteria schemes for multicentric Castleman disease in 75 cases. J Acquir Immune Defic Syndr 2014;65:e80-2.
59. Luppi M, Trovato R, Barozzi P, et al. Treatment of herpesvirus associated primary effusion lymphoma with intracavity cidofovir. Leukemia 2005; 19:473-6.
60. Ripamonti D, Marini B, Rambaldi A, Suter F. Treatment of primary effusion lymphoma with highly active antiviral therapy in the setting of HIV infection. AIDS 2008;22:1236-7.
61. Nakatsuka S, Yao M, Hoshida Y, Yamamoto S, Iuchi K, Aozasa K. Pyothorax-associated lymphoma: a review of 106 cases. J Clin Oncol 2002; 20:4255-60.
62. Nakamichi I, Takakuwa T, Tanio Y, Iuchi K, Aozasa K. Pyothorax-associated lymphoma: an unusual case with both T- and B-cell genotypes. Virchows Arch 2005;447:888-91.
63. Sasajima Y, Yamabe H, Kobashi Y, Hirai K, Mori S. High expression of the Epstein-Barr virus latent protein EB nuclear antigen-2 on pyothorax-associated lymphomas. Am J Pathol 1993;143:1280-5.
64. Miwa H, Takakuwa T, Nakatsuka S, Tomita Y, Iuchi K, Aozasa K. DNA sequences of the immunoglobulin heavy chain variable region gene in pyothorax-associated lymphoma. Oncology 2002;62:241-50.
65. Nishiu M, Tomita Y, Nakatsuka S, et al. Distinct pattern of gene expression in pyothorax-associated lymphoma (PAL), a lymphoma developing in long-standing inflammation. Cancer Sci 2004;95:828-34.
66. Taniguchi A, Hashida Y, Nemoto Y, Taguchi T, Iwahara Y, Daibata M. Pyothorax-associated lymphoma (PAL) with biclonal Epstein-Barr virus infection: characterization of a novel PAL cell line with unique features. Leuk Res 2013;37:1545-50.

67. Dunleavy K, Roschewski M, Wilson WH. Lymphomatoid granulomatosis and other Epstein-Barr virus associated lymphoproliferative processes. Curr Hematol Malig Rep 2012;7:208-15.
68. Nicholson AG, Wotherspoon AC, Diss TC, et al. Lymphomatoid granulomatosis: evidence that some cases represent Epstein-Barr virus-associated B-cell lymphoma. Histopathology 1996; 29:317-24.
69. Song JY, Pittaluga S, Dunleavy K, et al. Lymphomatoid granulomatosis—a single institute experience: pathologic findings and clinical correlations. Am J Surg Pathol 2015;39:141-56.
70. Wang F, Lan H. A case report on the effect of rituximab on pyothorax-associated lymphoma. Medicine (Baltimore) 2019;98:e18393.
71. Wittenberg KH, Swensen SJ, Myers JL. Pulmonary involvement with Erdheim-Chester disease: radiographic and CT findings. Amer J Roentgenol 2000;174:1327-31.
72. Brun AL, Touitou-Gottenberg D, Haroche J, et al. Erdheim-Chester disease: CT findings of thoracic involvement. Eur Radiol 2010;20:2579-87.
73. Mirmomen SM, Sirajuddin A, Nikpanah M, et al. Thoracic involvement in Erdheim-Chester disease: computed tomography imaging findings and their association with the $BRAF^{V600E}$ mutation. Eur Radiol 2018;28:4635-42.
74. Goyal G, Heaney ML, Collin M, et al. Erdheim-Chester disease: consensus recommendations for evaluation, diagnosis, and treatment in the molecular era. Blood 2020;135:1929-45.
75. Rush WL, Andriko JA, Galateau-Salle F, et al. Pulmonary pathology of Erdheim-Chester disease. Modern Pathol 2000;13:747-54.
76. Nicholson AG, Anderson E, Saha S, Indrajith M, Conry B, Hughes J. Progressive dyspnoea, pleural effusions and lytic bone lesions. Thorax 2008; 63:492, 554.
77. Diamond EL, Dagna L, Hyman DM, et al. Consensus guidelines for the diagnosis and clinical management of Erdheim-Chester disease. Blood 2014;124:483-92.
78. Vencio EF, Jenkins RB, Schiller JL, et al. Clonal cytogenetic abnormalities in Erdheim-Chester disease. Am J Surg Pathol 2007;31:319-21.
79. Chetritt J, Paradis V, Dargere D, et al. Chester-Erdheim disease: a neoplastic disorder. Hum Pathol 1999;30:1093-6.
80. Yahng SA, Kang HH, Kim SK, et al. Erdheim-Chester disease with lung involvement mimicking pulmonary lymphangitic carcinomatosis. Am J Med Sci 2009;337:302-4.
81. Protopapadakis C, Antoniou KM, Nicholson AG, et al. Erdheim-Chester disease: pulmonary presentation in a case with advanced systemic involvement. Respiration 2009;77:337-40.
82. Diamond EL, Subbiah V, Lockhart AC, et al. Vemurafenib for BRAF V600-mutant Erdheim-Chester disease and Langerhans cell histiocytosis: analysis of data from the histology-independent, phase 2, open-label VE-BASKET study. JAMA Oncol 2018;4:384-8.
83. Haroche J, Cohen-Aubart F, Emile JF, et al. Dramatic efficacy of vemurafenib in both multisystemic and refractory Erdheim-Chester disease and Langerhans cell histiocytosis harboring the BRAF V600E mutation. Blood 2013;121:1495-500.
84. Haroche J, Cohen-Aubart F, Emile JF, Donadieu J, Amoura Z. Vemurafenib as first line therapy in BRAF-mutated Langerhans cell histiocytosis. J Am Acad Dermatol 2015;73:e29-30.
85. Berti A, Cavalli G, Guglielmi B, et al. Tocilizumab in patients with multisystem Erdheim-Chester disease. Oncoimmunology 2017;6:e1318237.

7 SECONDARY MALIGNANT NEOPLASMS

Secondary malignant neoplasms arise at other sites and involve serosal surfaces via contiguous or noncontiguous spread. Since normal clearance of fluid can be compromised, these malignancies can cause effusions or ascites.

PLEURAL DISEASE

Sites of Origin

Lung carcinoma is the most common source of tumor in the pleural space. Nevertheless, carcinomas of various organs, including breast, ovary, gastrointestinal tract (stomach, colon, and pancreas), and kidney are also commonly encountered. Pleural involvement can be part of initial tumor presentation but can also occur later in the disease course. Of histologic types of carcinoma, adenocarcinoma is most frequent, followed by small cell carcinoma, with squamous carcinoma the least common.

Carcinomas are not the sole histologic type encountered. Secondary involvement of pleura by lymphoma is also common. Melanomas and sarcomas can also secondarily involve the pleura. In addition, although uncommon tumors, thymomas frequently involve the pleura either at presentation or more importantly as a site of recurrence.

Mechanism of Serosal Involvement

Direct extension of tumor into pleura is a common mechanism of involvement and is part of the staging of lung carcinoma (1). Invasion progresses from visceral to parietal pleura, or can further spread via pleural fluid and tumor seeding. Lymphatic and vascular involvement can occur from tumor of any origin. For tumors outside of the chest, thoracic space direct extension occurs through the diaphragm from abdominal or liver metastasis or from direct extension from the chest wall, as in breast cancer.

Pleural effusion occurs when lymphatic or nodal involvement impedes fluid clearance from the space. When malignant cells are found in the fluid, this is termed *malignant pleural effusion* in diagnosis and staging.

Patterns of Involvement

The patterns of pleural involvement include multiple nodules, diffuse involvement, or effusion. These patterns can appear together or singly, and certain tumor types have a predilection for one pattern over another.

Nodules. Pleural nodules are seen in carcinoma, melanoma, sarcoma, and thymoma, but less commonly in lymphoma. Diffuse pleural nodules reflect higher-stage disease and therefore poorer prognosis. In thymoma, "drop metastases" can occur at presentation or as a later complication of recurrence. These may be unilateral. Rarely, nodular involvement of the pleura by meningioma occurs via either extension from the spine or by hematogenous spread.

Diffuse. Secondary diffuse neoplastic pleural involvement can mimic malignant mesothelioma. This pattern of pseudomesotheliomatous carcinoma is most often, but not exclusively, from a lung source. Most often these are adenocarcinomas (2–5), although other histologies can produce this pattern including pleomorphic carcinoma, squamous cell carcinoma, and small cell carcinoma (2,6). Extrapulmonary primaries, such as from bladder, prostate gland, kidney, and pancreas, have all been reported to produce this pattern (2,7). Some mesenchymal tumors, especially those of vascular origin, have been reported to mimic mesothelioma as well (8–10), as has the systemic histiocytic disorder Erdheim-Chester disease (see chapter 6) (11).

Effusion. Clinically, rates of malignant pleural effusion increase with age, starting at 30 to 40 years, but reaching a peak in the 60s or 70s,

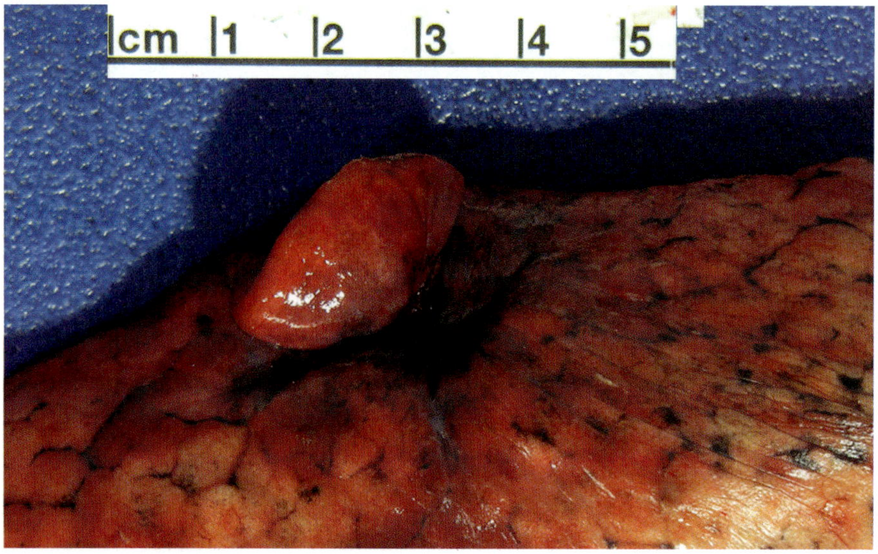

Figure 7-1

PLEURAL METASTASIS

A solitary pleural metastasis from a lung cancer (main tumor not shown) causes invagination of the visceral pleura.

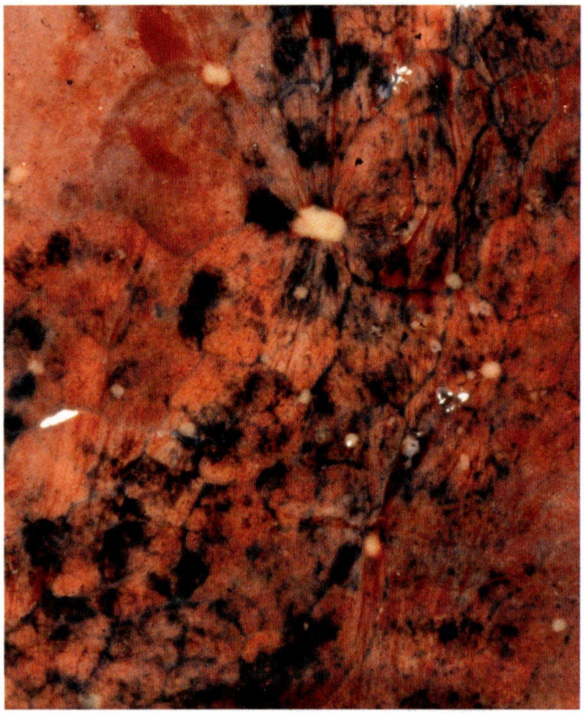

Figure 7-2

PLEURAL METASTASIS

Nodules of tumor are scattered across the visceral pleura.

reflecting the epidemiology of the underlying cancer. For example, patients with carcinomas have an average age over 60, while some hematologic malignancies are seen in younger patients. No sex predilection is reported, since this is linked to the epidemiology of the underlying malignancy.

While pleural effusions occur from a variety of causes, including heart failure, unilateral pleural effusion suggests malignancy. Lung carcinoma is the most common cause, followed by breast carcinoma, gastric carcinoma, and ovarian carcinoma (12–15). Often, the primary site is not known at the time of first effusion (12). Lymphomatous pleural involvement is more often an effusion rather than nodular involvement (16–18); of chylous effusions, roughly one sixth are due to lymphoma (19).

Gross Findings

Minimal pleural involvement frequently occurs in peripheral lung carcinoma with visceral pleura invasion (fig. 7-1). This can be seen in association with a pleural pucker or an area of adhesion on the pleural surface. Parietal pleural involvement is treated surgically, in some instances, after neoadjuvant therapy, with partial chest wall resection.

Once multiple nodules on the pleura are present, the disease is usually unresectable. Such nodules can be scattered (fig. 7-2) or extensive (fig. 7-3), and may be only millimeters in diameter. In other instances, there are fewer larger nodules (fig. 7-4). Nodules can be on the visceral or parietal pleura (fig. 7-5). Commonly, these are white to gray-white, but the color may be related to the underlying histology, such as melanoma (brown-black) or angiosarcoma

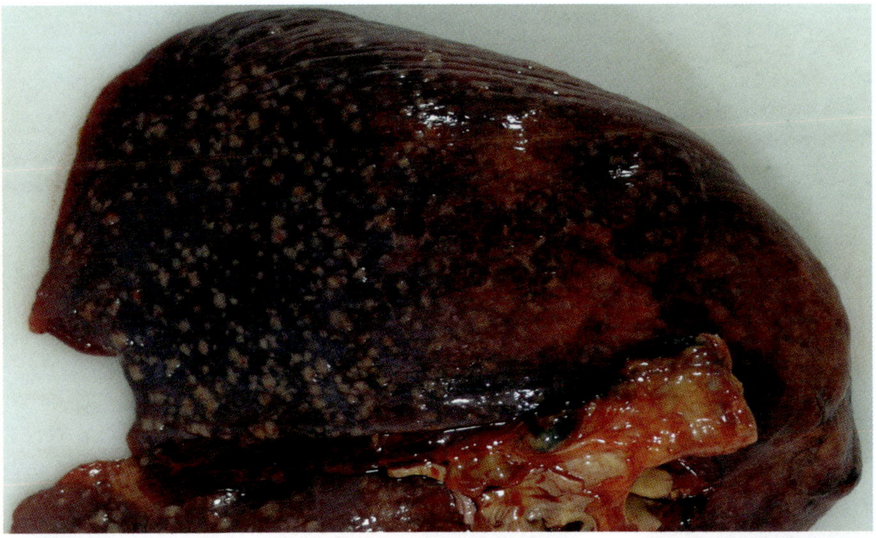

Figure 7-3

PLEURAL METASTASIS

Nodules of tumor are widely disseminated throughout the visceral pleura.

Figure 7-4

PLEURAL METASTASIS

Larger nodules of tumor stud the visceral pleura.

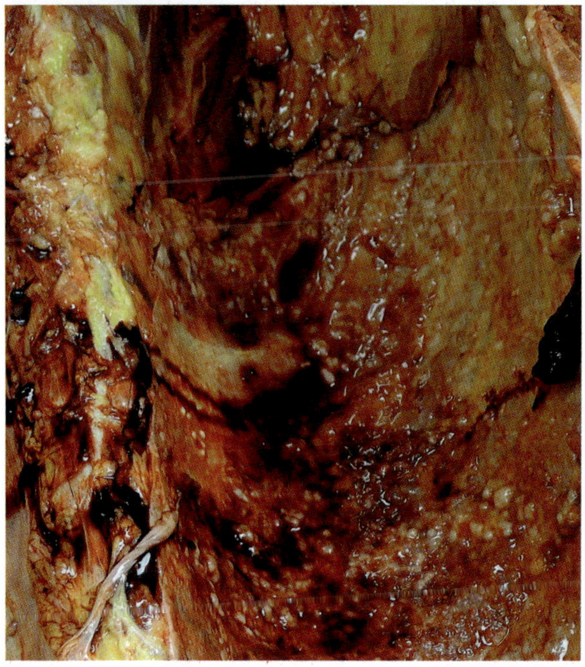

Figure 7-5

PLEURAL METASTASIS

Multiple, sometimes coalescent, nodules of tumor involve both visceral and parietal pleura.

(red). Involvement of pleural lymphatics on the visceral pleural of the lung may have the appearance of a meshwork (fig. 7-6) and highlight the lobular architecture of the lung.

Pseudomesotheliomatous carcinoma is a more diffuse visceral pleural thickening that can extend into the parietal pleura. In some instances, a lung mass, reflecting primary lung cancer or a metastatic tumor, is identified.

Microscopic Findings

As previously noted, the histology of secondary involvement can be quite varied, although adenocarcinoma predominates since tumors of lung,

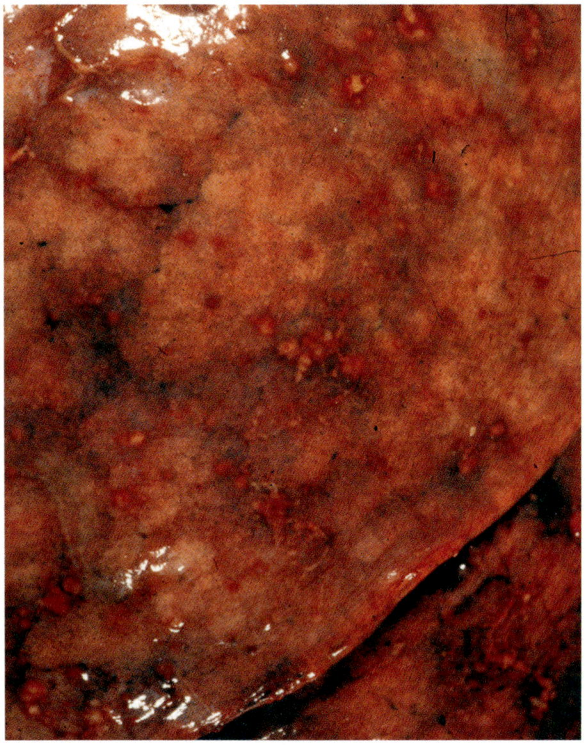

Figure 7-6

PLEURAL METASTASIS

Metastatic tumor follows the distribution of lymphatics within the visceral pleura, with a focally linear arrangement.

ovary, gastrointestinal tract, genitourinary tract, and gynecologic tract are generally adenocarcinomas (fig. 7-7). However, small cell carcinoma can involve the pleura (fig. 7-8), as can squamous cell carcinoma (fig. 7-9) and sarcomatoid histologies. The diagnostic specimen is often cytologic from an effusion, and cell blocks are generally used to undertake ancillary studies (13,15,20,21).

Malignant melanoma is an important pitfall as its histology can mimic carcinoma, not least because a history of previous tumor or an overt primary site may not be identifiable. For thymoma, while all types can be seen in the pleura, the most commonly encountered is World Health Organization (WHO) type B (fig. 7-10). Metastatic thymoma should not be confused with lymphomatous involvement of the pleura (fig. 7-11). Conversely, lymphoma should still be a consideration whenever dense nodular lymphoid infiltrates are present.

Pseudomesotheliomatous adenocarcinoma can cause significant sclerosis, mimicking a pleural plaque. A background of pleural fibrosis does not help in distinguishing secondary adenocarcinoma involvement from pleural mesothelioma (fig. 7-12).

As a result of the above microscopic findings, the clinical history and disease distribution are of

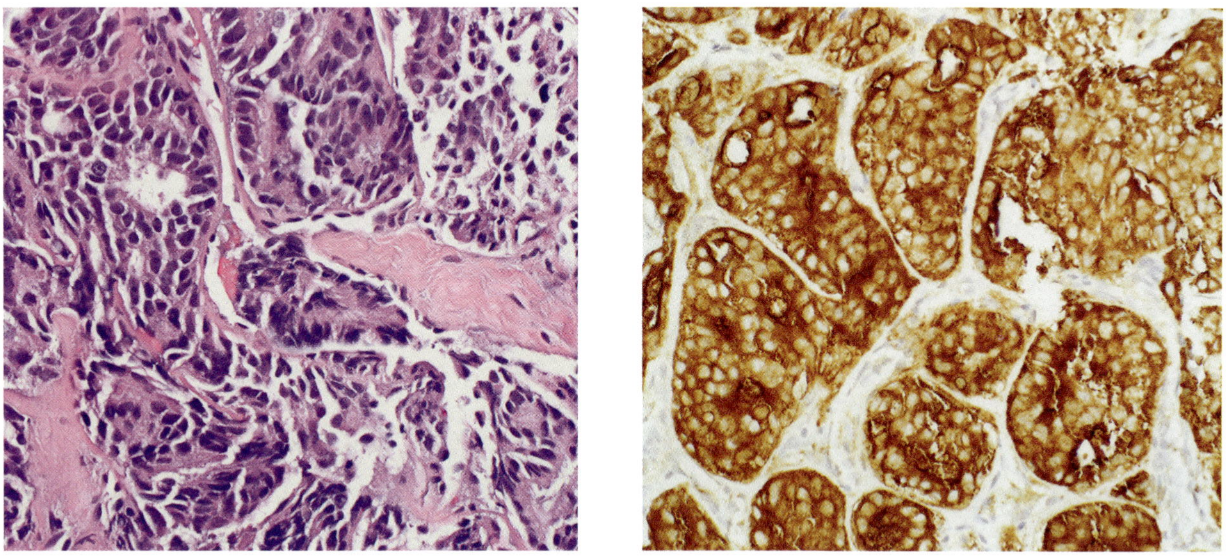

Figure 7-7

PLEURAL METASTASIS

The pleura is infiltrated by metastatic prostatic adenocarcinoma (left), which stains intensely for prostate-specific membrane antigen (right).

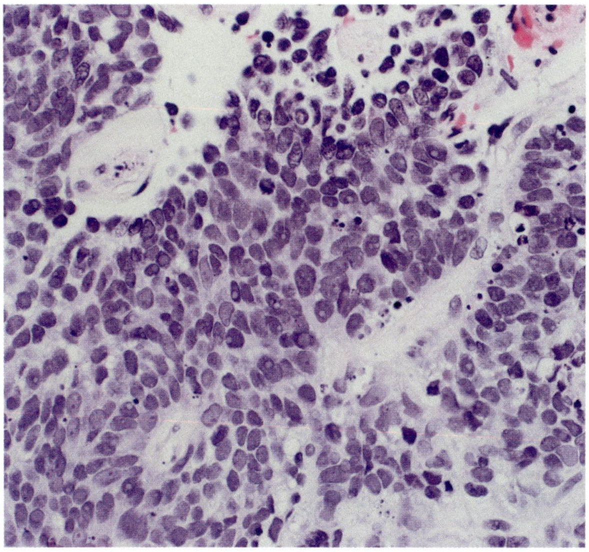

Figure 7-8

PLEURAL METASTASIS

The pleura is infiltrated by metastatic small cell carcinoma.

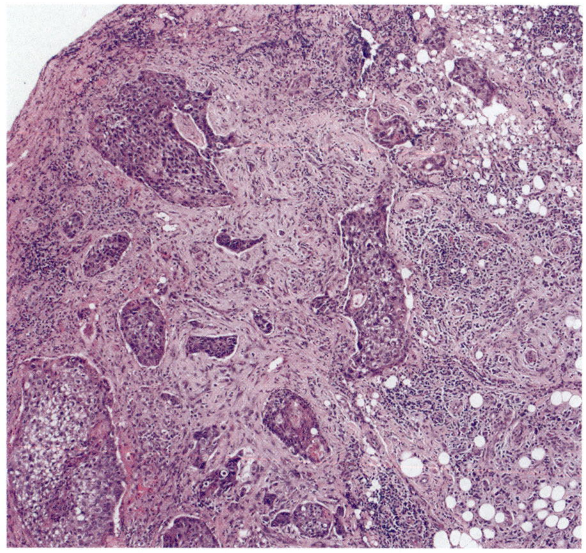

Figure 7-9

PLEURAL METASTASIS

The pleura is infiltrated by metastatic squamous cell carcinoma.

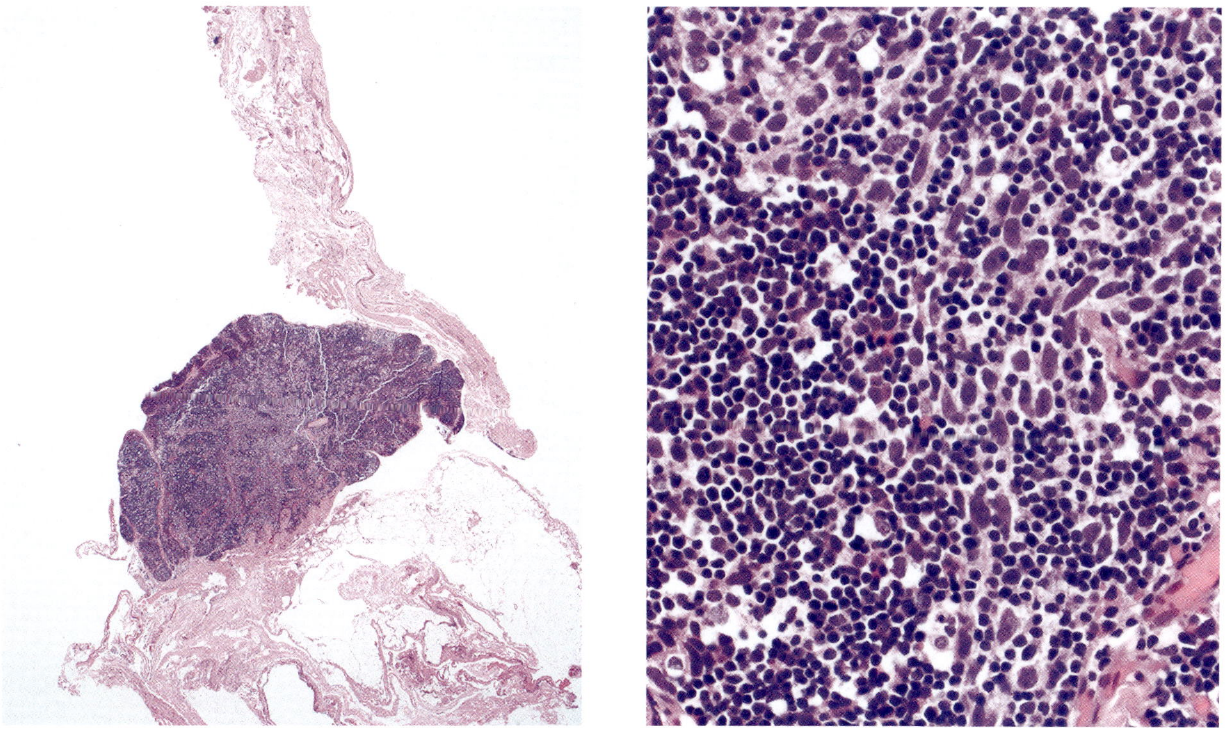

Figure 7-10

PLEURAL METASTASIS

Left: Metastatic thymoma forms a "drop metastasis" on the parietal pleural surface.
Right: At high power, the features are those of a WHO type B thymoma.

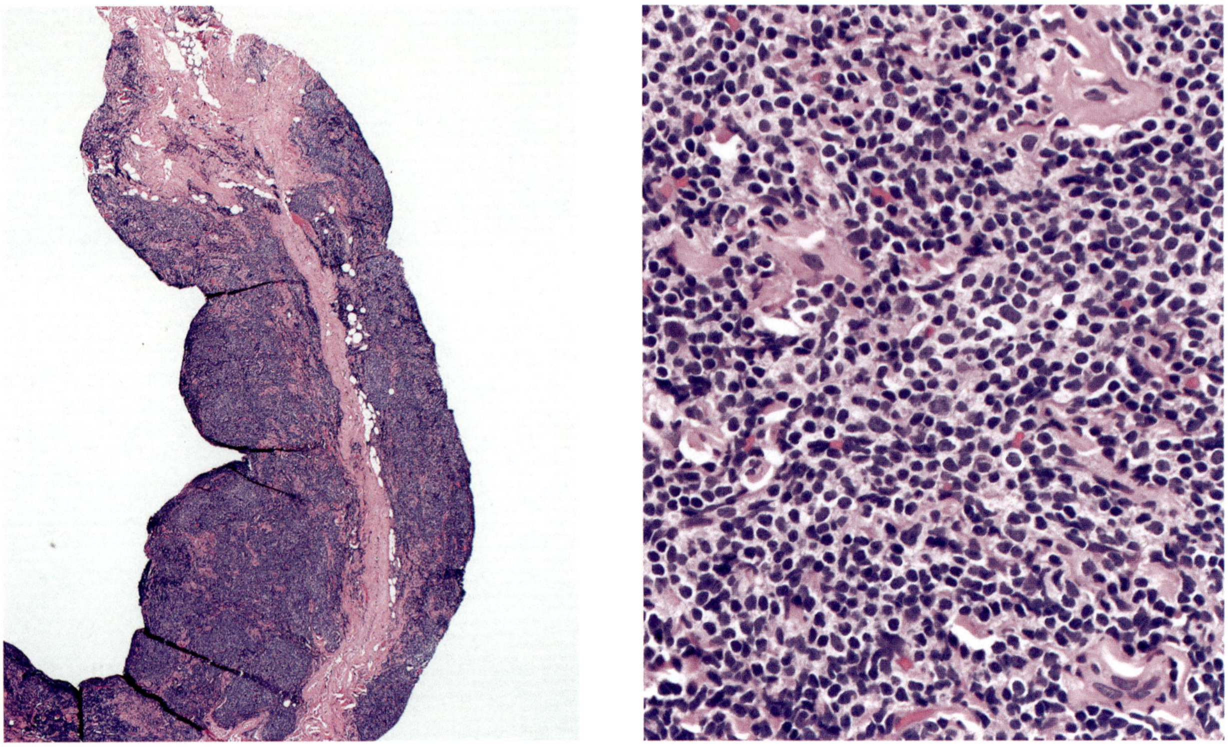

Figure 7-11

PLEURAL METASTASIS

The tumor is diffusely infiltrated by a non-Hodgkin lymphoma (left), consisting of a monotonous population of lymphocytes (right).

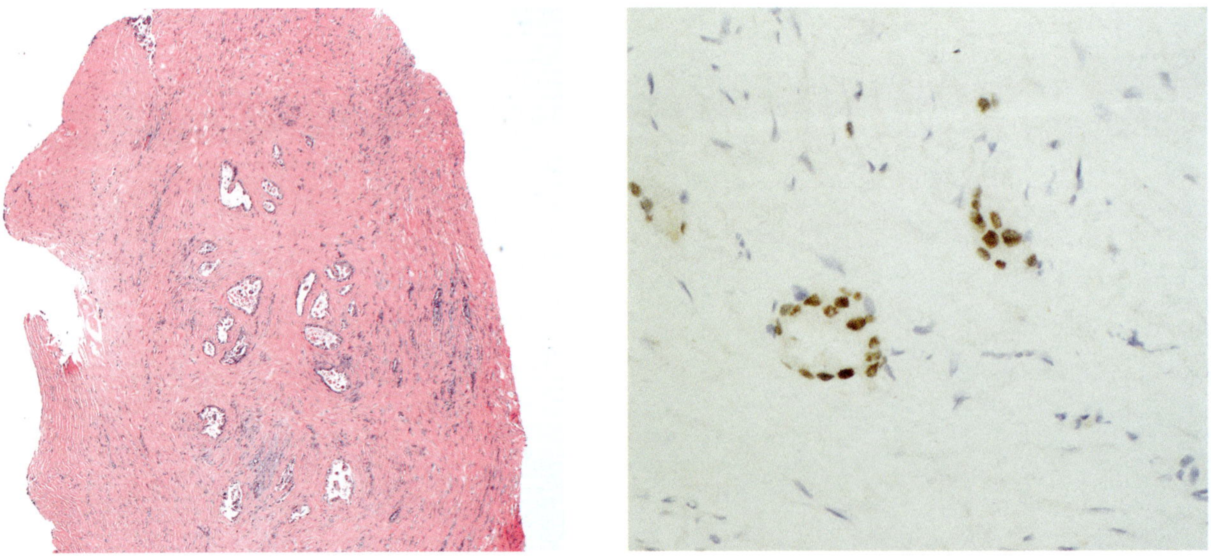

Figure 7-12

PLEURAL METASTASIS

Left: Metastatic adenocarcinoma mimics an epithelioid mesothelioma with diffuse pleural thickening.
Right: The tumor cells stain for TTF1, confirming pulmonary origin.

Table 7-1

IMMUNOHISTOCHEMISTRY IN SECONDARY TUMORS OF THE PLEURA[a]

Tumor Type	Immunohistochemical Markers
Lung adenocarcinoma	TTF1, Napsin A
Breast carcinoma	GATA3, mammaglobin, +/- estrogen receptor, progesterone receptor, GCDFP-15, SOX10
Ovarian serous carcinoma	PAX8, WT-1
Gynecologic (other than serous)	PAX8, estrogen receptor, progesterone receptor
Intestinal	CDX2, CK7, CK20, SATB2
Melanoma	S-100 protein, HMB45, SOX10, +/- melan A, MITF, tyrosinase
Sarcoma	CD34, CD31, ERG, SMA, desmin, S-100 protein, +/- MDM2, STAT6, INI1,TLE-1, SS18-SSX fusion specific antibody
Thymoma	P40/p63, PAX8 + evaluate lymphocytes as TDT, CD1a positive
Thymic carcinoma	PAX8, CD5 +/- CD117

[a]Data from references 22 through 31.

great importance in the evaluation of secondary pleural malignancies. However, valuable though this information can be, it is important to evaluate pleural biopsies based on histology and a broad differential diagnosis to avoid error. Immunohistochemical panels are an adjunct to this evaluation and should be used for the first diagnosis of malignancy to ensure accurate diagnosis and stage designation, given the broad differential of secondary pleural involvement (Table 7-1).

Differential Diagnosis

As well as the wide differential diagnosis caused by having multiple potential metastatic sites, another important consideration is differentiating primary tumors of the pleura, such as mesothelioma or solitary fibrous tumor. Panels for the diagnosis of mesothelioma are described elsewhere (see chapter 3), and in most instances, malignant solitary fibrous tumor presents as a single large mass rather than multifocal or diffuse pleural involvement (see chapter 5).

Non-neoplastic diseases also mimic pleural neoplasia. In most cases, nodular granulomatous disease and amyloidosis are easily distinguished from secondary malignancy on microscopy (see chapter 8).

Treatment and Prognosis

Metastatic tumor can occur in either visceral or parietal pleura, and treatment reflects a combination of local and systemic approaches. For example, recurrent pleural effusion may require drainage and pleurodesis; methods of drainage include thoracentesis as well as longer-term drainage solutions. For certain tumors, resection of solitary or oligometastatic disease is a consideration; this can include sarcomas and thymomas. Extrapleural pneumonectomy is a consideration in some instances (e.g., thymic tumors) (32), although this is an uncommon option. Chemotherapy and radiation therapy are options determined by tumor type, tumor stage, and patient symptoms.

For most tumors, pleural involvement reflects high stage and requires nonsurgical options. With the revolution in targeted and immunomodulatory therapies, the pleura is often a site where advanced disease is sampled to obtain tumor for molecular tests and screening for PD-L1 status. If the site of origin is already known, then limited tissue should not be wasted in "rediagnosing" the cancer. The prognosis of patients with pleural metastatic disease is primarily governed by treatment strategies for the primary cancer.

PERITONEUM

Sites of Origin

Secondary involvement of the peritoneal cavity by carcinoma, mostly adenocarcinomas, includes tumors of the ovary, endometrium, stomach, colon, pancreas, and breast (33–35). In

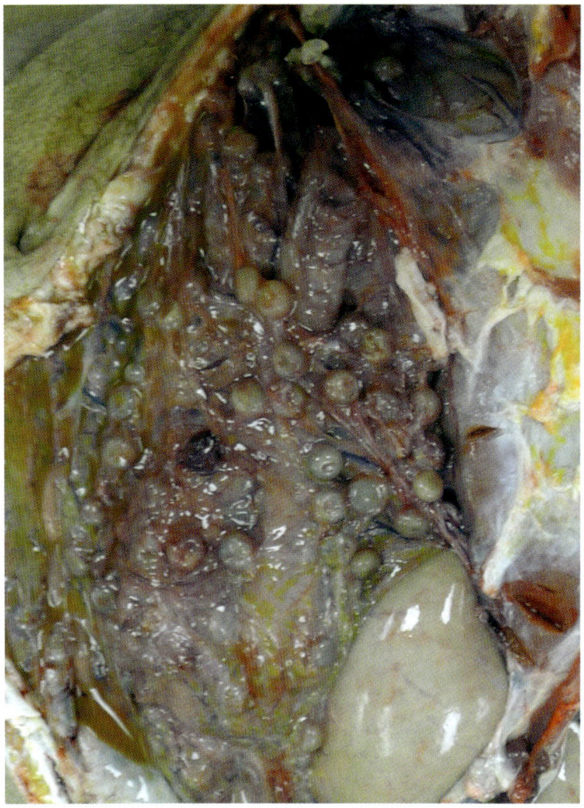

Figure 7-13

PERITONEAL METASTASIS

Variably sized tumor nodules stud the peritoneal surface.

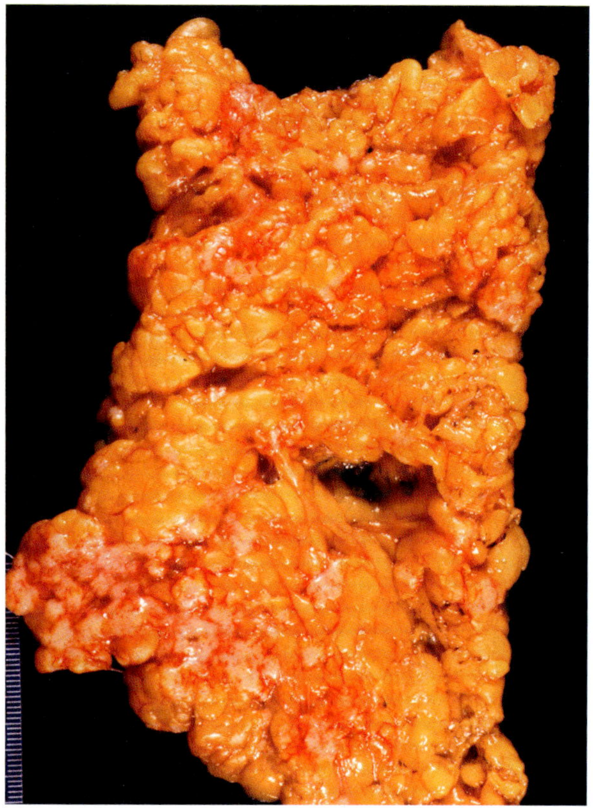

Figure 7-14

PERITONEAL METASTASIS

The tumor forms an "omental cake," in which diffuse confluent disease replaces the omentum.

some cases, the tumor gains access to this layer by direct invasion, as seen in high T-stage colorectal carcinoma, but seeding can occur though peritoneal fluid and angiolymphatic spread. Tumoral involvement can appear as small or medium-sized nodules (fig. 7-13), but these often become confluent as "omental cake" (fig. 7-14). Sometimes the distribution of disease is a clue to the primary mechanism of tumor spread. For example, accumulation of tumor nodules in dependent locations may indicate intracavitary spread via fluid. In some cases, serosal spread occurs along lymphatic routes, detectable microscopically.

Clinical Features

Symptoms from secondary disease include abdominal enlargement from tumor bulk or, more often, ascites. Survival with peritoneal involvement is poor, and this is true with involvement at initial presentation or relapse (36). Ascites is a poor prognosis factor (34). Pain can also occur. Obstruction of the bowel can lead to pain, nausea, and vomiting, and may be a life-threatening condition.

Microscopic Findings

The same principles of microscopy and immunohistochemistry, discussed in relation to the pleura, apply to metastatic carcinomas affecting the peritoneum. Lymphoma can also diffusely involve the abdomen, including visceral organs and serosal surfaces. It can be associated with lymphadenopathy, splenomegaly, and ascites (37,38). Sarcomas also spread to the peritoneum, although this is unusual for soft tissue sarcomas (39). Gastrointestinal stromal tumors, however, can spread on peritoneal surfaces (40), especially with recurrent disease.

Intraperitoneal spread of uterine sarcoma occurs, with leiomyosarcoma the most common,

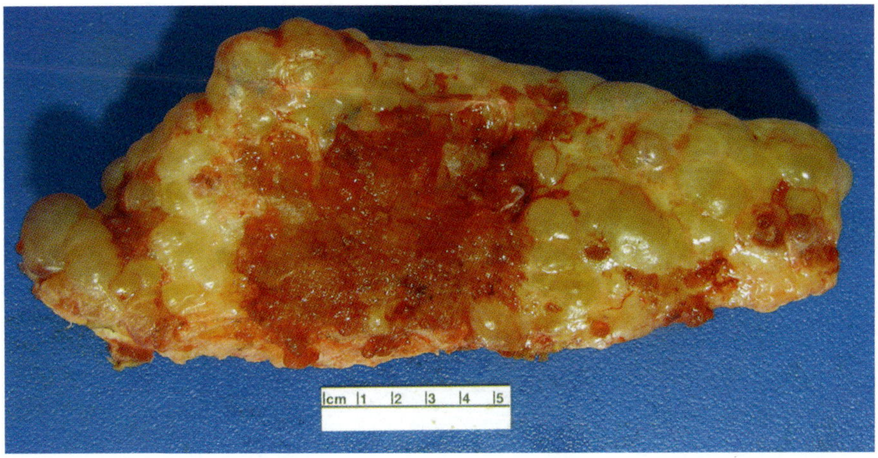

Figure 7-15

PSEUDOMYXOMA PERITONEI

The peritoneal surface is covered by mucinous tumor.

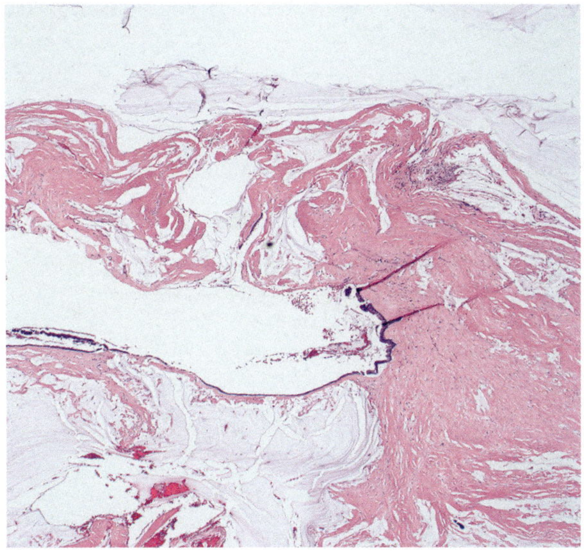

Figure 7-16

PSEUDOMYXOMA PERITONEI

Islands of mucoid material containing focal adenocarcinoma cells lie within the fibrous stroma.

followed by endometrial stromal sarcoma and adenosarcoma (41). Sarcomas are more often discrete nodules and less likely to result in ascites (38). Specifically, in pediatric patients, desmoplastic small round cell tumor and rhabdomyosarcoma cause peritoneal sarcomatosis (42).

There are also a few specific entities that merit further consideration in relation to this site because they are likely of uncertain or overlapping histogenesis. Most of these are discussed in chapter 8, although pseudomyxoma peritonei is discussed here.

Pseudomyxoma Peritonei. The peritoneal cavity can become filled predominantly with mucin, a phenomenon known as pseudomyxoma peritonei (fig. 7-15). This is often the result of a mucinous neoplasm of appendiceal origin (43,44), but can be derived from tumors from other parts of the gastrointestinal tract, or in some cases, nongastrointestinal extra-appendiceal sites (45). About 20 percent of mucinous neoplasms result in pseudomyxoma peritonei (46). Some series show a female predominance (47).

The nomenclature of such lesions has been a source of confusion. However, the term *pseudomyxoma peritonei* has been accepted and can be divided into low grade, high grade, and high grade with signet ring cells; these categories include identification of neoplastic mucinous epithelium, and the latter two categories are definitively carcinomatous. With these categories in mind, the histology of low-grade tumors (figs. 7-16, 7-17) includes uniform low-grade nuclei in cells with intracellular mucin within pools of extracellular mucin (48). High-grade lesions have a more complex architecture and cytologic atypia, or single cell invasion as signet ring cells.

Separation into low- and high-grade lesions has prognostic impact (49,50). The 5-year survival rate for patients with low-grade lesions is 75 percent, but is less than 10 percent for those with high-grade lesions. Tumors with signet ring cells may confer the poorest prognosis (51), specifically those with tissue invasive signet ring cells (52).

Treatment may impact the course of pseudomyxoma peritonei. Successful debulking is more likely in lower-grade diseases, and the

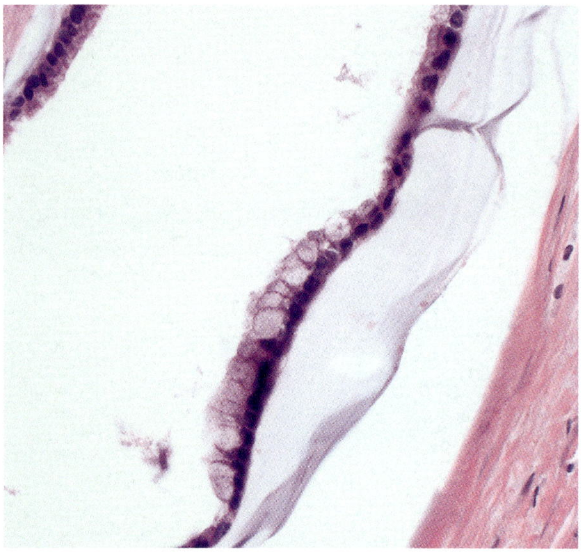

Figure 7-17

PSEUDOMYXOMA PERITONEI

The tumor cells show columnar mucinous morphology.

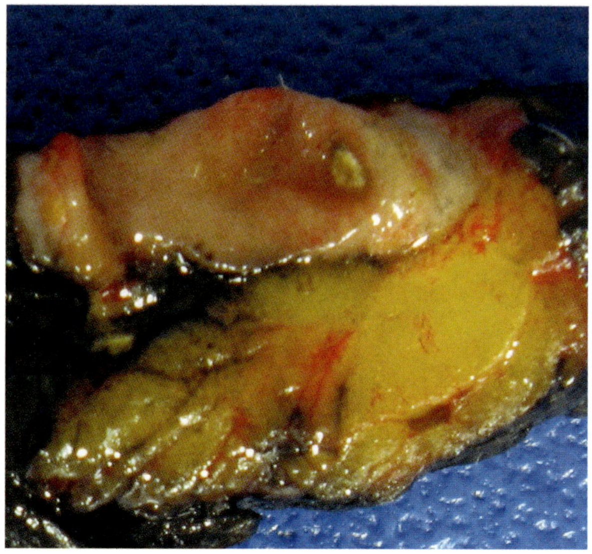

Figure 7-18

PRIMARY PERITONEAL CARCINOMA

A solitary tumor nodule involves mesenteric fat. The patient had normal-sized ovaries (not shown).

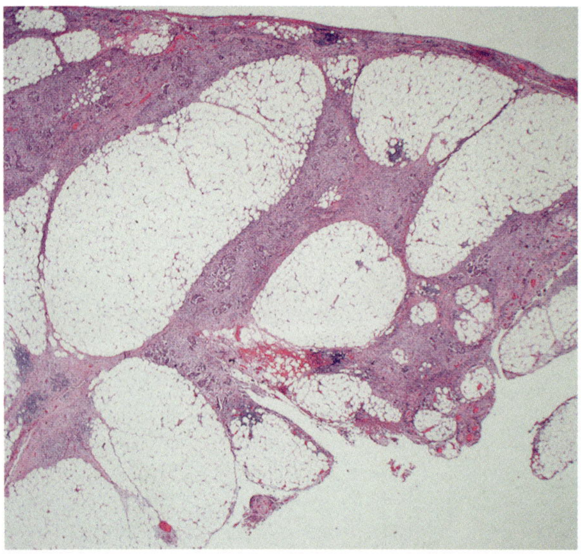

Figure 7-19

PRIMARY PERITONEAL CARCINOMA

Tumor diffusely involves peritoneal fat.

degree of cytoreduction is likely important in overall survival (53). Strategies of intraperitoneal chemotherapy have also been used (54,55). Nevertheless, the grade of the tumor remains the most important and independent prognostic variable in disease assessment.

Differential Diagnosis

Primary Peritoneal Carcinoma. While this term implies a primary neoplasm, it may be that these tumors are within the spectrum of tubo-ovarian neoplasms. In contrast to tumors of the ovary, this tumor can be present in the upper abdomen and mesentery (fig. 7-18), with little ovarian or pelvic involvement. The definition of this entity requires that both ovaries are of normal size, the bulk of the tumor is peritoneal, and the ovarian component is either not present or confined to surface involvement (56). Involvement can be nodular as well as diffuse (fig. 7-19). The histology of this tumor is usually serous, psammoma bodies may be present, and PAX8 and WT-1 are positive by immunohistochemistry. Other mesothelial markers are negative (figs. 7-20, 7-21).

Age at presentation is mainly in the 60s (57). It has been proposed that many so-called primary peritoneal carcinomas are derived from serous tubal in situ carcinoma (STIC) (58), but even extensive examination does not reveal STIC in all cases. It remains possible that clear cell and endometrioid patterns arise from endometriosis (59). Recent molecular analyses

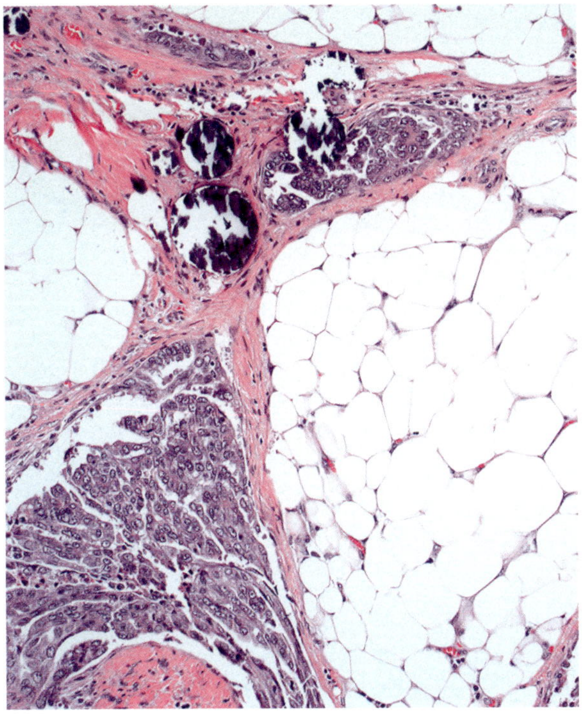

Figure 7-20

PRIMARY PERITONEAL CARCINOMA

Focal papillary islands of carcinoma cells are present, with psammoma body formation.

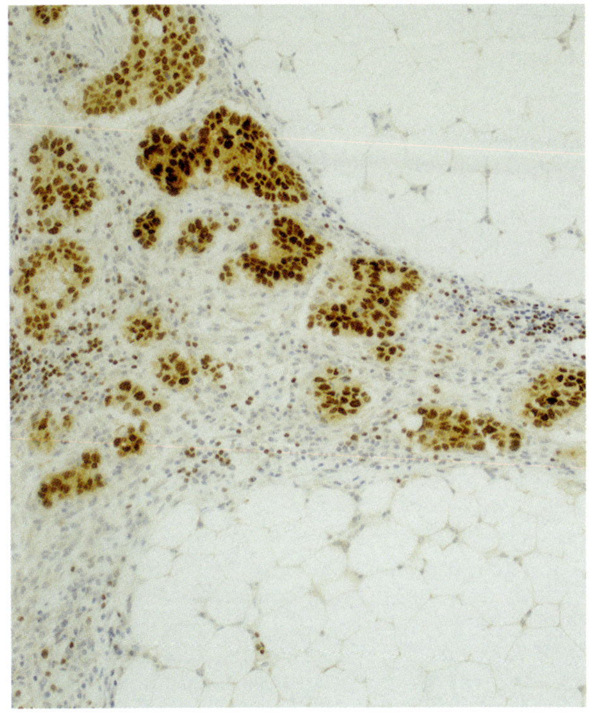

Figure 7-21

PRIMARY PERITONEAL CARCINOMA

The tumor cells stain for PAX8.

support common origin of serous carcinomas with and without STIC lesions (60).

Rare primary peritoneal mesenchymal tumors can occur. As an example, epithelioid hemangioendothelioma of the peritoneum can mimic mesothelioma (61,62), as it does in the pleura (63).

PERICARDIUM

Metastatic carcinoma of the lung, esophagus, and breast as well as sarcomas can involve the pericardium (fig. 7-22) (64–66). Pleural mesothelioma can involve the pericardium. Because of their anatomic location in the anterior mediastinum, thymomas commonly abut and invade the pericardium (67). Involvement of the pericardium by thymoma is part of staging (68,69).

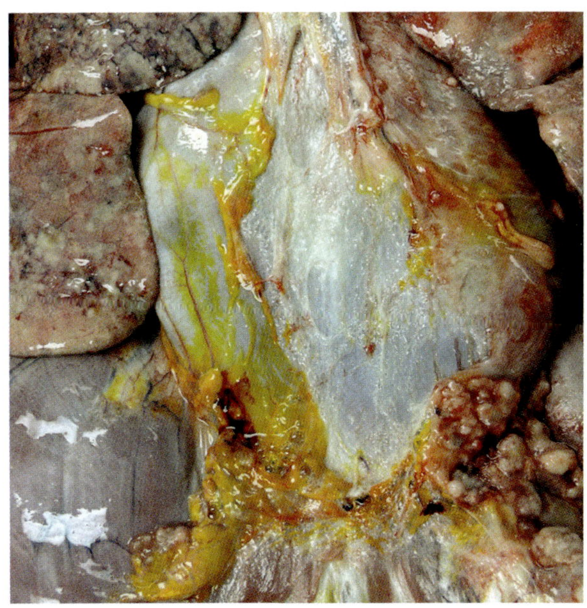

Figure 7-22

PERICARDIAL METASTASIS

The pericardium is studded with variably confluent nodules of tumor.

REFERENCES

1. Eberhardt WE, Mitchell A, Crowley J, et al. The IASLC Lung Cancer Staging Project: proposals for the revision of the M descriptors in the forthcoming eighth edition of the TNM classification of lung cancer. J Thorac Oncol 2015;10:1515-22.
2. Attanoos RL, Gibbs AR. 'Pseudomesotheliomatous' carcinomas of the pleura: a 10-year analysis of cases from the Environmental Lung Disease Research Group, Cardiff. Histopathology 2003;43:444-52.
3. Dessy E, Pietra GG. Pseudomesotheliomatous carcinoma of the lung. An immunohistochemical and ultrastructural study of three cases. Cancer 1991;68:1747-53.
4. Harwood TR, Gracey DR, Yokoo H. Pseudomesotheliomatous carcinoma of the lung. A variant of peripheral lung cancer. Am J Clin Pathol 1976;65:159-67.
5. Koss MN, Fleming M, Przygodzki RM, Sherrod A, Travis W, Hochholzer L. Adenocarcinoma simulating mesothelioma: a clinicopathologic and immunohistochemical study of 29 cases. Ann Diagn Pathol 1998;2:93-102.
6. Falconieri G, Zanconati F, Bussani R, Di Bonito L. Small cell carcinoma of lung simulating pleural mesothelioma. Report of 4 cases with autopsy confirmation. Pathol Res Pract 1995;191:1147-52.
7. Taylor DR, Page W, Hughes D, Varghese G. Metastatic renal cell carcinoma mimicking pleural mesothelioma. Thorax 1987;42:901-2.
8. Del Frate C, Mortele K, Zanardi R, et al. Pseudomesotheliomatous angiosarcoma of the chest wall and pleura. J Thorac Imaging 2003;18:200-3.
9. Falconieri G, Bussani R, Mirra M, Zanella M. Pseudomesotheliomatous angiosarcoma: a pleuropulmonary lesion simulating malignant pleural mesothelioma. Histopathology 1997;30:419-24.
10. Yousem SA, Hochholzer L. Unusual thoracic manifestations of epithelioid hemangioendothelioma. Arch Pathol Lab Med 1987;111:459-63.
11. Nicholson AG, Anderson E, Saha S, Indrajith M, Conry B, Hughes J. Progressive dyspnoea, pleural effusions and lytic bone lesions. Thorax 2008;63:492, 554.
12. Dermawan JK, Policarpio-Nicolas ML. Malignancies in pleural, peritoneal, and pericardial effusions. Arch Pathol Lab Med 2020;144:1086-91.
13. Siddiqui MT, Schmitt F, Churg A. Proceedings of the American Society of Cytopathology companion session at the 2019 United States and Canadian Academy of Pathology Annual meeting, part 2: effusion cytology with focus on theranostics and diagnosis of malignant mesothelioma. J Am Soc Cytopathol 2019;8:352-61.
14. Lepus CM, Vivero M. Updates in effusion cytology. Surg Pathol Clin 2018;11:523-44.
15. Sundling KE, Cibas ES. Ancillary studies in pleural, pericardial, and peritoneal effusion cytology. Cancer Cytopathol 2018;126(Suppl 8):590-8.
16. Hughes JC, Votaw ML. Pleural effusion in multiple myeloma. Cancer 1979;44:1150-4.
17. Leuallen EC, Carr DT. Pleural effusion—a statistical study of 436 patients. N Engl J Med 1955; 252:79-83.
18. Weick JK, Kiely JM, Harrison EG Jr, Carr DT, Scanlon PW. Pleural effusion in lymphoma. Cancer 1973;31:848-53.
19. Doerr CH, Allen MS, Nichols FC 3rd, Ryu JH. Etiology of chylothorax in 203 patients. Mayo Clin Proc 2005;80:867-70.
20. Brun C, Gay P, Cottier M, et al. Comparison of cytology, chest computed and positron emission tomography findings in malignant pleural effusion from lung cancer. J Thorac Dis 2018;10:6903-11.
21. Husain AN, Colby TV, Ordonez NG, et al. Guidelines for pathologic diagnosis of malignant mesothelioma 2017 update of the consensus statement from the International Mesothelioma Interest Group. Arch Pathol Lab Med 2018; 142:89-108.
22. Conner JR, Hornick JL. Metastatic carcinoma of unknown primary: diagnostic approach using immunohistochemistry. Adv Anat Pathol 2015;22:149-67.
23. Kalhor N, Moran CA. Primary thymic adenocarcinomas: a clinicopathological and Immunohistochemical study of 16 cases with emphasis on the morphological spectrum of differentiation. Hum Pathol 2018;74:73-82.
24. Ordonez NG. Value of PAX 8 immunostaining in tumor diagnosis: a review and update. Adv Anat Pathol 2012;19:140-51.
25. Ordonez NG. Value of melanocytic-associated immunohistochemical markers in the diagnosis of malignant melanoma: a review and update. Hum Pathol 2014;45:191-205.
26. Wei S, Henderson-Jackson E, Qian X, Bui MM. Soft tissue tumor immunohistochemistry update: illustrative examples of diagnostic pearls to avoid pitfalls. Arch Pathol Lab Med 2017; 141:1072-91.

27. Tozbikian GH, Zynger DL. A combination of GATA3 and SOX10 is useful for the diagnosis of metastatic triple-negative breast cancer. Hum Pathol 2019;85:221-7.
28. Baranov E, McBride MJ, Bellizzi AM, et al. A novel SS18-SSX fusion-specific antibody for the diagnosis of synovial sarcoma. Am J Surg Pathol 2020;44:922-33.
29. El Beaino M, Jupiter DC, Assi T, et al. Diagnostic value of TLE1 in synovial sarcoma: a systematic review and meta-analysis. Sarcoma 2020; 2020:7192347.
30. Asirvatham JR, Esposito MJ, Bhuiya TA. Role of PAX-8, CD5, and CD117 in distinguishing thymic carcinoma from poorly differentiated lung carcinoma. Appl Immunohistochem Mol Morphol 2014;22:372-6.
31. Dabir PD, Svanholm H, Christiansen JJ. SATB2 is a supplementary immunohistochemical marker to CDX2 in the diagnosis of colorectal carcinoma metastasis in an unknown primary. APMIS 2018;126:494-500.
32. Moser B, Fadel E, Fabre D, et al. Surgical therapy of thymic tumours with pleural involvement: an ESTS Thymic Working Group Project. Eur J Cardiothorac Surg 2017;52:346-55.
33. Levy AD, Shaw JC, Sobin LH. Secondary tumors and tumorlike lesions of the peritoneal cavity: imaging features with pathologic correlation. Radiographics 2009;29:347-73.
34. Chu DZ, Lang NP, Thompson C, Osteen PK, Westbrook KC. Peritoneal carcinomatosis in nongynecologic malignancy. A prospective study of prognostic factors. Cancer 1989;63:364-7.
35. Sadeghi B, Arvieux C, Glehen O, et al. Peritoneal carcinomatosis from non-gynecologic malignancies: results of the EVOCAPE 1 multicentric prospective study. Cancer 2000;88:358-63.
36. Glehen O, Osinsky D, Beaujard AC, Gilly FN. Natural history of peritoneal carcinomatosis from nongynecologic malignancies. Surg Oncol Clin N Am 2003;12:729-39, xiii.
37. Karaosmanoglu D, Karcaaltincaba M, Oguz B, Akata D, Ozmen M, Akhan O. CT findings of lymphoma with peritoneal, omental and mesenteric involvement: peritoneal lymphomatosis. Eur J Radiol 2009;71:313-7.
38. O'Neill AC, Shinagare AB, Rosenthal MH, Tirumani SH, Jagannathan JP, Ramaiya NH. Differences in CT features of peritoneal carcinomatosis, sarcomatosis, and lymphomatosis: retrospective analysis of 122 cases at a tertiary cancer institution. Clin Radiol 2014;69:1219-27.
39. Zanarini D, Sugarbaker PH. Extremity soft tissue sarcoma with metastases to abdominopelvic surfaces. J Surg Oncol 1997;64:68-72.
40. DeMatteo RP, Lewis JJ, Leung D, Mudan SS, Woodruff JM, Brennan MF. Two hundred gastrointestinal stromal tumors: recurrence patterns and prognostic factors for survival. Ann Surg 2000;231:51-8.
41. Sardi A, Sipok A, Baratti D, et al. Multi-institutional study of peritoneal sarcomatosis from uterine sarcoma treated with cytoreductive surgery and hyperthermic intraperitoneal chemotherapy. Eur J Surg Oncol 2017;43:2170-7.
42. Huh WW, Fitzgerald NE, Mahajan A, Hayes-Jordan A. Peritoneal sarcomatosis in pediatric malignancies. Pediatr Blood Cancer 2013;60:12-7.
43. Carr NJ, Cecil TD, Mohamed F, et al. A consensus for classification and pathologic reporting of pseudomyxoma peritonei and associated appendiceal neoplasia: the results of the Peritoneal Surface Oncology Group International (PSOGI) Modified Delphi Process. Am J Surg Pathol 2016;40:14-26.
44. Panarelli NC, Yantiss RK. Mucinous neoplasms of the appendix and peritoneum. Arch Pathol Lab Med 2011;135:1261-8.
45. Baratti D, Kusamura S, Milione M, et al. Pseudomyxoma peritonei of extra-appendiceal origin: a comparative study. Ann Surg Oncol 2016; 23:4222-30.
46. Smeenk RM, van Velthuysen ML, Verwaal VJ, Zoetmulder FA. Appendiceal neoplasms and pseudomyxoma peritonei: a population based study. Eur J Surg Oncol 2008;34:196-201.
47. Fernandez RN, Daly JM. Pseudomyxoma peritonei. Arch Surg 1980;115:409-14.
48. Misdraji J. Mucinous epithelial neoplasms of the appendix and pseudomyxoma peritonei. Mod Pathol 2015;28(Suppl 1):S67-79.
49. Ronnett BM, Zahn CM, Kurman RJ, Kass ME, Sugarbaker PH, Shmookler BM. Disseminated peritoneal adenomucinosis and peritoneal mucinous carcinomatosis. A clinicopathologic analysis of 109 cases with emphasis on distinguishing pathologic features, site of origin, prognosis, and relationship to "pseudomyxoma peritonei." Am J Surg Pathol 1995;19:1390-408.
50. Pai RK, Beck AH, Norton JA, Longacre TA. Appendiceal mucinous neoplasms: clinicopathologic study of 116 cases with analysis of factors predicting recurrence. Am J Surg Pathol 2009;33:1425-39.
51. Shetty S, Natarajan B, Thomas P, Govindarajan V, Sharma P, Loggie B. Proposed classification of pseudomyxoma peritonei: influence of signet ring cells on survival. Am Surg 2013;79:1171-6.
52. Sirintrapun SJ, Blackham AU, Russell G, et al. Significance of signet ring cells in high-grade mucinous adenocarcinoma of the peritoneum from appendiceal origin. Hum Pathol 2014;45: 1597-604.

53. Sugarbaker PH. Cytoreductive surgery and peri-operative intraperitoneal chemotherapy as a curative approach to pseudomyxoma peritonei syndrome. Eur J Surg Oncol 2001;27:239-43.
54. Baratti D, Kusamura S, Nonaka D, Cabras AD, Laterza B, Deraco M. Pseudomyxoma peritonei: biological features are the dominant prognostic determinants after complete cytoreduction and hyperthermic intraperitoneal chemotherapy. Ann Surg 2009;249:243-9.
55. Chan CH, Cusack JC, Ryan DP. A critical look at local-regional management of peritoneal metastasis. Hematol Oncol Clin North Am 2015;29:153-8.
56. Nik NN, Vang R, Shih Ie M, Kurman RJ. Origin and pathogenesis of pelvic (ovarian, tubal, and primary peritoneal) serous carcinoma. Annu Rev Pathol 2014;9:27-45.
57. Komiyama S, Nishijima Y, Kondo H, et al. Multicenter clinicopathological study of high-grade serous carcinoma presenting as primary peritoneal carcinoma. Int J Gynecol Cancer 2018;28:657-65.
58. Horn LC, Kafkova S, Leonhardt K, Kellner C, Einenkel J. Serous tubal in situ carcinoma (STIC) in primary peritoneal serous carcinomas. Int J Gynecol Pathol 2013;32:339-44.
59. Worley MJ, Welch WR, Berkowitz RS, Ng SW. Endometriosis-associated ovarian cancer: a review of pathogenesis. Int J Mol Sci 2013;14:5367-79.
60. Ducie J, Dao F, Considine M, et al. Molecular analysis of high-grade serous ovarian carcinoma with and without associated serous tubal intra-epithelial carcinoma. Nat Commun 2017;8:990.
61. Attanoos RL, Dallimore NS, Gibbs AR. Primary epithelioid haemangioendothelioma of the peritoneum: an unusual mimic of diffuse malignant mesothelioma. Histopathology 1997;30:375-7.
62. Posligua L, Anatelli F, Dehner LP, Pfeifer JD. Primary peritoneal epithelioid hemangioendothelioma. Int J Surg Pathol 2006;14:257-67.
63. Crotty EJ, McAdams HP, Erasmus JJ, Sporn TA, Roggli VL. Epithelioid hemangioendothelioma of the pleura: clinical and radiologic features. AJR Am J Roentgenol 2000;175:1545-9.
64. Chen KT. Extracardiac malignancy presenting with cardiac tamponade. J Surg Oncol 1983;23:167-8.
65. Fraser RS, Viloria JB, Wang NS. Cardiac tamponade as a presentation of extracardiac malignancy. Cancer 1980;45:1697-704.
66. Wilkes JD, Fidias P, Vaickus L, Perez RP. Malignancy-related pericardial effusion. 127 cases from the Roswell Park Cancer Institute. Cancer 1995;76:1377-87.
67. Batata MA, Martini N, Huvos AG, Aguilar RI, Beattie EJ Jr. Thymomas: clinicopathologic features, therapy, and prognosis. Cancer 1974;34:389-96.
68. Nicholson AG, Detterbeck FC, Marino M, et al. The IASLC/ITMIG Thymic Epithelial Tumors Staging Project: proposals for the T component for the forthcoming (8th) edition of the TNM classification of malignant tumors. J Thorac Oncol 2014;9(Suppl 2):S73-80.
69. Kondo K, Van Schil P, Detterbeck FC, et al. The IASLC/ITMIG Thymic Epithelial Tumors Staging Project: proposals for the N and M components for the forthcoming (8th) edition of the TNM classification of malignant tumors. J Thorac Oncol 2014;9(Suppl 2):S81-7.

8 NON-NEOPLASTIC ENTITIES AND ENTITIES OF UNCERTAIN OR OVERLAPPING HISTOGENESIS

Reactive conditions involving the serosal membranes are common and vary from diffuse predominantly inflammatory disorders to more specific and sometimes localized fibrotic disorders. The serosal membranes are also common sites for the presentation of certain systemic diseases. It is, therefore, unsurprising that a reporting pathologist is frequently faced with the diagnostic dilemma of whether a serosal proliferation is reactive or neoplastic. Although the specifics of dealing with the differential diagnosis of mesothelioma are covered in chapter 3, this chapter describes not only a variety of non-neoplastic conditions but also entities of uncertain histogenesis. There are also entities in the literature where knowledge of the histogenesis has increased since the last edition of this Fascicle, for example, pseudomyxoma peritonei is now discussed in chapter 7 under metastatic disease.

NEUTROPHILIC EFFUSIONS AND EMPYEMA

A purely *neutrophilic infiltrate* is commonly found in patients undergoing cytologic assessment for either a pleural, ascitic, or pericardial effusion. In most cases, a bacterial infection will be, or already has been, confirmed by microbiology, although the presence of an intra-abdominal abscess can elicit a neutrophil-rich pleural effusion. The samples are then sent to cellular pathology primarily to ensure that there is no evidence of an underlying malignancy.

In the pleural space, a diagnosis of *empyema* is made when there is persistent accumulation of purulent material within the space, which often leads to the development of a surrounding fibrous wall or cortex (fig. 8-1). The evacuation of purulent material and debridement of the cavity wall are common histologic and cytologic samples (fig. 8-2) and should be carefully assessed in order to ensure there is no underlying tumor (see Reactive Fibrous Pleuritis below). The incidence of empyema appears to be increasing in both children and young adults (1,2).

LYMPHOCYTIC EFFUSIONS AND REACTIVE FIBROUS PLEURITIS

A *lymphocytic effusion* is one of the most commonly encountered samples in thoracic pathology. Many cases are due to mycobacterial infection (3–5), with pleural effusions occurring in about one third of cases of tuberculosis. Other causes include malignancy (6), connective tissue disorders (7,8), iatrogenic (post-surgical) (9), pulmonary emboli (10), chylothorax, and lymphangioleiomyomatosis (11).

Effusions due to tuberculosis were once thought solely to be due to a delayed hypersensitivity reaction but are currently believed to

Figure 8-1

EMPYEMA

The pleural space is filled with hemorrhagic and fibrinopurulent debris. An enclosing pleural cortex is present.

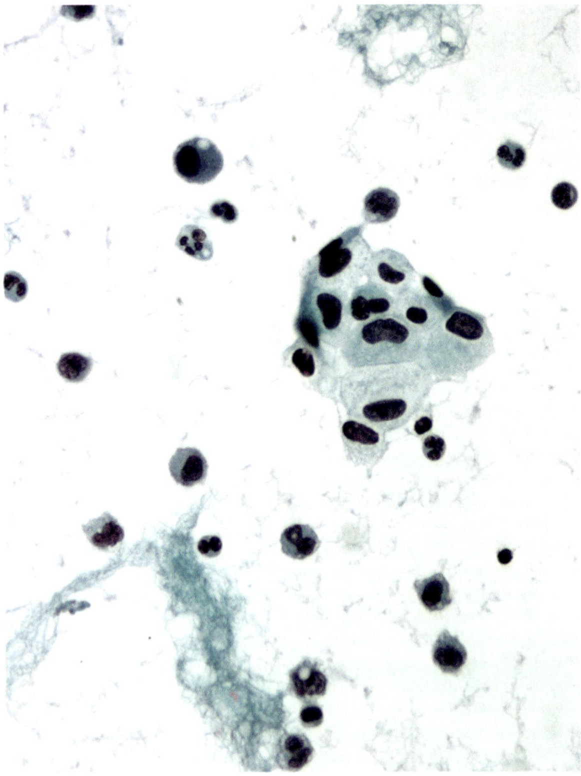

Figure 8-2

ACUTE AND CHRONIC EFFUSION

Both neutrophils and lymphocytes are present within proteinaceous debris. A cluster of mesothelial cells shows mild reactive atypia.

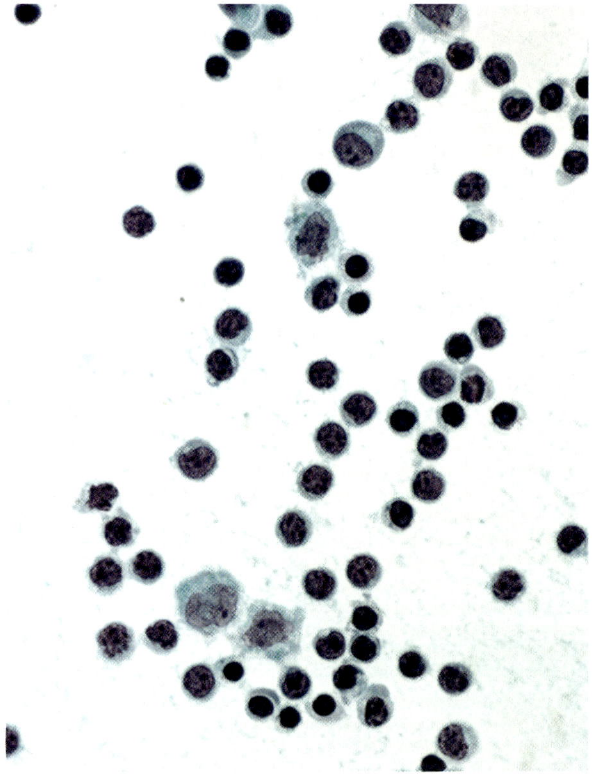

Figure 8-3

LYMPHOCYTIC EFFUSION

Most cells are small lymphocytes without atypia, mixed with a few plasma cells and histiocytes.

be in part secondary to direct infection. Special staining for acid-fast bacilli is rare in cytologic samples, since culture and identification using polymerase chain reaction (PCR) is more sensitive for identifying organisms (12,13). Necrotic material may also be seen within effusions in tuberculosis, although this does not distinguish it from other causes of effusions such as nonmycobacterial empyema or noninfective causes such as chronic effusions secondary to rheumatoid disease (14). A combination of a lymphocytic effusion and high adenosine deaminase (ADA) levels is used to infer a diagnosis of tuberculosis, although identification of organisms is the ideal (3–5). Correlation with other clinical and serologic parameters aid in distinguishing neoplastic from non-neoplastic causes (15,16).

In most cases, the majority of cells in a lymphocytic effusion are small lymphocytes without atypia, mixed with histiocytes and mildly atypical reactive mesothelial cells that can be either single or form small flat sheets or small clusters. Sometimes, the lymphoid infiltrate is more heterogeneous, and both plasma cells and larger blastic cells are seen (fig. 8-3), although this does not indicate that the infiltrate is malignant. Plasma cells may be present and, if abundant, IgG4-related disease should be considered (see below) (17). Chronic follicular pleuritis has also been described, although with no difference in clinical associations from reactive pleuritis without follicles (18).

If a lymphoid infiltrate is particularly dense and monomorphic, then a more thorough assessment may be required to determine whether the lymphoid cells are neoplastic. This typically means using immunohistochemistry, often on a cell block made from the effusion, to determine whether there is a predominance of B cells rather than T cells, as well as the presence

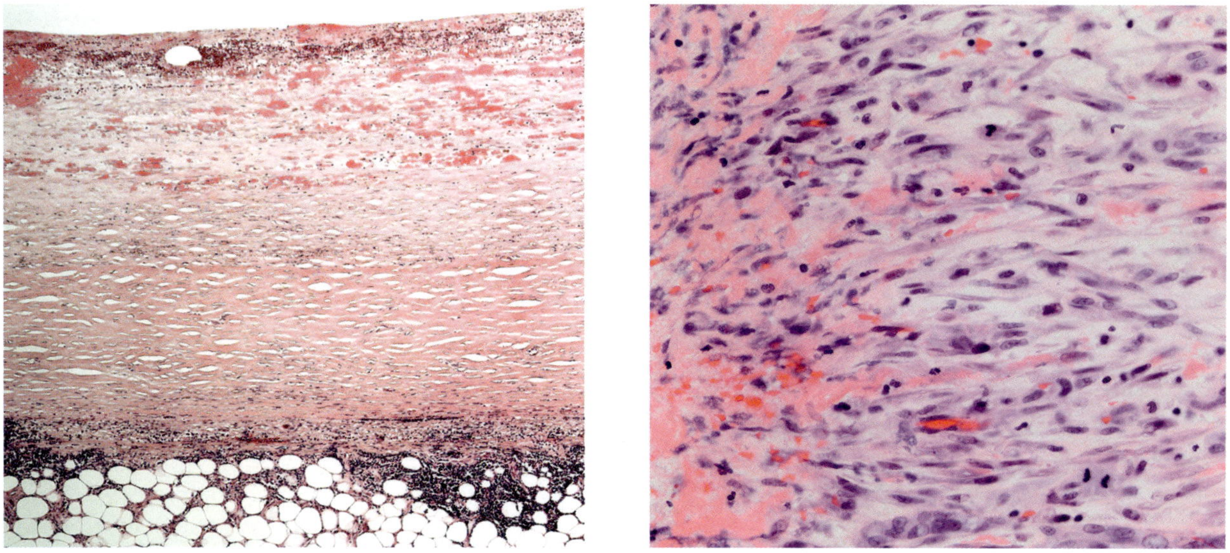

Figure 8-4

REACTIVE FIBROUS PLEURITIS

Left: The surface is replaced by chronically inflamed granulation tissue and fibrin, with loose fibroblastic tissue that matures to denser, more established fibrosis away from the surface. As is often the case, there is a linear chronic inflammatory cell infiltrate at the interface between the fibrosis and subpleural fat.

Right: Granulation tissue show moderate atypia and mitotic activity.

of light chain restriction. Flow cytometry and molecular studies such as PCR to look for clonality may also be of value. The clinical history should be sought, as a diagnosis of lymphoma at another site has often already been made, but that information has not been provided to the pathologist. Imaging is also important, especially in children with dense lymphocytic effusions, as mediastinal lymphomas produce alarmingly dense lymphocyte-rich samples.

Fibrous (Organizing) Pleuritis. Fibrous pleuritis reflects the organization and deposition of a variably collagenous stroma within the pleural space. This is also referred to as *pleurisy*, *fibrous pleurisy,* and *organizing pleurisy*. Prefixing as "reactive" fibrous pleuritis may be helpful in emphasizing the non-neoplastic nature of the sample. When fibrous pleuritis is marked and results in fibrosis surrounding and compressing the lung, it is referred to by some as *fibrothorax*.

Etiologically, reactive fibrous pleuritis may be part of empyema of any cause, although it is also seen in association with other disorders such as exposure to asbestos, connective tissue disorders, chronic infections lacking a neutrophilic component, iatrogenic (postsurgical), and drug-related causes. Rarely, bilateral idiopathic chronic pleuritis occurs (19).

The histologic features vary, depending on the extent of inflammatory activity. In active lesions, fibrin may cover the surface, with underlying granulation tissue. Inflammation within the granulation tissue may be both acute and chronic. Mesothelial cells are either absent or sparse. As the fibrosis extends away from the pleural surface, there is decreasing cellularity and increasing fibrous tissue. It is, therefore, especially important to orientate the specimens so that sections are taken perpendicular to the surface (fig. 8-4, left). This "maturation with depth" helps distinguish reactive fibrous pleuritis from desmoplastic mesothelioma, since mesotheliomas typically lack the zonation of a reactive process and are often more cellular at the deep, advancing tumor edge. Also, when properly orientated, capillaries may be prominent and lie perpendicular to the surface. Sometimes, hemosiderin-laden macrophages are present. PD-L1 staining has been shown to be of value, with reactive pleuritis being negative and sarcomatoid mesotheliomas generally being positive (20).

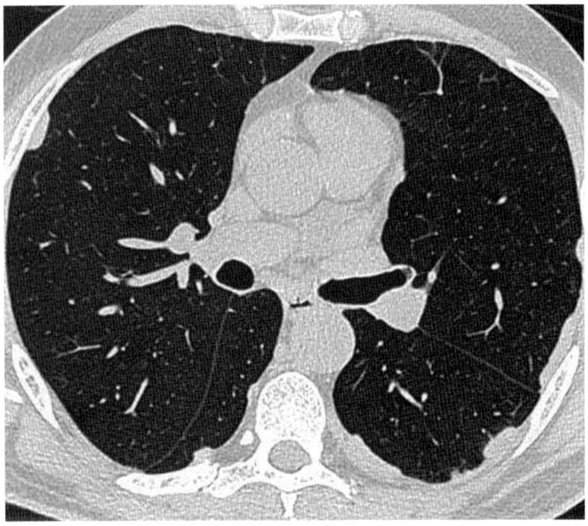

Figure 8-5

PLEURAL PLAQUE

A computerized tomography (CT) image shows localized fibrous plaque-like nodules within the pleura.

Cytologically, stromal cells immediately deep to the surface within the granulation tissue may show quite marked atypia but, in the context of the above architecture, this is reactive atypia rather than neoplastic in nature (fig. 8-4, right; see figs. 3-107, 3-108). Spindle cells within the granulation tissue are fibroblastic, often with plump nuclei. Toward the deeper aspect, the cells become more fibrocytic, thin, and spindled, situated between bands of collagen. In difficult cases, immunohistochemistry for broad-spectrum keratins highlights the maturation of the fibrosis, with stronger staining toward the surface within the granulation tissue and lesser intensity toward the deeper aspect. This staining pattern often separates desmoplastic mesotheliomas, where the staining is generally more diffuse and stronger, often with greater intensity at the advancing edge of the tumor. Reactive fibrous pleuritis also often stops abruptly at the junction between pleura and chest wall fat, where there may be a greater aggregation of chronic inflammatory cells. However, it can extend into subpleural fat, so its presence should not be used as a marker of malignancy. In general, reactive fibrous stroma within subpleural fat is paucicellular and cytokeratin negative (21–23).

A process morphologically identical to fibrous pleuritis occurs in hydroceles. There is similar zonation, with decreasing cellularity away from the lumen of the hydrocele and similar parallel layering to that seen in fibrous pleuritis. Organizing peritonitis may also occur but this is much less uncommon.

Fibrothorax is an extreme example of organizing pleuritis in which a prior inflammatory process results in more or less total replacement of the fat by dense paucicellular fibrosis, often with extension into chest wall muscle. This can cause functional restriction and contraction of the lung. Calcification may be present.

There are diagnostic pitfalls to be avoided. Chronic pleural effusions and fibrosis may be induced by malignancies and thorough sampling of submitted tissue should be undertaken as both fibrous pleuritis and malignancy may coexist, with malignancy only present in a few areas. This can be especially problematic when sarcomatoid mesothelioma and fibrous pleuritis coexist. Conversely, the presence of mesothelial cells within the fibrous stroma does not necessarily indicate malignant infiltration, as native reactive mesothelial cells may become entrapped within fibrous pleuritis. When this occurs, the reactive mesothelial cells generally have a linear arrangement parallel to the surface, as opposed to a more irregular infiltrative architecture when neoplastic.

PLEURAL PLAQUES

Pleural plaques are localized scars typically found on the parietal pleura. They may also be found on the surface of the diaphragm or posterolaterally, running along the direction of the ribs. Microscopically, they are sharply circumscribed, white to gray firm lesions of variable thickness, with smooth surfaces. Sometimes they are nodular, an appearance that has been compared to dripping candle wax. Pleural plaques are often calcified or even ossified, and are widely visible on chest radiographs or computerized tomography (CT) (figs. 8-5; 8-6, left). They are associated with exposure to asbestos in most cases.

Histologically, pleural plaques consist of acellular hyalinized collagen arranged in a "basket-weave" pattern (fig. 8-6, right). Sometimes, there is focal chronic inflammation at the interface between the pleural plaque and adjacent

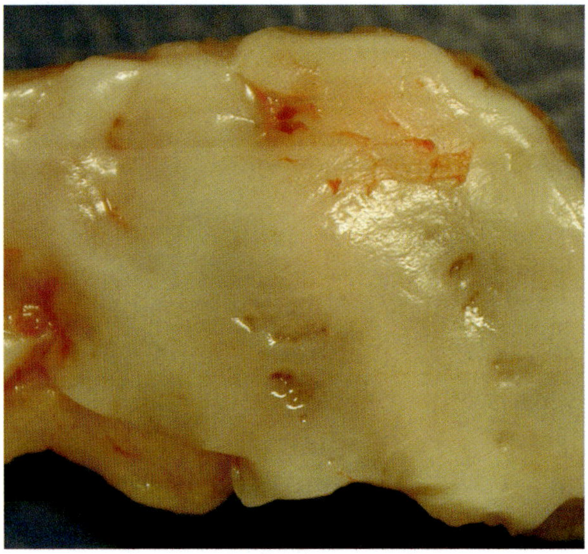

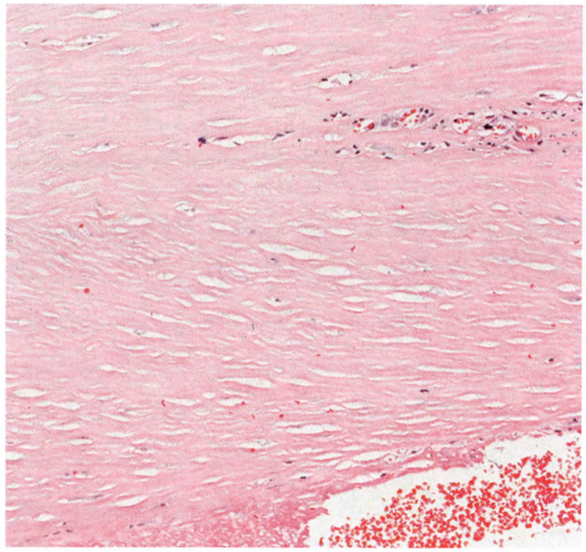

Figure 8-6

PLEURAL PLAQUE

Left: A sharply demarcated, hard, almost waxy plaque lies on the parietal pleural surface.
Right: The fibrosis shows a lamellar "basket-weave" pattern.

fibrosis or fat. Rarely, spindle cells are positive for cytokeratins. Pleural plaques are usually readily distinguishable from the desmoplastic variant of sarcomatoid mesothelioma since they are much less cellular and the fibrosis has a more regular architecture, being basket-weave rather than storiform or haphazard.

Pleural plaques have no malignant potential. Occasionally, they become sufficiently rounded that they form thoracoliths, which may detach from the pleural surface (fig. 8-7).

Figure 8-7

THORACOLITH

A smooth, partly calcified nodule was removed from the pleural cavity, which showed the characteristic fibrous pattern of a pleural plaque.

MESOTHELIAL HYPERPLASIA

Normal mesothelium is described in chapter 1 and consists of flat inconspicuous cells covering serosal surfaces. When hyperplastic, the cells become much more prominent and cuboidal on serosal surfaces and more rounded in effusions. The nuclear features, however, remain bland and mostly show no more than mild atypia.

Simple hyperplasia consists of only a single layer on the surface that is regularly spaced (fig. 8-8A,B). This is common when the patient has benign ascites, for example, in disorders such as endometriosis, hernia sacs, and hydroceles, or is part of a reaction to pneumothorax. In pneumothorax, reactive mesothelial cells may be mixed with palisading histiocytic inflammation and eosinophils.

Atypical mesothelial hyperplasia is a term used when reactive mesothelial proliferations are more florid, with greater atypia in single or aggregated cells on the pleural surface (fig. 8-8C,D; see figs. 3-98–3-100) or in pleural effusions. As opposed to mesothelioma with a tubulopapillary architecture, these aggregates lack stromal cores.

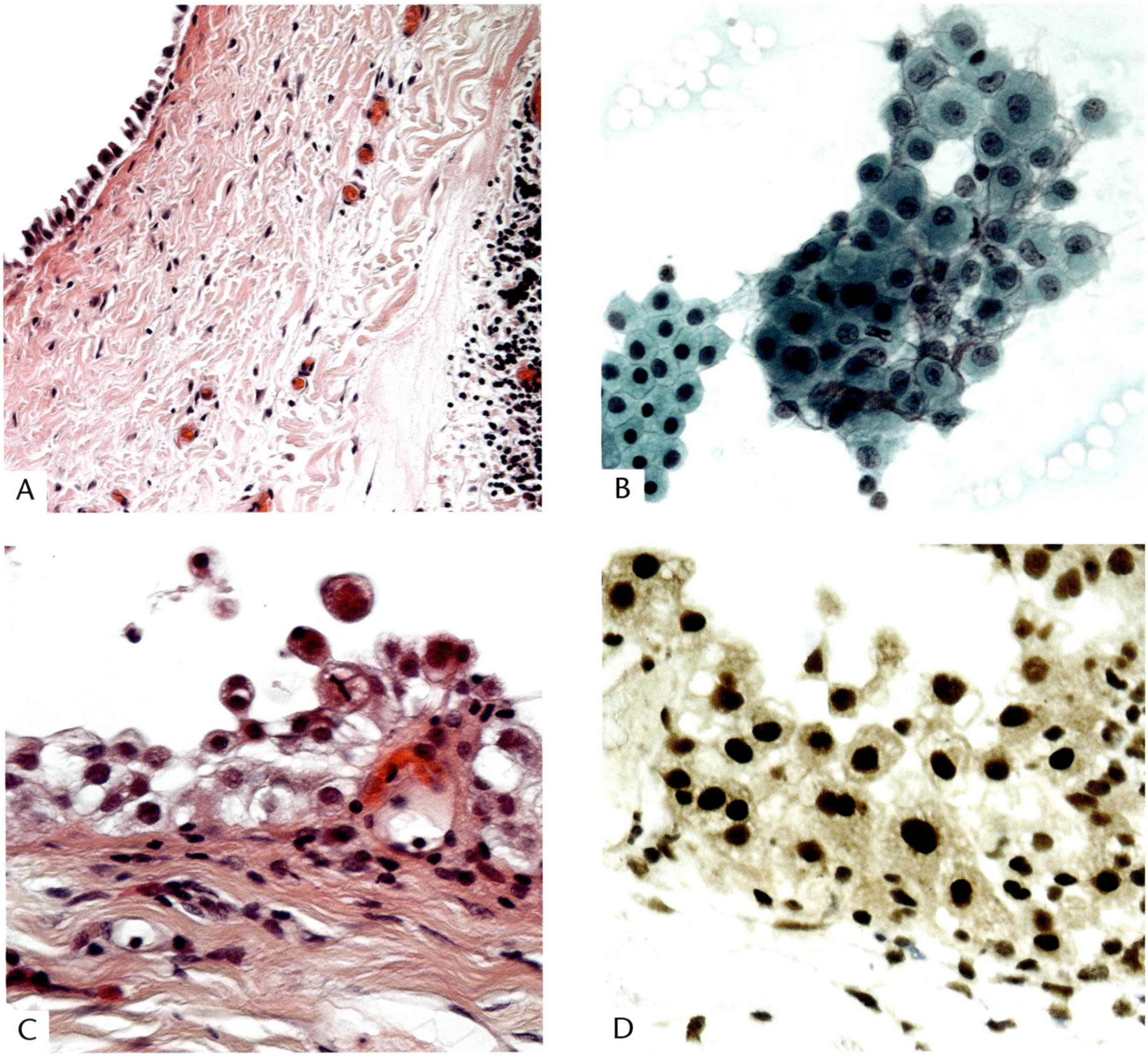

Figure 8-8

SIMPLE AND ATYPICAL MESOTHELIAL HYPERPLASIA

A,B: Simple: A single layer of mildly atypical mesothelial cells lines chronically inflamed fibrotic pleura (A). An effusion contains an aggregate of histiocytes to the right of a small sheet of benign hyperplastic mesothelial cells (B).

C,D: Atypical: The peritoneal surface is covered by a population of atypical mesothelial cells. There is mitotic activity but mitotic figures are not atypical. Tubulopapillary architecture is lacking (C) and there is no loss of BAP1 staining (D).

Cytologically, the cells may have prominent nucleoli and enlarged nuclei with anisocytosis. The distinction between atypical mesothelial hyperplasia from epithelioid mesothelioma is discussed in chapter 3 (21–24). In cases where distinction between atypical mesothelial hyperplasia and mesothelioma cannot be made, the term *atypical mesothelial proliferation* is often used.

Occasionally, mixed nodular aggregates of histiocytes and mesothelial cells are found, for which many terms are used: *nodular histiocytic/mesothelial hyperplasia*, *nodular histiocytic hyperplasia*, and *mesothelial/monocytic incidental cardiac excrescences* (MICE) (25,26). Nodular proliferative lesions within hernia sacs also likely fall into this category. These nodules may

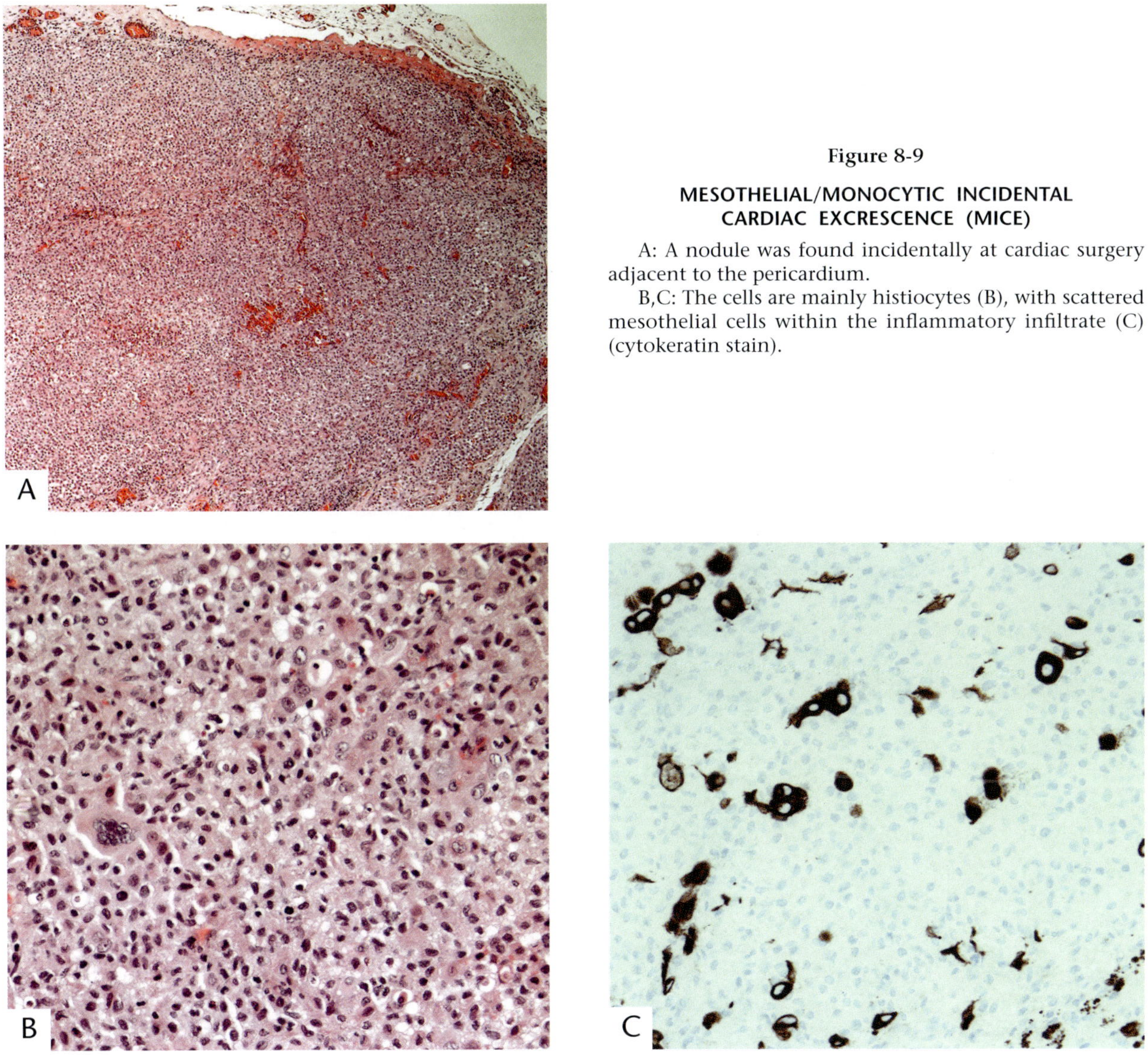

Figure 8-9

MESOTHELIAL/MONOCYTIC INCIDENTAL CARDIAC EXCRESCENCE (MICE)

A: A nodule was found incidentally at cardiac surgery adjacent to the pericardium.

B,C: The cells are mainly histiocytes (B), with scattered mesothelial cells within the inflammatory infiltrate (C) (cytokeratin stain).

be identified at surgery and mimic a diffuse nodular neoplastic proliferation when they form grossly visible masses (27). Their presence may be associated with previous surgical instrumentation and sometimes they coexist with a true neoplasm.

The histologic findings are those of a nodular aggregate comprising monomorphic cytologically bland cells with pale cytoplasm and round to oval nuclei. Mitotic activity may be present, although atypical variants are not seen. These cells stain mainly for histiocytic markers such as CD68. Staining for cytokeratins and mesothelial markers usually shows a few scattered mesothelial cells lying within the histiocytic infiltrate (fig. 8-9).

CYSTIC TUMOR OF THE ATRIOVENTRICULAR NODE

Cystic tumor of the atrioventricular (AV) node is considered to be an endodermally derived embryologic rest (28–30) rather than a proliferation of mesothelial cells. This is the recommended term within the 2021 World Health Organization (WHO) classification and the term mesothelioma of the atrioventricular node is no longer recommended (31). There is no evidence that this is a clonal neoplastic

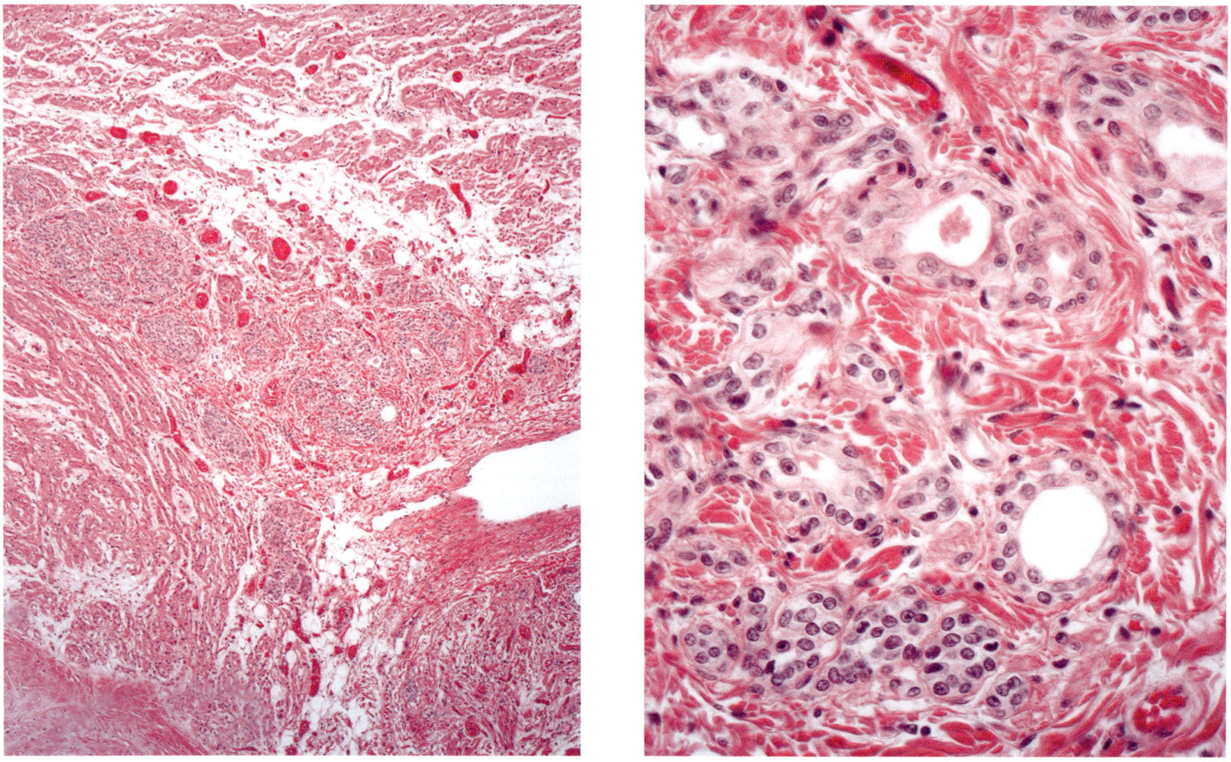

Figure 8-10

CYSTIC TUMOR OF THE ATRIOVENTRICULAR NODE

Left: Within bundles of myocardium, nests of an epithelioid proliferation are seen.
Right: This lesion is composed of bland tubules and solid nests. (Courtesy of Dr. J. Graham, New York, NY)

proliferation. Synonyms include *endodermal/ectopic heterotopia/inclusion/cyst* and *mesothelioma of the atrioventricular node*.

Sudden death can occur as a result of this lesion, since it is associated with the conduction system and can result in electrical disruption and arrhythmia (32). As a result, it is usually detected postmortem. However, surgical excision has been successful when the lesion is detected antemortem (33,34), with lesional size likely contributing to detection. Presentation is usually secondary to AV conduction heart block. Cystic lesions may be visible on CT or magnetic resonance imaging (MRI) (34).

This is generally a small lesion, 2 mm to up to 2 cm, localized in the triangle of Koch in the AV node, at the base of the atrial septum. Histologically, the cells of this lesion are epithelioid, and in this regard, resemble mesothelial cells. They are flat or cuboidal, nonciliated, and can be part of a single layer or a multilayered solid cell nest. It has been confirmed that the cells are epithelial (35) (fig. 8-10).

Immunohistochemistry supports epithelial and sometimes neuroendocrine differentiation (35). There is positive staining for CK5 but staining for other mesothelial markers, such as calretinin, is negative. Electron microscopy does not show the prominent microvilli of mesothelial differentiation (36).

Rarely simple foregut cysts can occur within the heart and should be differentiated. Teratomas also require exclusion.

EOSINOPHILIC PLEURITIS

Eosinophilic pleuritis is a common finding in surgical pathology, primarily due to the submission of pleural samples postpneumothorax (fig. 8-11, top). It may also be seen with hemothorax, although this likely reflects the fact that there is a coexistent pneumothorax. It has been hypothesized that the reason for eosinophilic

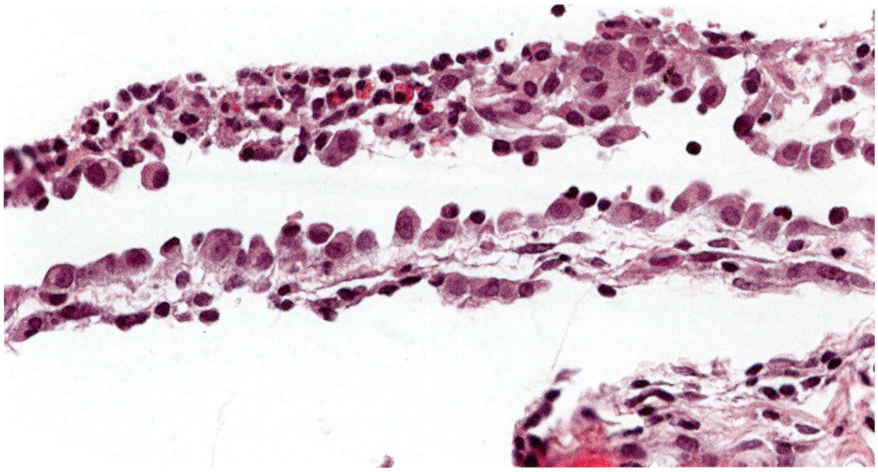

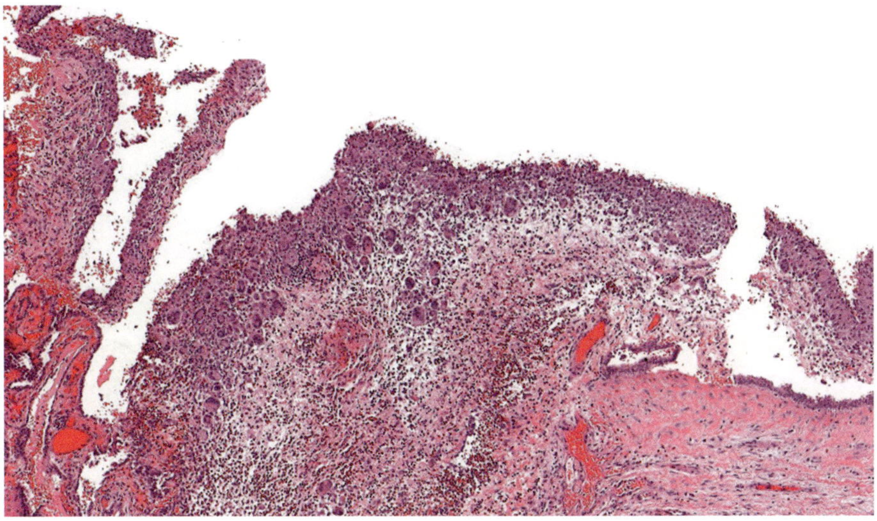

Figure 8-11

EOSINOPHILIC/HISTIOCYTIC PLEURITIS SECONDARY TO PNEUMOTHORAX

Top: The pleura is covered by a layer of palisading histiocytic cells mixed with variable numbers of eosinophils. Reactive mesothelial cells form a layer deep to the inflammation.

Bottom: The pleural surface is partly replaced by palisading granulomatous pleuritis containing abundant histiocytic giant cells.

infiltration is an inflammatory reaction to the presence of air in the pleural space, although it is incongruous that neither pneumoperitoneum nor interstitial emphysema induce significant eosinophilia within soft tissues or lung parenchyma.

Eosinophilia may also be seen in effusions: an eosinophilic effusion is defined as greater than 10 percent eosinophils, with a variety of reported causes including air or blood in the pleural spaces and drug reactions (37–40). If secondary to chemotherapeutic agents, then there may be significant reactive atypia within the mesothelial cells, and knowledge of previous or current therapies is useful in helping avoid a misdiagnosis of malignancy. Infections with parasites and fungus are also sometimes associated with eosinophilia (41,42).

GRANULOMATOUS PLEURITIS

Granulomatous pleuritis may present microscopically with a variety of patterns. The most commonly seen is palisading histiocytic inflammation replacing the pleural surface, a feature associated with pneumothorax when there may be coexistent eosinophilic pleuritis (fig. 8-11, bottom). Granulomatous inflammation may also be discrete or confluent, and either necrotizing or non-necrotizing. When present, infection always requires thorough exclusion, specifically mycobacterial infection (4). The presence or absence of necrosis is not a sensitive marker for distinguishing infectious from noninfectious causes in this situation. It is good practice to always use Ziehl-Nielsen and Grocott (or other fungal) stains to assess for infection when granulomas are present (fig. 8-12).

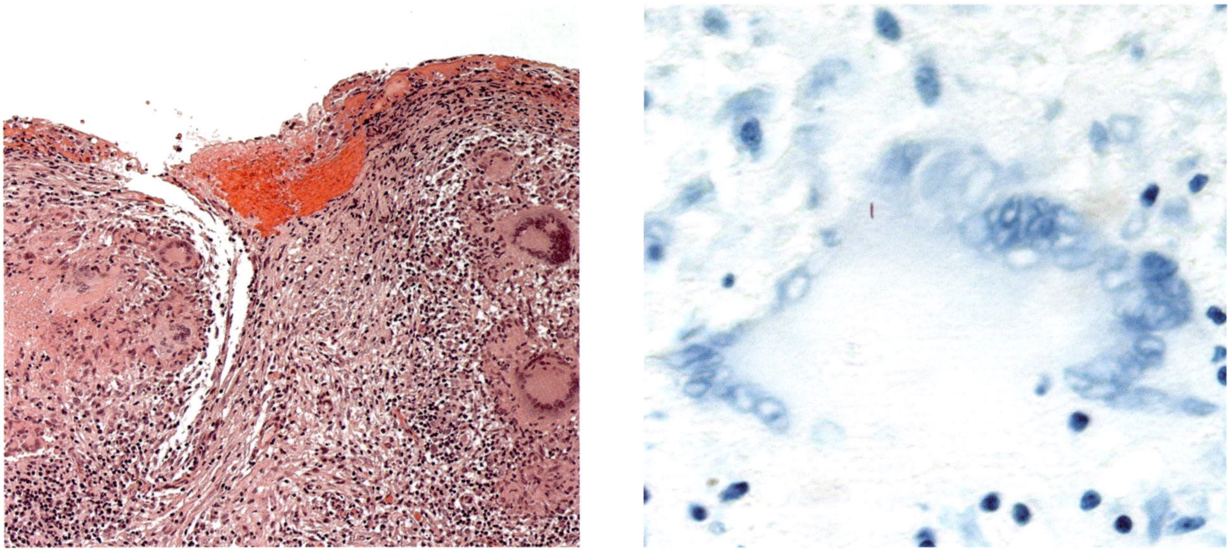

Figure 8-12

GRANULOMATOUS PLEURITIS DUE TO TUBERCULOSIS

Left: The pleura contains both necrotizing and non-necrotizing granulomas.
Right: An acid-fast bacillus is seen with a Ziehl-Nielsen stain.

Other noninfectious causes of serosal granulomatous inflammation do occur (43), such as sarcoidosis. Fistulation of granulomatous lung cavities, for example, secondary to rheumatoid nodules or pulmonary granulomatosis with polyangiitis (44), may also rarely lead to granulomatous changes in the pleura. Foreign body type granulomatous inflammation may occur, most commonly seen in relation to talc pleurodesis (fig. 8-13). Serosal involvement by histiocytic diseases may occur, with rare cases of Rosai-Dorfman disease described involving the pleura, pericardium, and peritoneum (45,46). Erdheim-Chester disease more commonly involves the serosal membranes, but is now classified as a neoplasm and is discussed in chapter 6, although involvement is nearly always part of multiorgan disease (47).

HEMOTHORAX

Hemothorax is arbitrarily defined as a pleural effusion with a hematocrit of 50 percent of the patient's peripheral blood hematocrit, although a degree of hemorrhage is often seen in many effusions. Its presence should heighten suspicion of malignancy. Abundant red blood cells are seen within the effusion and, dependent on chronicity, hemosiderin deposition may also be seen. Indeed, hemosiderin accumulation may be seen within the cytoplasm of reactive mesothelial cells, indicating that they have the capacity to phagocytose (see fig. 1-8). Aggregated hemosiderin particles are referred to as Heinz bodies. Hemothorax is usually due to trauma, although it may be due to bleeding secondary to causes such as tumor deposition, iatrogenic injury (fig. 8-14), and endometriosis.

ROUNDED ATELECTASIS

Rounded atelectasis is a localized area of collapsed lung parenchyma entrapped by overlying fibrotic pleura. It is also known as *folded lung* and *Blesovsky syndrome*. Presentation is often asymptomatic, with identification as an incidental finding on imaging. Patients may, however, present with chest pain. CT shows a characteristic "comet-tail" sign, composed of a bronchovascular bundle extending from the mass and curving toward the hilum of the lung (fig. 8-15) (48–50).

As discussed earlier, there are numerous conditions that can cause pleural fibrosis, with exposure to asbestos a common cause of this disorder (48–50). Pleural plaques and/or calcifications may also be present.

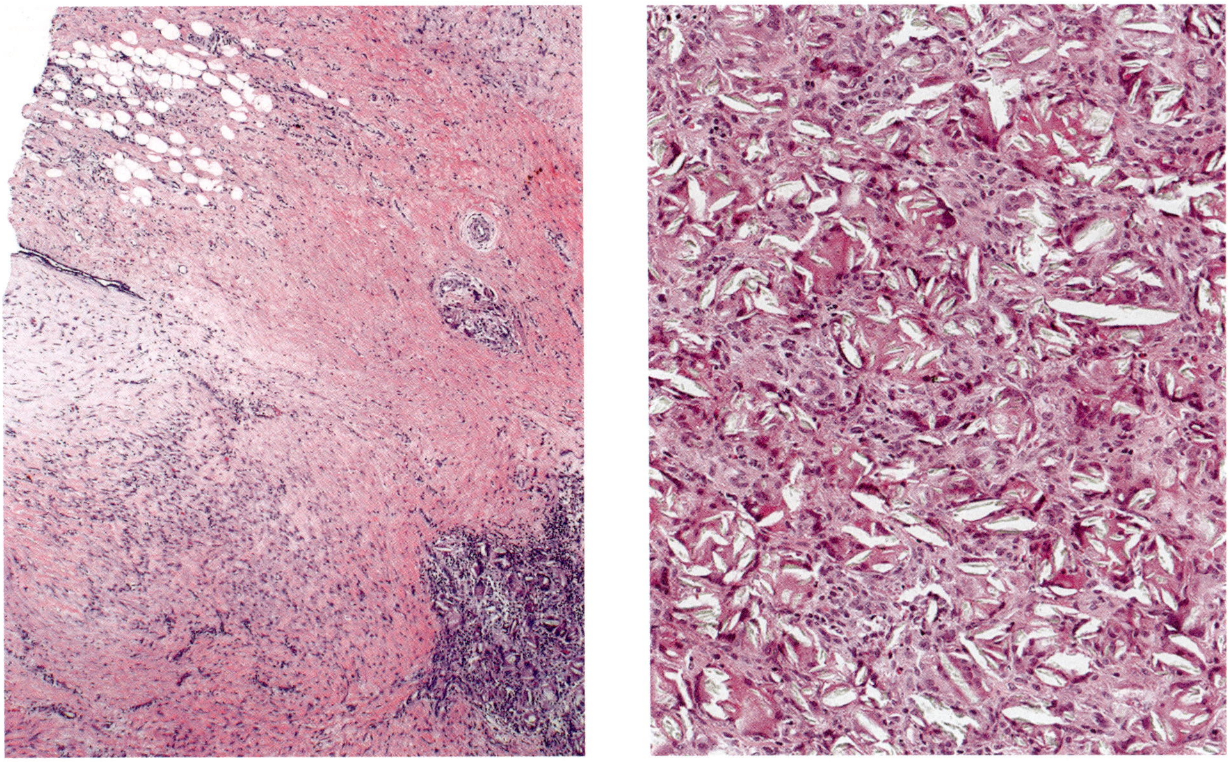

Figure 8-13

TALC PLEURODESIS

Left: Foreign body-type granulomatous inflammation induces dense fibrosis.
Right: Giant cells contain talc crystals.

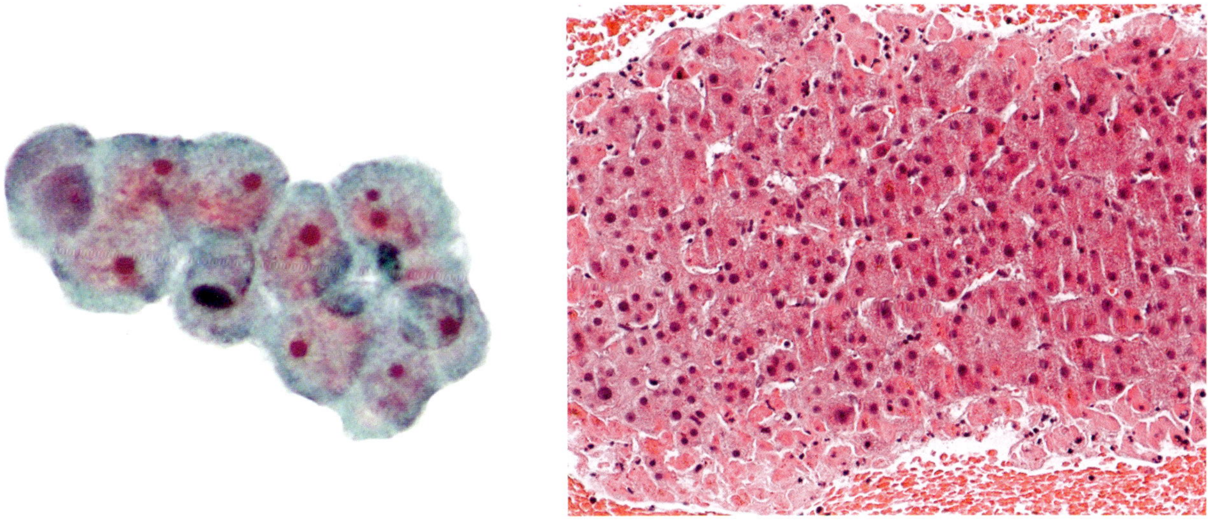

Figure 8-14

HEMOTHORAX DUE TO IATROGENIC INJURY

Left: Atypical epithelioid cells were seen in a blood-stained pleural effusion, and a cell block was subsequently created.
Right: The cell block shows clusters of histologically normal liver parenchyma, presumably secondary to injury during tube insertion.

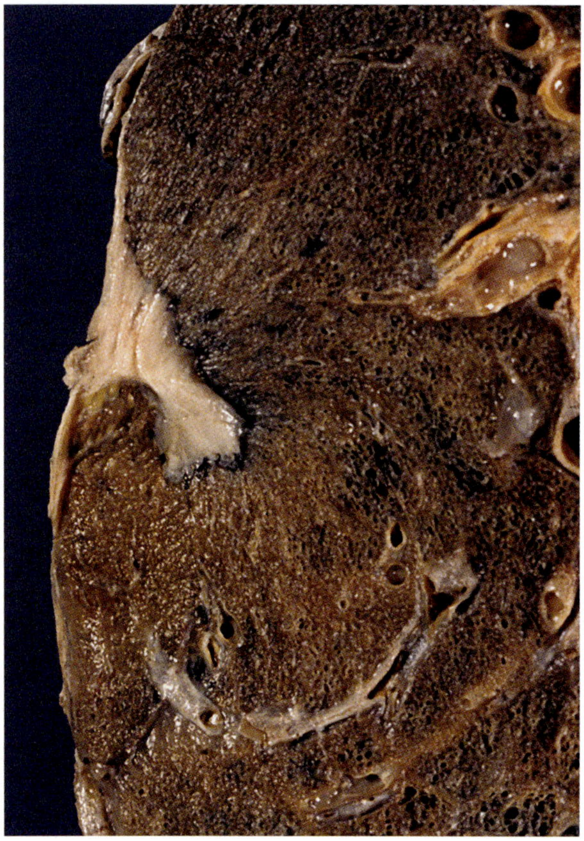

Figure 8-15

ROUNDED ATELECTASIS

There is a localized area of visceral pleural fibrosis that appears to be retracted into the lung parenchyma. The adjacent lung is atelectatic. The patient had prior radiotherapy for Hodgkin lymphoma.

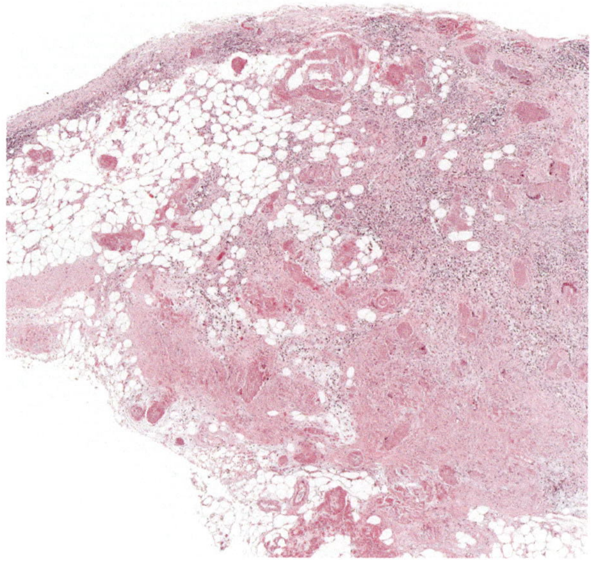

Figure 8-16

PLEURAL AMYLOIDOSIS

The pleura contains acellular eosinophilic material that is associated with a patchy chronic inflammatory cell infiltrate and elicits a foreign-body type giant cell reaction.

Because they form peripheral mass-like lesions, they are often mistaken clinically for a neoplasm. However, positron emission tomography (PET) scans are typically negative or show low uptake. On occasion, foci of rounded atelectasis may be resected, with a localized area of visceral pleural fibrosis often forming a rounded scar, with adjacent atelectatic lung. Sometimes, the surgeon will simply dissect the pleural fibrosis and the mass disappears without the need for dissection. If unresected, most areas either decrease in size or remain stable (51).

AMYLOIDOSIS

Pleural presentation of *amyloidosis* is rare and usually occurs in the context of primary systemic disease (52). When involving the pleura, patients develop a persistent pleural effusion in 1 to 2 percent of cases (53) and present with dyspnea and chest pain. There is often involvement of underlying lung parenchyma when the pleura is involved. Microscopically, there may be diffuse pleural thickening that mimics mesothelioma (54–56) or multiple small nodules within the serosal membranes (57,58).

The diagnosis is made by the identification of the characteristic acellular eosinophilic material that is deposited within the tissues, which is sometimes associated with foreign body-type granulomatous inflammation and a mild patchy lymphoplasmacytic infiltrate (fig. 8-16). Congo red staining shows characteristic apple-green birefringence and electron microscopy shows the diagnostic fibrillary appearance. Rare cases have been diagnosed on cytology (59).

If amyloidosis is diagnosed within the serosal membranes, the patient should be assessed for extent of systemic disease as involvement of other organ systems, particularly the heart, portend a poorer prognosis (60). Material should also be sent for subtyping of amyloid because management differs for some variants. Light chain deposition disease also involves the pleura in similar fashion, although is considerably rarer (61).

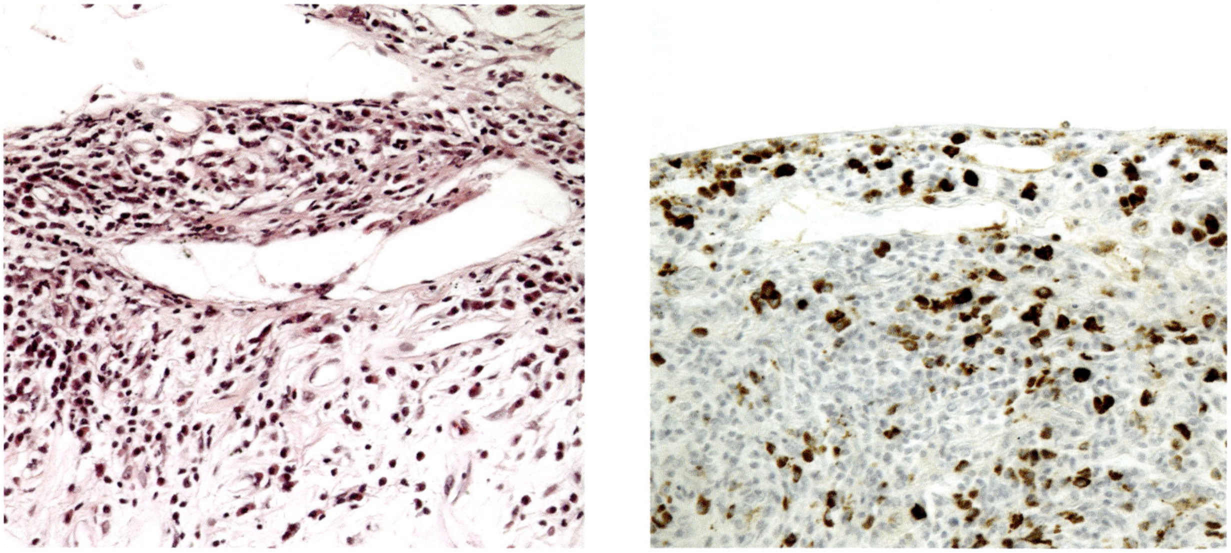

Figure 8-17

PLEURAL INVOLVEMENT BY IgG4-RELATED DISEASE

The pleura is infiltrated by abundant plasma cells (left), and the majority stain for IgG4 (right).

IgG4-RELATED DISEASE

IgG4-related disease is a recently described disorder, first recognized in the pancreas and now known to occur at many sites. It is rare within the serosal membranes, where there may be large pleural effusions as well as ascites and pericardial effusions (17,62–65). Rarely, do clinical findings mimic mesothelioma (66).

Morphologically, the features are similar to those seen at other sites: a predominance of IgG4-positive plasma cells (fig. 8-17) (67,68). However, an excess of IgG4-positive plasma cells is present in other disorders, and clinical correlation is required before definitive assignment of diagnosis (68,69).

ENDOMETRIOSIS

Endometriosis is a condition that can involve many organ sites, with the serosal membranes often involved by ectopic endometrial stroma in females of reproductive age. Endometriosis is considered to be a benign inflammatory lesion, but has some cancer-like features and may harbor somatic cancer driver mutations (70,71).

The peritoneum is the most common serosal site, although endometriosis involves the pleural surfaces occasionally, more frequently (over 90 percent) on the right side, believed due to extension through fenestrations in the diaphragm. Thoracic endometriosis tends to occur about 5 years later than pelvic disease and presentation peaks at a slightly older age (30 to 34 years versus 25 to 29 years) (72–74).

Pneumothorax is the most common manifestation of endometriotic involvement of the pleura, followed by hemothorax (72–74). When involving the pleura, patients may present with catamenial pneumothorax. Effusions may occur (72). Right-sided pneumothorax, a history of pelvic endometriosis, an age of 31 years or more, and a no smoking history suggest endometriosis rather than primary spontaneous pneumothorax (75). CT features are not specific although MRI may highlight hemorrhage within the lesions (76). When involving the peritoneum, abdominal/pelvic pain is the most common symptom.

Macroscopically, endometriosis appears as blue to brown nodules on the serosal surface that may be either cystic or hemorrhagic. Fibrosis may also occur, leading sometimes to extensive adhesions that can cause obstruction of visceral organs when severe.

At microscopy, endometriosis is characterized by benign endometrial glandular epithelial structures lying within endometrial-type stroma (fig. 8-18). The stromal elements are always

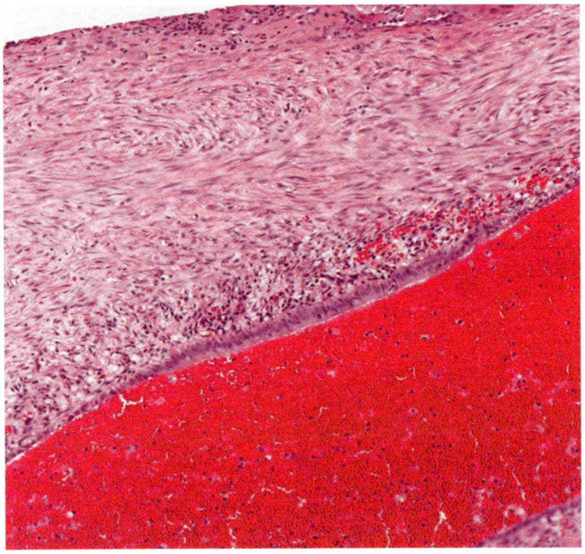

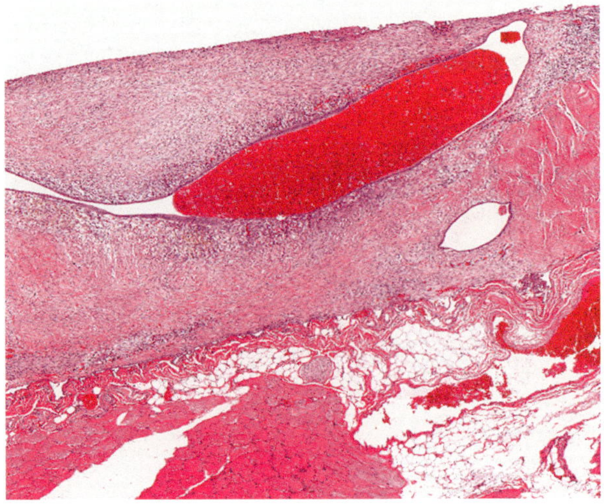

Figure 8-18

PLEURAL ENDOMETRIOSIS

Left: The pleura contains an endometrial gland with hemorrhage in its lumen.
Right: Endometrial stroma surrounds the gland.

present although epithelial elements are only seen in about 25 percent of cases (77). These foci may show alterations similar to those of cycling endometrium, and may show atrophic changes in postmenopausal women. Hyperplasia and metaplasia also occur. Occasionally, there is coexistent prominent smooth muscle proliferation, sometimes termed *endomyometriosis*. Rarely, carcinomas arise from foci of endometriosis. The spindle cells stain for hormone receptors and typically for CD10, with about 50 percent of cases staining for smooth muscle actin (SMA) (77,78).

For management of pneumothorax, surgical pleural abrasion is superior to hormonal treatment in the long term. Earlier diagnosis and effective therapy can decrease the morbidity of this disease in women during their reproductive period (74).

Endosalpingiosis is a similar metaplastic process, mainly presenting in the pelvic peritoneum, omentum, and on the ovarian surface. There are no reports of endosalpingiosis arising in the pleura, although mullerian cysts are described occasionally in the mediastinum (79). As with endometriosis, presentation may be incidental or associated with pelvic pain (80). Microscopically, the lesions are usually small and sometimes cystic, but rarely extensive enough that neoplasia is mimicked (81). At microscopy, glands lined by a single layer of cytologically bland ciliated epithelial cells are associated with fibrotic stroma that lacks features of endometrial-type stroma. Endocervical-type glands may also be encountered and coexistent psammoma bodies may be seen. Cells may stain for PAX8 (80,82).

Another rare finding in the peritoneal cavity is *deciduosis*, in which there is an ectopic decidual reaction that presents in pregnancy or following hormonal therapy (83,84). Within the submesothelial stroma, and more rarely in the abdominal sites removed from the pelvis, stromal cells show abundant pale granular eosinophilic cytoplasm with enlarged polygonal cells (fig. 8-19). Nuclei are cytologically bland, but there may be necrosis and mild nuclear atypia. When extensive, the features may be sufficiently florid to mimic malignancy (85), and morphologically may be misinterpreted as deciduoid mesothelioma. However, deciduoid mesotheliomas stain for cytokeratins and mesothelioma-associated markers (86–88), while true decidua is negative and also stains for hormone

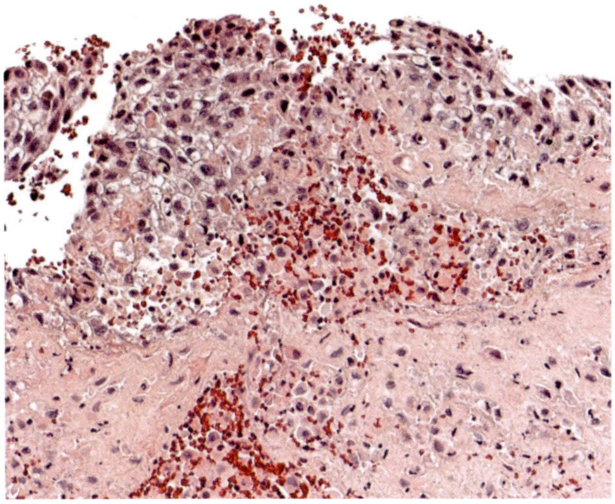

Figure 8-19

DECIDUOSIS

The peritoneal surface is replaced by decidual cells in a patient with an ectopic pregnancy.

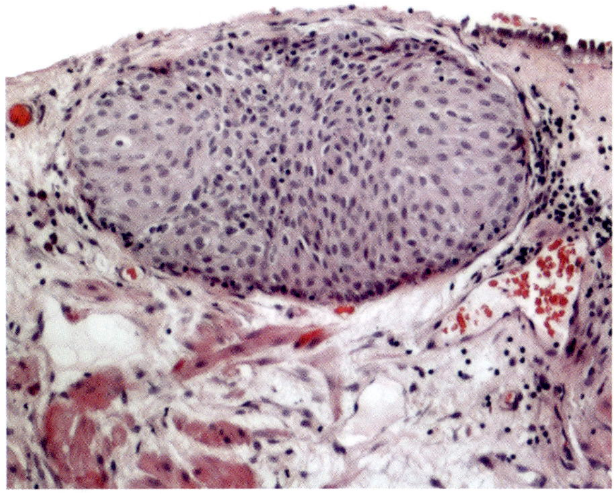

Figure 8-20

WALTHARD REST

An island of transitional looking cells is present deep to the peritoneal surface.

receptors (84). Rare cases of deciduosis causing pneumothorax are reported (89,90).

TROPHOBLASTIC IMPLANTS

Trophoblastic implants occur secondary to postoperative complications of a tubal pregnancy, with implantation of trophoblastic and villous elements within the peritoneum (91–93). Patients may present with intra-abdominal hemorrhage. Raised levels of beta-human chorionic gonadotrophin may be present (94).

WALTHARD RESTS

Walthard rests are incidental nodules that arise in the peritoneum within the serosa of the fallopian tubes or the ovarian hila. Macroscopically, they are small, glistening, solid, or cystic nodules, sometimes mistaken for disseminated malignancy. On microscopy, they consist of aggregates of cuboidal to attenuated cytologically bland epithelial cells with transitional features (fig. 8-20). Rarely, Brenner tumors are reported arising from Walthard rests (95).

GLIOMATOSIS PERITONEI

Proliferation of benign glial tissue has been reported on the peritoneal surfaces (fig. 8-21), sometimes in association with previous teratomas (96–98). This usually occurs in females. On other occasions, it is seen in association with either reactive conditions such as endometriosis (99) or iatrogenically in relation to the presence of ventriculo-peritoneal shunts. Its etiology, therefore, seems multifactorial, with some suggesting its presence may be metaplastic rather than representing implantation of neoplastic deposits (97,100). As such, the behavior of the lesion in reports is variable, although generally favorable (96,97), ranging from disappearance, persistence, to even malignant transformation (101), likely reflecting the multiple conditions associated with its presence.

DISSEMINATED PERITONEAL LEIOMYOMATOSIS

Disseminated peritoneal leiomyomatosis (DPL) is a condition associated with typical uterine leiomyomas (102). There is also a strong association with pregnancy and oral contraceptive use. Rarely, DPL occurs in association with endometriosis (103). The disorder is viewed by some as metaplasia within the submesothelial mesenchyme, although there also appears to be an increased risk of DPL after uterine morcellation (104,105).

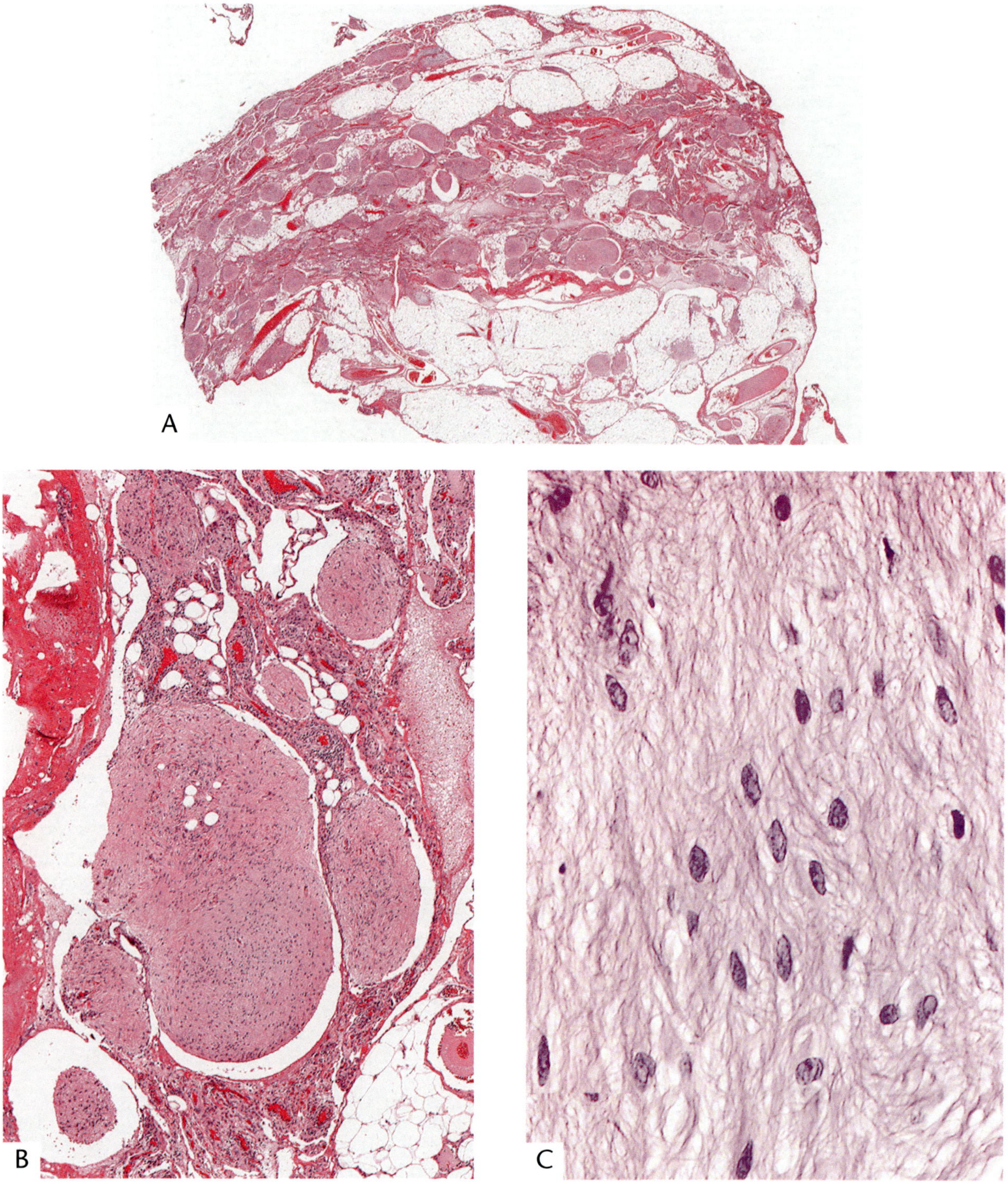

Figure 8-21

GLIOMATOSIS PERITONEI

A: At low power, abundant nodules of glial cells are present within peritoneal fat.

B: Intermediate power shows nests of glial cells within the peritoneum.

C: Higher magnification shows glial tissue consisting of spindle cells in a fibrillary matrix. (Courtesy of Dr. S. Robboy, Durham, NC)

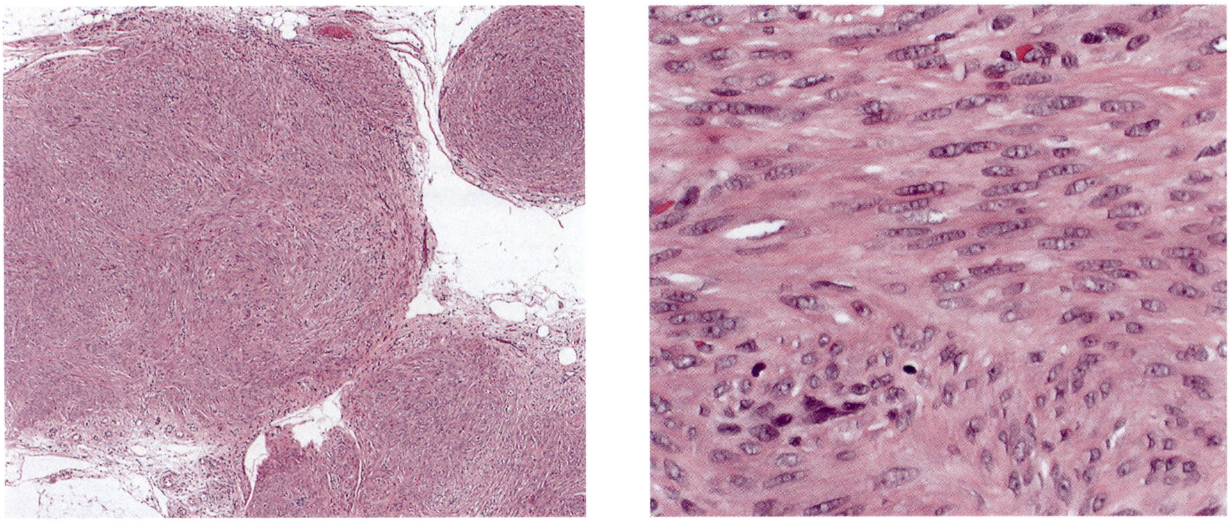

Figure 8-22

DISSEMINATED PERITONEAL LEIOMYOMATOSIS

Left: Low power shows nodules of eosinophilic spindle cells in a background of omental adipose tissue. Right: High power shows smooth muscle cells with bland nuclear features and an absence of mitoses.

Microscopically, there are well-demarcated firm white nodules that, on microscopy, consist of cytologically bland smooth muscle cells with a fascicular architecture that stain for desmin, SMA, and muscle-specific actin (fig. 8-22). There is also staining for hormone receptors. The clinical behavior is usually benign, with spontaneous regression in most cases, although recurrence and malignant transformation have been reported (106).

SPLENOSIS

Splenosis is the result of implantation of splenic tissue after either traumatic rupture or splenectomy. Patients are generally asymptomatic and lesions are incidental findings at either laparotomy or autopsy. The nodules may range considerably in size and there can be coexistent intrathoracic disease (fig. 8-23) (107–110).

MELANOSIS

Benign peritoneal *melanosis* is a rarely described phenomenon that seemingly has varied origins, with some cases associated with mature cystic teratomas and serous cystadenoma, and others associated with enteric duplication cysts (111–114). Macroscopically, there is black pigmentation of the peritoneal membranes. Microscopy shows histiocytes containing abundant melanin.

CHONDROID AND OSSEOUS METAPLASIA

Osseous or *chondroid metaplasia* may rarely involve the peritoneal cavity, usually as an incidental finding at laparotomy. Most patients have a history of previous surgery (115–118), with ossification also described after rupture of an aneurysm (116), abdominal trauma (119), and in association with peritoneal dialysis (120). Chondroid metaplasia is more common in women and mesenteric ossification is more common in men. Microscopically, there are well-circumscribed nodules of either chondroid or osseous stroma, lacking cytologic atypia. The lesions are presumed to represent metaplasia within the submesothelial mesenchyme.

OMENTAL-MESENTERIC MYXOID HAMARTOMA

Omental-mesenteric myxoid hamartoma is composed of multiple omental and mesenteric nodules that macroscopically may resemble metastatic implants (121–123). On microscopy, they consist of highly vascular loose myxoid stroma containing plump mesenchymal cells. Follow-up has shown a benign clinical course without recurrence.

It remains uncertain whether these reflect a persistent reactive process, a variant

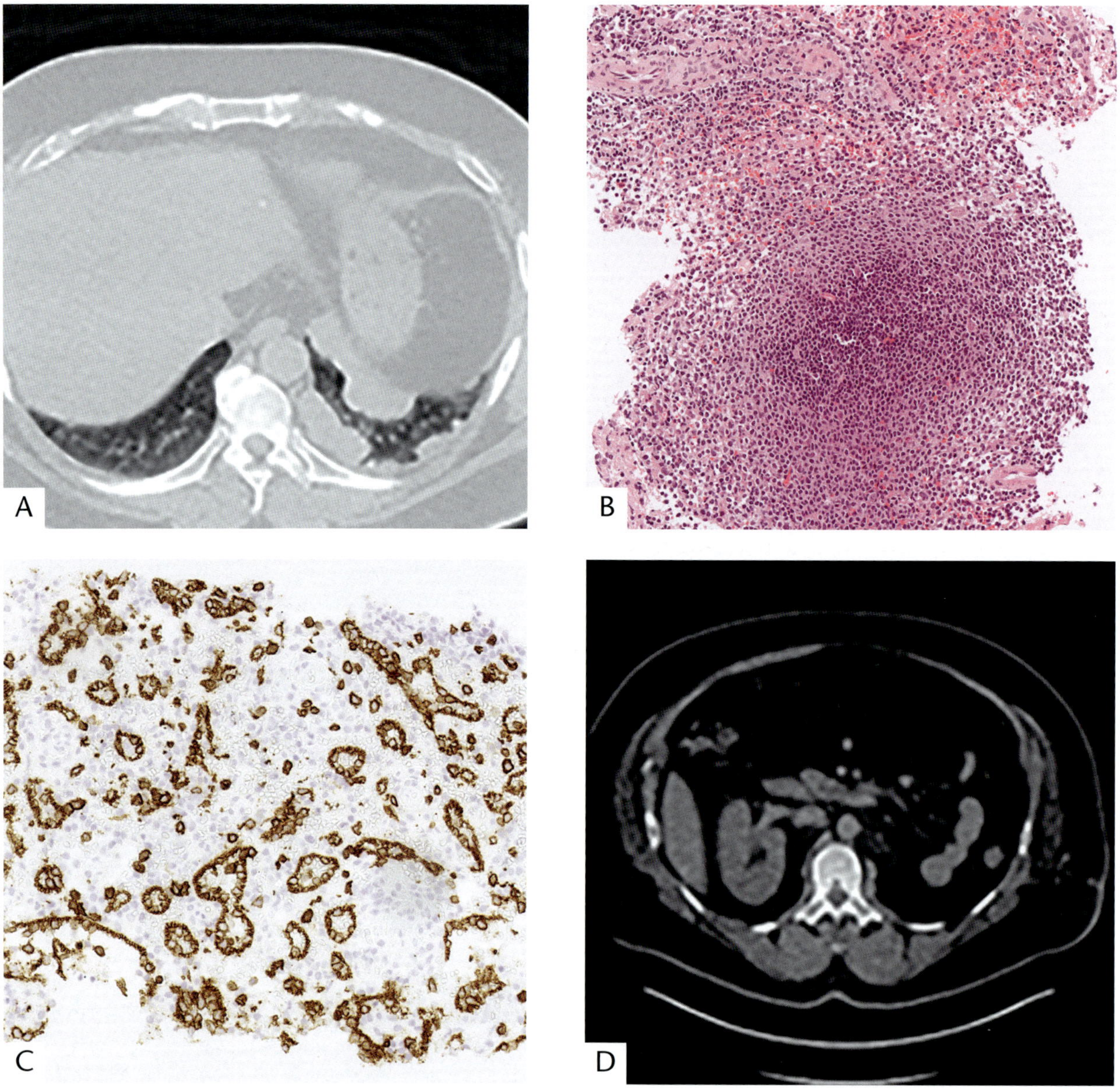

Figure 8-23

SPLENOSIS

A: Multiple nodules are present in the left pleural space.
B: Biopsy shows predominantly lymphoid stroma.
C: CD8 staining highlights splenic vasculature.
D: Abdominal imaging review revealed that the spleen was absent, and a history of splenic trauma in a road traffic accident was found on further history taking.

of inflammatory myofibroblastic tumor, or a metaplastic process within the submesothelial mesenchyme. One recent case showed an *ALK* gene rearrangement, providing supportive evidence of an inflammatory myofibroblastic tumor (124).

REFERENCES

1. Stankey CT, Spaulding AB, Doucette A, et al. Blood culture and pleural fluid culture yields in pediatric empyema patients - a retrospective review, 1996-2016. Pediatr Infect Dis J 2018; 37:952-4.
2. Burgos J, Falco V, Pahissa A. The increasing incidence of empyema. Curr Opin Pulm Med 2013; 19:350-6.
3. Zhai K, Lu Y, Shi HZ. Tuberculous pleural effusion. J Thorac Dis 2016;8:E486-94.
4. Vorster MJ, Allwood BW, Diacon AH, Koegelenberg CF. Tuberculous pleural effusions: advances and controversies. J Thorac Dis 2015;7:981-91.
5. Porcel JM. Advances in the diagnosis of tuberculous pleuritis. Ann Transl Med 2016;4:282.
6. Dixon G, Bhatnagar R, Zahan-Evans N, et al. A prospective study to evaluate a diagnostic algorithm for the use of fluid lymphocyte subset analysis in undiagnosed unilateral pleural effusions. Respiration 2018;95:98-105.
7. Lama A, Ferreiro L, Toubes ME, et al. Characteristics of patients with pseudochylothorax-a systematic review. J Thorac Dis 2016;8:2093-101.
8. Chung A, Wilgus ML, Fishbein G, Lynch JP 3rd. Pulmonary and bronchiolar involvement in Sjogren's syndrome. Semin Respir Crit Care Med 2019;40:235-54.
9. Light RW. Pleural effusions following cardiac injury and coronary artery bypass graft surgery. Semin Respir Crit Care Med 2001;22:657-64.
10. Light RW. Pleural effusion due to pulmonary emboli. Curr Opin Pulm Med 2001;7:198-201.
11. Lama A, Ferreiro L, Golpe A, et al. Characteristics of patients with lymphangioleiomyomatosis and pleural effusion: a systematic review. Respiration 2016;91:256-64.
12. Reechaipichitkul W, Lulitanond V, Sungkeeree S, Patjanasoontorn B. Rapid diagnosis of tuberculous pleural effusion using polymerase chain reaction. Southeast Asian J Trop Med Public Health 2000;31:509-14.
13. Antonangelo L, Faria CS, Sales RK. Tuberculous pleural effusion: diagnosis & management. Expert Rev Respir Med 2019;13:747-59.
14. Avnon LS, Abu-Shakra M, Flusser D, Heimer D, Sion-Vardy N. Pleural effusion associated with rheumatoid arthritis: what cell predominance to anticipate? Rheumatol Int 2007;27:919-25.
15. Kim CH, Oh HG, Lee SY, et al. Differential diagnosis between lymphoma-associated malignant pleural effusion and tuberculous pleural effusion. Ann Transl Med 2019;7:373.
16. Gonzalez A, Fielli M, Ceccato A, Luna C. Score for differentiating pleural tuberculosis from malignant effusion. Med Sci (Basel) 2019;7:36.
17. Kita T, Araya T, Ichikawa Y, et al. IgG4-related pleuritis with no other organ involvement. Am J Med Sci 2018;356:487-91.
18. Yousem SA. Chronic follicular pleuritis: a B cell-rich form of nonspecific pleuritis/fibrosis. Hum Pathol 2019;90:14-9.
19. Buchanan DR, Johnston ID, Kerr IH, Hetzel MR, Corrin B, Turner-Warwick M. Cryptogenic bilateral fibrosing pleuritis. Br J Dis Chest 1988; 82:186-93.
20. Derakhshan F, Ionescu D, Cheung S, Churg A. Use of programmed death ligand-1 (PD-L1) staining to separate sarcomatoid malignant mesotheliomas from benign mesothelial reactions. Arch Pathol Lab Med 2020;144:185-8.
21. Husain AN, Colby TV, Ordonez NG, et al. Guidelines for pathologic diagnosis of malignant mesothelioma 2017 update of the consensus statement from the International Mesothelioma Interest Group. Arch Pathol Lab Med 2018; 142:89-108.
22. Churg A, Sheffield BS, Galateau-Salle F. New markers for separating benign from malignant mesothelial proliferations: are we there yet? Arch Pathol Lab Med 2016;140:318-21.
23. Churg A, Colby TV, Cagle P, et al. The separation of benign and malignant mesothelial proliferations. Am J Surg Pathol 2000;24:1183-200.
24. Cagle PT, Churg A. Differential diagnosis of benign and malignant mesothelial proliferations on pleural biopsies. Arch Pathol Lab Med 2005;129:1421-7.
25. Jiao N, Zhang W, Wang W, et al. Mesothelial/monocytic incidental cardiac excrescence: a case report and review of literature. Int J Clin Exp Pathol 2014;7:6219-24.
26. Suarez-Vilela D, Izquierdo-Garcia FM. Mesothelial/monocytic incidental cardiac excrescence: a process mediated by adhesion molecules? Am J Surg Pathol 2005;29:1545.
27. Mallick S, Arava S, Muthukumaran S, Sharma B, Choudhary SK, Ray R. Mesothelial/monocytic incidental cardiac excrescence mimicking cardiac tumor. Asian Cardiovasc Thorac Ann 2016;24:42-4.
28. Luc JG, Phan K, Tchantchaleishvili V. Cystic tumor of the atrioventricular node: a review of the literature. J Thorac Dis 2017;9:3313-8.

29. Linder J, Shelburne JD, Sorge JP, Whalen RE, Hackel DB. Congenital endodermal heterotopia of the atrioventricular node: evidence for the endodermal origin of so-called mesotheliomas of the atrioventricular node. Hum Pathol 1984;15:1093-8.
30. Fine G, Morales AR. Mesothelioma of the atrioventricular node. Arch Pathol 1971;92:402-8.
31. WHO Classification of Tumours Editorial Board. Thoracic tumours, 5th ed. Lyon: IARC; 2021.
32. Lie JT, Lufschanowski R, Erickson EE. Heterotopic epithelial replacement (so-called "mesothelioma") of the atrioventricular node, congenital heart block, and sudden death. Am J Forensic Med Pathol 1980;1:131-7.
33. Guo J, Zuo S, Lin C, Ji Y. Surgical treatment of a giant cystic tumor of the atrioventricular nodal region. Interact Cardiovasc Thorac Surg 2009;8:592-3.
34. Paniagua JR, Sadaba JR, Davidson LA, Munsch CM. Cystic tumour of the atrioventricular nodal region: report of a case successfully treated with surgery. Heart 2000;83:E6.
35. Cameselle-Teijeiro J, Santias RR, Nallib IA, Forteza J, Barreiro NA. Cystic tumor of the atrioventricular node: a rare cardiac pseudoneoplastic lesion. Arch Pathol Lab Med 2010;134:1584-6.
36. Evans CA, Suvarna SK. Cystic atrioventricular node tumour: not a mesothelioma. J Clin Pathol 2005;58:1232.
37. Cudzilo C, Aragaki A, Guitron J, Benzaquen S. Methotrexate-induced pleuropericarditis and eosinophilic pleural effusion. J Bronchology Interv Pulmonol 2014;21:90-2.
38. Oba Y, Abu-Salah T. The prevalence and diagnostic significance of eosinophilic pleural effusions: a meta-analysis and systematic review. Respiration 2012;83:198-208.
39. Krenke R, Light RW. Drug-induced eosinophilic pleural effusion. Eur Respir Rev 2011;20:300-1.
40. Krenke R, Nasilowski J, Korczynski P, et al. Incidence and aetiology of eosinophilic pleural effusion. Eur Respir J 2009;34:1111-7.
41. Akaba T, Takeyama K, Toriyama M, et al. Pulmonary paragonimiasis: the detection of a worm migration track as a diagnostic clue for uncertain eosinophilic pleural effusion. Intern Med 2016;55:503-6.
42. Aktogu Ozkan S, Erer OF, Y AY, Yuncu G, Aydogdu Z. Hydatid cyst presenting as an eosinophilic pleural effusion. Respirology 2007;12:462-4.
43. Ferreiro L, Alvarez-Dobano JM, Valdes L. Systemic diseases and the pleura. Arch Bronconeumol 2011;47:361-70.
44. Koyama S, Murakami K, Sakakibara T, et al. Massive pleural effusion and bronchopleural fistula in Wegener's granulomatosis. Sarcoidosis Vasc Diffuse Lung Dis 2010;27:76-9.
45. Dhillon N, Skinnider B, Darabian S, Carruthers M. Histiocytosis masquerading in the mesentery and pleura. BMJ Case Rep 2019;12:e232165.
46. Chen J, Tang H, Li B, Xiu Q. Rosai-Dorfman disease of multiple organs, including the epicardium: an unusual case with poor prognosis. Heart Lung 2011;40:168-71.
47. Ozkaya N, Rosenblum MK, Durham BH, et al. The histopathology of Erdheim-Chester disease: a comprehensive review of a molecularly characterized cohort. Mod Pathol 2018;31:581-97.
48. Lynch DA, Gamsu G, Ray CS, Aberle DR. Asbestos-related focal lung masses: manifestations on conventional and high-resolution CT scans. Radiology 1988;169:603-7.
49. Hillerdal G. Rounded atelectasis. Clinical experience with 74 patients. Chest 1989;95:836-41.
50. McHugh K, Blaquiere RM. CT features of rounded atelectasis. AJR Am J Roentgenol 1989;153:257-60.
51. Azour L, Billah T, Salvatore MM, et al. Causative factors, imaging findings, and CT course of round atelectasis. Clin Imaging 2018;50:250-7.
52. Berk JL. Pleural effusions in systemic amyloidosis. Curr Opin Pulm Med 2005;11:324-8.
53. Mansalis KA, Klein DA, Demartini SD, Powers JF, Danielson DS. Pleural findings in a patient with persistent pulmonary effusions from systemic amyloidosis. Amyloid 2011;18:29-31.
54. Coolbear F, Bilawich AM, Tongson J, Adamo J, Churg A. Pleural amyloidosis imitating pleural malignancy. Respir Med Case Rep 2017;20:195-7.
55. Nakano T, Endo S, Tetsuka K, Fukushima N. Asymptomatic localized pleural amyloidosis mimicking malignant pleural mesothelioma: report of a case. J Thorac Dis 2016;8:E157-60.
56. Adams AL, Castro CY, Singh SP, Moran CA. Pleural amyloidosis mimicking mesothelioma: a clinicopathologic study of two cases. Ann Diagn Pathol 2001;5:229-32.
57. Kavuru MS, Adamo JP, Ahmad M, Mehta AC, Gephardt GN. Amyloidosis and pleural disease. Chest 1990;98:20-3.
58. Knapp MJ, Roggli VL, Kim J, Moore JO, Shelburne JD. Pleural amyloidosis. Arch Pathol Lab Med 1988;112:57-60.
59. Manur R, Lamzabi I. Amyloidosis diagnosed in cytology specimen of pleural effusion: a case report. Diagn Cytopathol 2018;46:522-4.
60. Shah PL, Gillmore JD, Copley SJ, et al. The importance of complete screening for amyloid fibril type and systemic disease in patients with amyloidosis in the respiratory tract. Sarcoidosis Vasc Diffuse Lung Dis 2002;19:134-42.
61. Mahto M, Mohapatra S, Sumitra G, Kaushik S, Mishra TK, Saxena A. A case of light chain deposition disease (LCDD) in a young patient. Indian J Clin Biochem 2011;26:420-2.

62. Lee HJ, Sul H, Kim HW, et al. IgG4-related systemic disease presenting with pleural effusion, ascites and dilated cardiomyopathy. Tumori 2016;102(Suppl 2).
63. Kojima M, Nakazato Y, Kaneko Y, Sugihara S, Masawa N, Nakamura N. Cytological findings of IgG4-related pleural effusion: a case report. Cytopathology 2013;24:338-40.
64. Rossi G, Marchioni A, Guicciardi N, Cadioli A, Cavazza A. Recurrent pleural and pericardium effusions in a white woman with IgG4-related syndrome. Am J Surg Pathol 2009;33:802-3.
65. Murata Y, Aoe K, Mimura Y. Pleural effusion related to IgG4. Curr Opin Pulm Med 2019;25:384-90.
66. Choi IH, Jang SH, Lee S, Han J, Kim TS, Chung MP. A case report of IgG4-related disease clinically mimicking pleural mesothelioma. Tuberc Respir Dis (Seoul) 2014;76:42-5.
67. Zen Y, Inoue D, Kitao A, et al. IgG4-related lung and pleural disease: a clinicopathologic study of 21 cases. Am J Surg Pathol 2009;33:1886-93.
68. Wallace ZS, Naden RP, Chari S, et al. The 2019 American College of Rheumatology/European League Against Rheumatism classification criteria for IgG4-related disease. Ann Rheum Dis 2020; 79:77-87.
69. Shrestha B, Sekiguchi H, Colby TV, et al. Distinctive pulmonary histopathology with increased IgG4-positive plasma cells in patients with autoimmune pancreatitis: report of 6 and 12 cases with similar histopathology. Am J Surg Pathol 2009;33:1450-62.
70. Suda K, Nakaoka H, Yoshihara K, et al. Clonal expansion and diversification of cancer-associated mutations in endometriosis and normal endometrium. Cell Rep 2018;24:1777-89.
71. Anglesio MS, Papadopoulos N, Ayhan A, et al. Cancer-associated mutations in endometriosis without cancer. N Engl J Med 2017;376:1835-48.
72. Bobbio A, Canny E, Mansuet Lupo A, et al. Thoracic endometriosis syndrome other than pneumothorax: clinical and pathological findings. Ann Thorac Surg 2017;104:1865-71.
73. Ghigna MR, Mercier O, Mussot S, et al. Thoracic endometriosis: clinicopathologic updates and issues about 18 cases from a tertiary referring center. Ann Diagn Pathol 2015;19:320-5.
74. Joseph J, Sahn SA. Thoracic endometriosis syndrome: new observations from an analysis of 110 cases. Am J Med 1996;100:164-70.
75. Haga T, Kataoka H, Ebana H, et al. Thoracic endometriosis-related pneumothorax distinguished from primary spontaneous pneumothorax in females. Lung 2014;192:583-7.
76. Rousset P, Rousset-Jablonski C, Alifano M, Mansuet-Lupo A, Buy JN, Revel MP. Thoracic endometriosis syndrome: CT and MRI features. Clin Radiol 2014;69:323-30.
77. Haga T, Kumasaka T, Kurihara M, Kataoka H, Miura M. Immunohistochemical analysis of thoracic endometriosis. Pathol Int 2013;63:429-34.
78. Guedj N, Cote JF, Lepimpec-Barthes F, et al. [Immuno-histochemical study contribution in thoracic endometriosis: about an analysis of eight cases.] Ann Pathol 2009;29:475-80. [French]
79. Karpathiou G, Da Cruz V, Patoir A, et al. Mediastinal cyst of mullerian origin: evidence for developmental endosalpingiosis. Pathology 2017;49:83-4.
80. Heinig J, Gottschalk I, Cirkel U, Diallo R. Endosalpingiosis-an underestimated cause of chronic pelvic pain or an accidental finding? A retrospective study of 16 cases. Eur J Obstet Gynecol Reprod Biol 2002;103:75-8.
81. Singh N, Murali S, Zangmo R. Florid cystic endosalpingiosis, masquerading as malignancy in a young patient: a brief review. BMJ Case Rep 2014;2014.
82. Sneige N, Dawlett MA, Kologinczak TL, Guo M. Endosalpingiosis in peritoneal washings in women with benign gynecologic conditions: thirty-eight cases confirmed with paired box-8 immunohistochemical staining and correlation with surgical biopsy findings. Cancer Cytopathol 2013;121:582-90.
83. Markou GA, Goubin-Versini I, Carbunaru OM, Karatzios C, Muray JM, Fysekidis M. Macroscopic deciduosis in pregnancy is finally a common entity. Eur J Obstet Gynecol Reprod Biol 2016;197:54-8.
84. Flieder DB, Moran CA, Travis WD, Koss MN, Mark EJ. Pleuro-pulmonary endometriosis and pulmonary ectopic deciduosis: a clinicopathologic and immunohistochemical study of 10 cases with emphasis on diagnostic pitfalls. Hum Pathol 1998;29:1495-503.
85. Cruz DB, Dhamer T, da Rocha VW, Dupont RF. Diffuse peritoneal deciduosis mimicking metastatic lesions. BMJ Case Rep 2014;2014.
86. Ordonez NG. Deciduoid mesothelioma: report of 21 cases with review of the literature. Mod Pathol 2012;25:1481-95.
87. Shanks JH, Harris M, Banerjee SS, et al. Mesotheliomas with deciduoid morphology: a morphologic spectrum and a variant not confined to young females. Am J Surg Pathol 2000;24:285-94.
88. Nascimento AG, Keeney GL, Fletcher CD. Deciduoid peritoneal mesothelioma. An unusual phenotype affecting young females. Am J Surg Pathol 1994;18:439-45.
89. Dudek W, Schreiner W, Strehl J, Sirbu H. Spontaneous pneumothorax due to ectopic deciduosis: a case report. Thorac Cardiovasc Surg Rep 2014; 3:58-60.

90. Kim YD, Min KO, Moon SW. Thoracoscopic treatment of recurrent pneumothorax in a pregnant woman: a case of ectopic deciduosis. Thorac Cardiovasc Surg 2010;58:429-30.
91. Bucella D, Buxant F, Anaf V, Simon P, Fayt I, Noel JC. Omental trophoblastic implants after surgical management of ectopic pregnancy. Arch Gynecol Obstet 2009;280:115-7.
92. Pal L, Parkash V, Rutherford TJ. Omental trophoblastic implants and hemoperitoneum after laparoscopic salpingostomy for ectopic pregnancy. A case report. J Reprod Med 2003;48:57-9.
93. Ben-Arie A, Goldchmit R, Dgani R, et al. Trophoblastic peritoneal implants after laparoscopic treatment of ectopic pregnancy. Eur J Obstet Gynecol Reprod Biol 2001;96:113-5.
94. Seo HW, Jeong MJ, Namkung J, Kim CJ, Kwon JY. Persistent low-level elevation of serum human chorionic gonadotropin after termination of pregnancy: a rare case of peritoneal trophoblastic implant. Obstet Gynecol Sci 2017;60:616-20.
95. Roma AA, Masand RP. Ovarian Brenner tumors and Walthard nests: a histologic and immunohistochemical study. Hum Pathol 2014;45:2417-22.
96. Wang D, Jia CW, Feng RE, Shi HH, Sun J. Gliomatosis peritonei: a series of eight cases and review of the literature. J Ovarian Res 2016;9:45.
97. Liang L, Zhang Y, Malpica A, et al. Gliomatosis peritonei: a clinicopathologic and immunohistochemical study of 21 cases. Mod Pathol 2015;28:1613-20.
98. Alexander M, Cope N, Renninson J, Hong A, Simpson RH, Hirschowitz L. Relationship between endometriosis, endometrioid adenocarcinoma, gliomatosis peritonei, and carcinoid tumor in a patient with recurrent ovarian teratoma. Int J Gynecol Pathol 2011;30:151-7.
99. Muller AM, Sondgen D, Strunz R, Muller KM. Gliomatosis peritonei: a report of two cases and review of the literature. Eur J Obstet Gynecol Reprod Biol 2002;100:213-22.
100. Ferguson AW, Katabuchi H, Ronnett BM, Cho KR. Glial implants in gliomatosis peritonei arise from normal tissue, not from the associated teratoma. Am J Pathol 2001;159:51-5.
101. Shefren G, Collin J, Soriero O. Gliomatosis peritonei with malignant transformation: a case report and review of the literature. Am J Obstet Gynecol 1991;164(Pt 1):1617-20.
102. Bisceglia M, Galliani CA, Pizzolitto S, et al. Selected case from the Arkadi M. Rywlin International Pathology Slide Series: leiomyomatosis peritonealis disseminata: report of 3 cases with extensive review of the literature. Adv Anat Pathol 2014;21:201-15.
103. Toriyama A, Ishida M, Amano T, et al. Leiomyomatosis peritonealis disseminata coexisting with endometriosis within the same lesions: a case report with review of the literature. Int J Clin Exp Pathol 2013;6:2949-54.
104. Ma Y, Wang S, Liu Q, Lu B. A clinicopathological and molecular analysis in uterine leiomyomas and concurrent/metachronous peritoneal nodules: new insights into disseminated peritoneal leiomyomatosis. Pathol Res Pract 2020;216:152938.
105. Lu B, Xu J, Pan Z. Iatrogenic parasitic leiomyoma and leiomyomatosis peritonealis disseminata following uterine morcellation. J Obstet Gynaecol Res 2016;42:990-9.
106. Zyla MM, Dzieniecka M, Kostrzewa M, et al. Leiomyomatosis peritonealis disseminata of unusual course with malignant transformation: case report. Acta Obstet Gynecol Scand 2015;94:220-3.
107. Ferrer Marrero TM, Prieto-Centurion V, Jaffe HA. Thoracic splenosis: history is the key. Respir Med Case Rep 2017;22:251-3.
108. Chemouni D, Laas O, Caporossi JM, Coze S, Panuel M, Chaumoitre K. A rare cause of pleural nodules: thoracic splenosis. Diagn Interv Imaging 2014;95:1103-4.
109. Khan AM, Manzoor K, Malik Z, Avsar Y, Shim C. Thoracic splenosis: know it—avoid unnecessary investigations, interventions, and thoracotomy. Gen Thorac Cardiovasc Surg 2011;59:245-53.
110. Syed S, Zaharopoulos P. Thoracic splenosis diagnosed by fine-needle aspiration cytology: a case report. Diagn Cytopathol 2001;25:321-4.
111. Lim CS, Thompson JF, McKenzie PR, McCarthy SW, Scolyer RA. Peritoneal melanosis associated with metastatic melanoma involving the omentum. Pathology 2012;44:255-7.
112. Kim SS, Nam JH, Kim SM, Choi YD, Lee JH. Peritoneal melanosis associated with mucinous cystadenoma of the ovary and adenocarcinoma of the colon. Int J Gynecol Pathol 2010;29:113-6.
113. Kim NR, Suh YL, Song SY, Ahn G. Peritoneal melanosis combined with serous cystadenoma of the ovary: a case report and literature review. Pathol Int 2002;52:724-9.
114. Jaworski RC, Boadle R, Greg J, Cocks P. Peritoneal "melanosis" associated with a ruptured ovarian dermoid cyst: report of a case with electron-probe energy dispersive X-ray analysis. Int J Gynecol Pathol 2001;20:386-9.
115. Fadare O, Bifulco C, Carter D, Parkash V. Cartilaginous differentiation in peritoneal tissues: a report of two cases and a review of the literature. Mod Pathol 2002;15:777-80.

116. Honjo H, Kumagai Y, Ishiguro T, et al. Heterotopic mesenteric ossification after a ruptured abdominal aortic aneurism: case report with a review of literatures. Int Surg 2014;99:479-84.
117. Patel RM, Weiss SW, Folpe AL. Heterotopic mesenteric ossification: a distinctive pseudosarcoma commonly associated with intestinal obstruction. Am J Surg Pathol 2006;30:119-22.
118. Wilson JD, Montague CJ, Salcuni P, Bordi C, Rosai J. Heterotopic mesenteric ossification ('intraabdominal myositis ossificans'): report of five cases. Am J Surg Pathol 1999;23:1464-70.
119. Como JJ, Yowler CJ, Malangoni MA. Extensive heterotopic mesenteric ossification after penetrating abdominal trauma. J Trauma 2008;65:1567.
120. Di Paolo N, Sacchi G, Lorenzoni P, Sansoni E, Gaggiotti E. Ossification of the peritoneal membrane. Perit Dial Int 2004;24:471-7.
121. Gonzalez-Crussi F, deMello DE, Sotelo-Avila C. Omental-mesenteric myxoid hamartomas. Infantile lesions simulating malignant tumors. Am J Surg Pathol 1983;7:567-78.
122. Shukla S, Singh SK, Pujani M, Pujani M, Chowdhury SR. Omental myxoid hamartomas—a case report with review of literature. Trop Gastroenterol 2009;30:49-50.
123. Vyas MC, Mathur DR, Ramdeo IN, Rohitasvadana. Omento-mesentral myxoid hamartoma—a case report. Indian J Cancer 1994;31:212-4.
124. Ludwig K, Alaggio R, Dall'Igna P, Lazzari E, d'Amore ES, Chou PM. Omental mesenteric myxoid hamartoma, a subtype of inflammatory myofibroblastic tumor? Considerations based on the histopathological evaluation of four cases. Virchows Arch 2015;467:741-7.

INDEX

A

Adenomatoid tumor, 104
Amyloidosis, 168
Anatomy, normal, 1
  Blood supply, 1
  Innervation, 2
  Lymphatic drainage, 1
  Microscopic anatomy, 2
  Sites, 1
Angiolipoma, 123
Angiosarcoma, 124
Asbestos and mesothelioma, 15
Askin tumor, 119
Atypical mesothelial proliferation, 162

B

Benign cystic mesothelioma, 101
Biphasic mesothelioma, 47
Blesovsky syndrome, 166

C

Calcifying fibrous tumor/pseudotumor, 122
Chondroid metaplasia, 173
Chondrosarcoma, 125
Classification of tumors, 11
  World Health Organization, 11
Clear cell sarcoma, 124
Cystic tumor of the atrioventricular node, 163

D

Deciduosis, 170
Desmoid-type fibromatosis, 122
Desmoplastic mesothelioma, 45
Desmoplastic small round cell tumor, 119
Diffuse large B-cell lymphoma associated with chronic inflammation, 136
Diffuse malignant mesothelioma, *see* Diffuse mesothelioma
Diffuse mesothelioma, 13
  Biphasic mesothelioma, 47
  Clinical findings, 17
  Cytologic findings, 50
  Desmoplastic mesothelioma, 45
  Diagnostic specimens, 18
  Differential diagnosis, 65
    Benign versus malignant mesothelial proliferations, 65
  Epidemiology, 13
  Epithelioid mesothelioma versus metastatic carcinoma, 72
  Epithelioid subtype, 28, *see also* Epithelioid diffuse mesothelioma
  Etiology, 15
  Gross findings, 19
  Histologic Patterns, 26
  Immunohistochemical findings, 48
  Incidence, 13
  Malignant features, 53
  Mesothelioma in situ, 23
  Microscopic findings, 20
    Biphasic mesothelioma, 47
    Diffuse mesothelioma: epithelioid mesothelioma, 28
    Sarcomatoid mesothelioma, 42
  Molecular genetic findings, 62
  Prognosis, 79
  Reactive atypia, 53
  Sarcomatoid mesothelioma, 42, 50, *see also* Sarcomatoid mesothelioma
  Sarcomatoid mesothelioma versus other spindle cell malignancies, 74
  Staging, 80, 81
  Treatment, 77
  Ultrastructural findings, 64
Disseminated peritoneal leiomyomatosis, 171

E

Elastofibroma, 123
Embryonic development, normal, 5
Empyema, 157
Endometriosis, 169
Endosalpingiosis, 170
Eosinophilic pleuritis, 164
Epithelioid diffuse mesothelioma, 28, *see also* Diffuse mesothelioma
  Architectural patterns, 28
  Cytologic variants, 34, 37
  Differential diagnosis, 65; differentiation from reactive hyperplasia, 65; from metastatic lung carcinoma, 72
  Grading, 41
  Immunohistochemical findings, 48
  Stromal variations, 40
Epithelioid hemangioendothelioma, 114

Erdheim-Chester disease, 136
Ewing sarcoma, 119
Extracavitary primary effusion lymphoma, 133

F

Fibrothorax, 159
Fibrous (organizing) pleuritis, 159
Folded lung, 166
Function, normal, 8

G

Gliomatosis peritonei, 171
Granulomatous pleuritis, 165

H

Hemothorax, 166

I

IgG4-related disease, 169

L

Leiomyosarcoma, 124
Lipoblastoma, 123
Lipoma, 123
Liposarcoma, 124
Localized mesothelioma, 99
Lung carcinoma metastatic to pleura, 143
Lymphocytic effusion, 157
Lymphoproliferative neoplasms, 133, *see also* under individual entities

M

Malignant peripheral nerve sheath tumor, 124
Malignant pleural effusion, 56, 143
Malignant solitary fibrous tumor, 110
Melanosis, 173
Mesenchymal neoplasms, 109, *see also* under individual entities
Mesothelial hyperplasia, 161
  Atypical hyperplasia, 161
  Simple hyperplasia, 161
Mesothelial/monocytic incidental cardiac excrescences (MICE), 162
Mesothelioma, *see* under individual entities
Mesothelioma in situ, 23
Metastatic neoplasms, *see* Secondary malignant neoplasms
Multicystic mesothelioma, 101

N

Neutrophilic effusions/infiltrates, 157
Nodular histiocytic hyperplasia, 162
Nodular histiocytic/mesothelial hyperplasia, 162

O

Omental-mesenteric myxoid hamartoma, 173
Osseous metaplasia, 173
Osteosarcoma, 125

P

Pericardium, normal anatomy, development, and function, 1
Peritoneal inclusion cyst, 101
Peritoneum, normal anatomy, development, and function, 1
Pleura, normal anatomy, development, and function, 1
Pleural plaques, 160
Pleurisy, 159
Primary effusion lymphoma, 133
Primary effusion lymphoma-like lymphoma, 133
Primary peritoneal carcinoma, 152
Pseudomyxoma peritonei, 151
Pyothorax-associated lymphoma, 136

R

Reactive fibrous pleuritis, 157
Rounded atelectasis, 166

S

Sarcoma, 124
Sarcomatoid mesothelioma, 42, *see also* Diffuse mesothelioma
  Differential diagnosis, 65; differentiation from fibrous pleurisy, 69; from other spindle cell malignancies, 74
  Immunohistochemical findings, 48, 50
  Variants, 37, 42
    Shared with epithelioid mesothelioma, 37
Schwannoma, 123
Secondary malignant neoplasms, 143
  Pericardium, 153
  Peritoneum, 149
    Clinical features, 150
    Differential diagnosis, 152
    Microscopic findings, 150
    Pseudomyxoma peritonei, 151
    Sites of origin, 149

Pleura, 143
Differential diagnosis, 149
Gross findings, 144
Immunohistochemistry, 149
Mechanisms of serosal involvement, 143
Microscopic findings, 145
Patterns of involvement, 143
Sites of origin, 143
Treatment and prognosis, 149
Simple hyperplasia, 161
Solitary fibrous tumor, 109
Splenosis, 173
Synovial sarcoma, 117

T

Talc pleurodesis, 167
Trophoblastic implants, 171
Tunica vaginalis, normal anatomy, development, and function, 1

W

Walthard rests, 171
Well-differentiated papillary mesothelial tumor, 97
Well-differentiated papillary mesothelioma, 97
World Health Organization classification of mesothelial tumors, 11